Daniel B. Dix

Large-Time Behavior of Solutions of Linear Dispersive Equations

Springer

Author

Daniel B. Dix
Department of Matheamtics
University of South Carolina
Columbia, SC 29208, USA
E-mail: dix@math.sc.edu

Cataloging-in-Publication Data applied for

Die Deutsche Bibliothek - CIP-Einheitsaufnahme

Dix, Daniel B.:
Large time behavior of solutions of linear dispersive equations /
Daniel B. Dix. - Berlin ; Heidelberg ; New York ; Barcelona ;
Budapest ; Hong Kong ; London ; Milan ; Paris ; Santa Clara ;
Singapore ; Tokyo : Springer, 1997
    (Lecture notes in mathematics ; 1668)
    ISBN 3-540-63434-7

Mathematics Subject Classification (1991):
35B40, 35B65, 35C20, 35E15, 35G10, 35Q53, 35Q55, 41A60

ISSN 0075-8434
ISBN 3-540-63434-7 Springer-Verlag Berlin Heidelberg New York

This work is subject to copyright. All rights are reserved, whether the whole or part
of the material is concerned, specifically the rights of translation, reprinting, re-use
of illustrations, recitation, broadcasting, reproduction on microfilms or in any other
way, and storage in data banks. Duplication of this publication or parts thereof is
permitted only under the provisions of the German Copyright Law of September 9,
1965, in its current version, and permission for use must always be obtained from
Springer-Verlag. Violations are liable for prosecution under the German Copyright
Law.

© Springer-Verlag Berlin Heidelberg 1997
Printed in Germany

The use of general descriptive names, registered names, trademarks, etc. in this
publication does not imply, even in the absence of a specific statement, that such
names are exempt from the relevant protective laws and regulations and therefore
free for general use.

Typesetting: Camera-ready TEX output by the author
SPIN: 10553322   46/3142-543210 - Printed on acid-free paper

CONTENTS

INTRODUCTION

This work concerns the large-time asymptotic behavior of solutions of the initial-value problem for the linear dispersive equation

$$u_t + iR(D)u = 0,$$
$$u(x, 0) = u_0(x),$$

where the dispersive symbol $R(k)$ is homogeneous of degree $r > 1$ and is generally of the form

$$R(k) = \begin{cases} \rho_+ k^r & k > 0, \\ \rho_-(-k)^r & k < 0, \end{cases}$$

where the constants ρ_+, ρ_- are real numbers. We define $[R(D)u]\hat{}(k) = R(k)\hat{u}(k)$ where $\hat{u}(k) = \int_{-\infty}^{\infty} e^{-ikx} u(x)\, dx$ is the Fourier transform of u. Different choices for r, ρ_+, ρ_- allow us to incorporate the following special cases:

Equation	$iR(D)u$	$R(k)$	r	ρ_+	ρ_-
Linear Schrödinger	$-i\rho u_{xx}$	ρk^2	2	ρ	ρ
Linearized Benjamin-Ono	$-\rho \mathcal{H} u_{xx}$	$-\rho k|k|$	2	$-\rho$	ρ
Linearized KdV	ρu_{xxx}	$-\rho k^3$	3	$-\rho$	ρ

KdV is an abbreviation for Korteweg-de Vries, $\rho \in \mathbb{R}$ and $\mathcal{H}$ denotes the Hilbert transform. The corresponding nonlinear dispersive wave equations

$$u_t \pm i|u|^2 u - i\rho u_{xx} = 0, \qquad \text{Nonlinear Schrödinger}$$
$$u_t + uu_x - \rho \mathcal{H} u_{xx} = 0, \qquad \text{Benjamin-Ono}$$
$$u_t + uu_x + \rho u_{xxx} = 0, \qquad \text{Korteweg-de Vries}$$

have been the subjects of intensive investigation in recent years because of their importance to physical applications (wave motions), because of the many fascinating properties of their solutions (such as the existence of solitons), and because there are methods (inverse scattering transforms) by which exact solutions may be computed and the pure initial-value problem solved in the same manner that the Fourier transform allows us to solve the above linear equations; see Newell [N]. It has also been found that solutions of more general nonlinear dispersive equations possess many of the same properties as solutions of these three *integrable* equations; however, the proofs of these properties cannot be based on any special structure of integrability. Frequently the proofs are based on an analysis of solutions of the linear dispersive equation, coupled with a perturbative argument allowing one to deal with the nonlinearity. In regard to large-time behavior of solutions, no comprehensive account exists in the literature dealing with the solutions of constant coefficient linear dispersive equations (although limited and scattered results for special cases have appeared). This work is an attempt to fill this gap. One of our main goals has been to provide a reference work which researchers and applied scientists in this area could consult for the detailed large-time asymptotic expansions (and their proofs) of solutions of these linear dispersive equations. It is hoped that these results might be a stimulus to further progress in the analysis of the large-time asymptotic behavior of solutions of nonlinear dispersive equations.

Before proceeding, we offer the following as motivation for the detailed study of the large-time asymptotic behavior of solutions of linear or nonlinear wave equations. In any physical situation where wave motion is encountered after sufficient time has elapsed for the transient and incidental features of the initial disturbance to have faded, one observes the "large-time" behavior of the wave motion. Actually, the elapsed time between the initial disturbance and the point at which the transient features have become relatively insignificant need not be that large. Thus large-time asymptotic expansions can be an effective way of approximating the wave motion over even intermediate time scales. Of course there is a definite limit to the accuracy of such an approximation to the solution at a fixed time. Greater accuracy can usually only be obtained by numerical calculations. In contrast to the results of numerical computations, the large-time asymptotic expansion can frequently be determined as a closed form expression. As a consequence of this closed form expression, one gains significant and deep insight into the solution; e.g. how it decays in various regions, how various functionals of the solution behave, where the solution oscillates and where it does not, which features of the initial disturbance are preserved and which ones are not, etc.. Such knowledge is often impractical to obtain by means of numerical calculations alone. Hence the large-time asymptotic behavior is an important and frequently accessible component of knowledge about the solution which is complementary to knowledge obtained by direct numerical computations.

We make the assumption that $u_0 = |D|^\gamma (v_0 + \mathcal{H}v_1)$, where v_0, v_1 are integrable complex-valued functions with compact support, and the real number γ satisfies $\gamma > -1$. We assume that $|\hat{v}_0(0)| + |\hat{v}_1(0)| \neq 0$. These assumptions will allow us, under different choices for γ, v_0, v_1 to understand several different types of solutions.

(1) **Derivatives of solutions**: $u_0 = \partial_x^j U_0$, where j is a nonnegative integer. These can be obtained by letting $\gamma = j$ and

$$(v_0, v_1) = \begin{cases} (U_0, 0) & j \equiv 0 \mod 4, \\ (0, -U_0) & j \equiv 1 \mod 4, \\ (-U_0, 0) & j \equiv 2 \mod 4, \\ (0, U_0) & j \equiv 3 \mod 4. \end{cases}$$

The understanding of the large-time behavior of derivatives is relevant to the problem of the asymptotic balance in solutions of nonlinear dispersive equations (see section 5.3).

(2) **Fractional Derivatives or Integrals of solutions**: $u_0 = |D|^\gamma U_0$, where $\gamma > -1$ is a real number. Understanding of the case $\gamma > 0$ has been found to be useful in the study of the large-time behavior of nonlinear dispersive equations. The case $-1 < \gamma < 0$ exposes some of the variety of large-time asymptotic behaviors that can result from initial data in L^p spaces (see section 5.2).

(3) **Solutions which decay slowly as** $|x| \to \infty$: see section 5.4.

The compact support condition is convenient for us, but not absolutely necessary; our results could be generalized somewhat. Also, it should be noted that many interesting initial data which decay as $|x| \to \infty$ like a negative power of $|x|$ can be put into the above form. It is not our purpose to prove results about the solution arising from an arbitrary initial datum in some commonly used function classes (such as $L^p(\mathbb{R})$); such an assumption on u_0 is compatible with a large variety of different asymptotic behaviors. It is our purpose to prove sharp results for initial conditions of the above type because such results map out the types of asymptotic behavior which can occur, with a minimum amount of extraneous complications. In fact, our results show why it is so hard to state or prove a sharp result valid for all initial data in spaces such as $L^p(\mathbb{R})$, $1 < p < \infty$. For more details, consult chapter five.

The solution of this initial-value problem can be expressed as the following Fourier integral

$$u(x,t) = \frac{1}{2\pi} \int_{-\infty}^{\infty} e^{ikx} e^{-iR(k)t} |k|^\gamma [\hat{v}_0(k) - i \operatorname{sgn}(k)\hat{v}_1(k)] \, dk.$$

This integral for $u(x,t)$ breaks naturally into the sum of two integrals:

$$u(x,t) = \frac{1}{2\pi} \int_0^{\infty} e^{ikx} e^{-i\rho_+ k^r t} k^\gamma [\hat{v}_0(k) - i\hat{v}_1(k)] \, dk$$
$$+ \frac{1}{2\pi} \int_0^{\infty} e^{-ikx} e^{-i\rho_- k^r t} k^\gamma [\hat{v}_0(-k) + i\hat{v}_1(-k)] \, dk.$$

Thus it suffices to analyze the large-time behavior of the single integral

$$I(x,t) = I(x,t;r,\rho,\gamma,V) \overset{\text{def}}{=} \int_0^{\infty} e^{ikx - i\rho k^r t} k^\gamma V(k) \, dk,$$

where $V(k)$ is an entire function of $k \in \mathbb{C}$. In fact, since

$$I(x,t;r,\rho,\gamma,V) = \overline{I(-x,t;r,-\rho,\gamma,\tilde{V})}$$

where $\tilde{V}(k) = \overline{V(\bar{k})}$, we see that it suffices to consider this integral when $\rho > 0$. Rather than define this integral as a limit of a family of absolutely convergent artificially regularized integrals, we will exploit the freedom we have (by Cauchy's theorem) to change the contour of integration, so that on the new contour the integral will converge absolutely, and to the same value. The exact nature of these new contours will be discussed later.

Although our ultimate goal is to analyze the solution $u(x, t)$, we set as an intermediate goal the complete asymptotic description of the integral $I(x, t; r, \rho, \gamma, V)$. This endeavor will occupy the first three chapters. In chapters four and five we will return to the analysis of $u(x, t)$. We approach the study of the integral $I(x, t)$ first from the standpoint of matched asymptotic expansions (chapters one and two) and then we obtain an expansion which is uniformly valid as $t \to \infty$ and $(-1)^\alpha x \geq 0$ is arbitrary (chapter three). Both of these approaches have significant redeeming features and we view them as complementary.

All this is necessary because the asymptotic expansion as $t \to \infty$, $b = xt^{-1} \in \mathbb{R}$ fixed, of the function $I(bt, t)$ is essentially different from the asymptotic expansion as $t \to \infty$, $\xi = xt^{-1/r} \in \mathbb{R}$ fixed, of $I(\xi t^{1/r}, t)$. The asymptotic expansions of $I(bt, t)$ are called the *outer expansions* (studied in chapter one); there are two of them, one for $b < 0$ and the other for $b > 0$. The terms of the outer expansions involve only elementary functions and the derivatives of $V(k)$. The derivation of the outer expansions and control of the associated error terms when b is fixed can be done using the classical method for Laplace contour integrals as discussed in Olver [O1], Bleistein and Handelsman [BH], Wong [W], and other places as well. We outline this method in section 1.1. In fact, in that same section we derive some general formulae for the coefficients which appear in Laplace expansions; these formulae are useful when one is studying (as we are) the dependence of the expansion on a parameter (such as b). The study of the outer expansions is rather lengthy because there is a different expansion for each connected component of the *steepest descent contour* which the initial path of integration is deformed into. Several authors have pointed out that the complete determination of the steepest descent contours is unnecessary in order to compute the Laplace expansion as well as to control the error in the classical case where b is constant. Nevertheless, in section 1.2 we have given a fairly complete study of these contours, and a proof that the original contour can be deformed into a union of these special contours. Our error estimates must be proved under significantly weaker assumptions on b than in the classical case (for reasons which will be explained below) and the proofs (sections 1.4 and 1.7) use our detailed knowledge of the steepest descent contours. In order that we might prove sharp error estimates, it is essential that we know exactly how the derivatives of $V(k)$ behave as $k \to \infty$ on rays and in sectors in the k plane. Thus in section 1.5 we study of the leading-order behavior of $V(k) = \hat{v}(k)$, under carefully formulated conditions on a function $v(x)$ with compact support. These assumptions on $v(x)$, which we call the *standard assumptions*, are invoked extensively in chapters four and five.

Unfortunately, the asymptotic character of the outer expansions break down and the error estimates obscure the "true" leading-order terms if b is allowed to tend to 0 too rapidly as $t \to \infty$. Indeed this must happen, since as we said above, there is an *inner expansion* of $I(\xi t^{1/r}, t)$, completely distinct from the outer expansions, which holds as $t \to \infty$, $\xi \in \mathbb{R}$ fixed. The inner expansion (studied in chapter two) is in descending powers of t with coefficients which involve the derivatives of V at $k = 0$ and certain special functions of ξ. These special functions appear naturally, and cannot be avoided. The inner expansion is derived by expanding $V(k)$ in a Taylor series at $k = 0$, and then integrating term-by-term (see section 2.1). The proofs of the sharp error estimates (sections 2.2 and 2.3) again utilize our detailed knowledge of the steepest descent contours.

The asymptotic character of the inner expansion breaks down as $|\xi| \to \infty$; hence we must live with three distinct expansions, each valid in its own asymptotic region. A complete matched asymptotic description of $I(x, t)$ is obtained by proving the validity of the outer and inner expansions in *extended* asymptotic regions ([KC]) in the (x, t) plane—the outer expansions in the disjoint regions R_+ and R_-, and the inner expansion in the inner region R_0. These extended regions will be defined precisely in sections 1.1 and 2.1, but they have the following properties. $R_\pm$ eventually (i.e for sufficiently large t) contains every ray $x = bt$, where $\pm b > 0$ is a constant. The inner region, R_0 eventually contains every curve of the form $x = \xi t^{1/r}$, where ξ is a real constant. R_+ and R_0 overlap for all $t \geq 1$; also R_- and R_0 overlap for all $t \geq 1$. Furthermore, we must prove that in the overlap of two regions the two expansions which are valid there can be matched to one another. We perform this matching in section 2.4.

The necessity of concerning oneself with carefully defined asymptotic regions and matching different expansions in an overlap region is the disagreeable aspect of the matched asymptotic description. This could be avoided if one had a single expansion which retains its asymptotic character independently of the value of x. Such an expansion is said to be *uniformly valid*. Methods for obtaining uniformly valid expansions have been studied for some time, see Wong [W] for references. When r is an integer, one finds that the phase function $kb - \rho k^r$ has exactly $r - 1$ complex stationary points, some of which contribute to the asymptotic expansion of $I(x, t)$, all of which are simple and distinct for $b \neq 0$, but which coalesce as $b \to 0$. This coalescence is the opportunity for the Laplace expansions to become nonuniformly valid. The usual approach to this problem of non-uniformity, as pioneered by Chester Friedman Ursell [CFU], and developed by many others, most notably Bleistein [B], is to write $V(k)$ as the sum of a polynomial in k which agrees with $V(k)$ at *all* the problem points (stationary points, endpoints, singularities, etc.) and a remainder. Thus $I(x, t)$ breaks into two integrals: the integral involving the polynomial approximation can be expressed in terms of special functions, and provides the first term(s) of an uniformly-valid expansion of $I(x, t)$; the integral involving the remainder decays more rapidly in t than the first term(s), as one discovers by integrating by parts, whereupon a factor of t^{-1} appears multiplied by an integral of the exact same appearance as $I(x, t)$ but with a different function substituted for $V(k)$. This process can then be repeated any number of times to generate as many terms as desired. See §3.1 for further discussion.

When r is not an integer, it is not exactly clear how this procedure is to be carried out. One problem when r is irrational is that there are now *infinitely many* stationary points on the Riemann surface forming the domain of the phase function, and it is not clear how to construct an approximating function to $V(k)$. Even when r is rational, when there are still finitely many stationary points, it is not clear how to construct the approximating function so that the remainder has good enough behavior so that the process can be repeated indefinitely, thereby generating a complete asymptotic expansion. When r is not an integer, it is not clear that the integration by parts procedure can be generalized.

Thus in section 3.1 we use a different procedure to generate a complete uniformly-valid asymptotic expansion of $I(x,t)$, an approach which will work for all real $r \geq 2$. The basic idea is simple. We separate $I(x,t)$ as a sum of integrals over the connected steepest descent contours employed in the analysis of the outer region. Then we expand $V(k)$ in a Taylor series about the saddle point; each connected contour passes through at most one saddle point. When this Taylor expansion is integrated term-by-term, certain special functions of the variable ξ emerge because of the scaling properties of the phase function. This yields an uniformly-valid expansion of the integral over each steepest descent contour. These expansions are then added together to obtain an uniformly-valid expansion of $I(x,t)$. These expansions are written down in full detail, together with their sharp error estimates, in section 3.2. The method of integrating Taylor expansions (or other expansions closely related to Taylor expansions) for generating uniformly-valid asymptotic expansions was used in the original paper of Chester, Friedmann, and Ursell [CFU], and on pages 352–357 in the book of Olver [O1]. As far as we can tell, the idea of applying this method to each connected contour of steepest descent is new. Since the nature of the steepest descent contours depends on whether $b > 0$ or $b < 0$, the resulting uniformly-valid expansions look different in these two cases. But they retain their asymptotic character even as $x \to 0$. In the inner region they are shown to agree (in section 3.1) with the inner expansion, which is the same for both $x > 0$ and $x < 0$. We obtain full asymptotic expansions as $|\xi| \to \infty$ of the special functions involved, and show (in section 3.1) that the uniformly-valid expansions reduce to the outer expansions in the outer regions.

The only disadvantage of the uniform expansions is that they involve a significant number of special functions, which are themselves defined by integrals. When $r = 2$ (section 4.1) or $r = 3$ (section 4.5) these can be identified with or expressed in terms of well-known special functions, such as confluent hypergeometric functions, or Airy functions; but for other values of r these special functions have not been studied as much. We need the fundamental special function

$$F_n(r,\delta;y) \overset{\text{def}}{=} \int_0^\infty \exp(-\tfrac{1}{r}\sigma^r + y\sigma)\sigma^{\delta-1}(\sigma - y^{\frac{1}{r-1}})^n \, d\sigma,$$

where $n \geq 0$ is an integer, $\delta > 0$ is a real number, and y lies on the Riemann surface of the function $y^{\frac{1}{r-1}}$, i.e. $y \in S\{y^{\frac{1}{r-1}}\}$. When $r \geq 3$ is an odd integer we also need two incomplete forms of this function. All the other special functions we need can be expressed in terms of these. When $r = 2$ or $r = 3$ we also derive recursion relations for these functions in the variable n, which allow us to reduce the number of special functions to a small number (i.e. r) of truly essential functions. These recursion relations also allow us to compare our uniformly-valid expansions to the uniformly-valid expansions which can be obtained (when $r = 2, 3$) by the integration-by-parts procedure mentioned above. Especially in the case $r = 3$ (in section 4.5) we discuss the relative merits of these two types of uniformly-valid expansions. As mentioned above, an expansion generated by an integration-by-parts procedure is not available when r is not an integer.

It turns out that even the leading-order terms of uniformly-valid asymptotic expansions are not uniquely determined. Since the higher-order terms depend very sensitively on the exact form of the leading-order term, they are also not uniquely determined. Despite this nonuniqueness it is worthwhile to compute all the terms of the asymptotic expansion. Knowledge of the higher terms in the expansion allows us to formulate sharp error estimates for finite term asymptotic approximations. Also, as we have already mentioned, asymptotic expansions can be used to compute the solution $u(x, t)$, and a variable number of terms of the expansion are needed in such calculations, depending on the particular point (x, t). We will not discuss the realistic computable *a priori* error bounds needed to decide if the desired accuracy has been obtained with a given number of terms.

In chapter four, in addition to the above topics, we also return to the original subject of the solution of the linear dispersive equation—but only in the special cases $r = 2$ and $r = 3$. We devote an entire section to each of the three linear equations listed as examples earlier, the Linear Schrödinger (section 4.3), the Linearized Benjamin-Ono (section 4.4), and the Linearized Korteweg-de Vries (section 4.6) equations. In each of these three cases we write down the complete uniformly-valid asymptotic expansions, together with the sharp error estimates. We also give the leading-order terms of the outer and inner expansions, and their error estimates. In the case $r = 2$ the outer expansion (and our uniformly-valid expansion) can lose its asymptotic character in the limit $|x| \to \infty$, $t \geq 1$ fixed. Thus we find the correct leading-order term in this limit, and prove an error estimate in section 4.2. This enables us to give the correct leading-order term of the asymptotics of the solutions of the Linear Schrödinger and Linearized Benjamin-Ono equations as $|x| \to \infty$, $t \geq 1$ fixed, when the initial data is only of moderate smoothness.

In chapter five we begin to discuss some applications of our results in the previous chapters to solutions of the linear dispersive equation with general $r \geq 2$. Although we restrict our discussion to equations possessing real-valued solutions, such as Linearized Benjamin-Ono and Linearized KdV, most of what we say can be reformulated to apply to Schrödinger-type equations, i.e. to $u_t + i\rho |D|^r u = 0$. Although we do not formulate or prove any results about solutions of nonlinear dispersive equations, everything in chapter five is directed toward the understanding of the nonlinear case. We have tried to make it possible to read chapter five without having to digest chapters one through four, although we have probably not succeeded completely. We make frequent references to results in the literature on nonlinear dispersive equations, and attempt to describe succinctly the relevant features of the theorems appearing in those papers, but we are not attempting a survey of such results. We hope our discussion will be helpful to both the expert and the newcomer to the field. It will be apparent that there is a great deal which is still unknown about the nonlinear case, and one of our main goals is to formulate some reasonable conjectures about when we can expect solutions of a nonlinear dispersive equation to behave to the leading-order as $t \to \infty$ like solutions of the corresponding linear dispersive equation. We do not completely reach even this goal; but we do identify some of the main unresolved issues which we feel are of central importance.

We have tried to keep the prerequisites to a minimum for the reader. A reasonable knowledge of single variable complex analysis (for example on the level of the book of Ahlfors, [Ahl]) should be adequate preparation for chapters one through three. We refer to Riemann surfaces several times, but we only use the fundamental idea of them as natural domains of certain analytic functions, especially when we wish to avoid a detailed discussion of branches of multivalued functions. In chapters four and five we also assume a certain familiarity with Fourier analysis, and Fourier multiplier operators, such as $|D|^\gamma$ or $\mathcal{H} = -i\operatorname{sgn}(D)$. Whenever a line of argument is presented for the first time, we try to write it out in detail. Considerable effort has been expended to make our expansions explicit, useful in computations, and reliable. In particular, we have endeavored to relate our special functions as much as possible to those discussed in Olver [O1] and Abramowitz and Stegun [AS]. However, it is inevitable that some errors are present. The author would appreciate being made aware of any errors the reader discovers in the text or formulae.

The author would like to acknowledge many helpful conversations with John Albert over a period of several years about the subject of this work. The author also greatly appreciates the invaluable assistance of F. W. J. Olver in directing him to the most recent references pertaining to chapters 1-3 of this work. Thanks to Doug Meade for his help in generating the figures using Maple. Thanks to Jerry Bona and John Albert for looking at preliminary versions of this work and making helpful suggestions. Thanks to Matania Ben-Artzi for references and discussion on smoothing effects relevant to section 5.2. The author acknowledges support of the Office of Naval Research, grant N00014-94-1-1163 during the summers of 1994 and 1995 while this work was in progress. Thanks are extended to the author's wife, Jean, and to his children, Amy and Nathan, for their love, patience, and prayers. Thanks also be to the Creator, Sustainer, and Lord of the universe, whose knowledge and understanding surpasses that of the entire human race more than the sun surpasses a spark. His unmerited favor, extended through His Son, Jesus Christ, has given me strength throughout this work.

CHAPTER 1

LAPLACE EXPANSIONS, OUTER REGIONS

1.1 Generalities, Notation

We define $b = x/t$ and $\varphi(k; b, r, \rho) = ikb - i\rho k^r$. We will usually suppress whatever parameters from our notation that are understood from the context. With our new notation we have

$$I(x, t; r, \rho, \gamma, V) = \int_0^\infty e^{\varphi(k; b)t} k^\gamma V(k)\, dk.$$

$V(k)$ is an entire function, the Fourier transform of an integrable function with compact support. Techniques for computing asymptotic expansions as $t \to \infty$ for this integral when b is fixed can be found in [O1], [BH], [W]. The outer regions $R_\pm$ are defined as follows:

$$R_\pm = \{(x, t) \in \mathbb{R} \times [1, \infty) \mid \pm x t^{-\beta} > \epsilon\},$$

where $\frac{1}{r} < \beta < 1$ and $0 < \epsilon < 1$ are constants. The principal contributions in this case arise from the singularity (or zero) and endpoint of integration at $k = 0$ and from the relevant saddle points $k_j \in \mathbb{C}$, where $\varphi'(k_j) = 0$. Since $\varphi'(k) = ib - ir\rho k^{r-1}$, we have that

$$k_j(b, r) = B^{\frac{1}{r-1}} e^{iS_j^\alpha}$$

where $\mathrm{sgn}(b) = (-1)^\alpha$, $\alpha \in \{0, 1\}$, $B = \left|\frac{b}{r\rho}\right|$, $S_j^\alpha = \frac{\pi}{r-1}(2j + \alpha)$, and $j \in \mathbb{Z}$. Not all of these saddle points are asymptotically relevant. Exactly which saddle points are relevant will be revealed when we deform the original contour of integration into a union of *steepest descent contours* (SDCs) starting at $k = 0$, (maybe) through some saddle points, and proceeding to infinity in the nearest *valley* to the ray $\arg k = 0$. A *valley* is a region in the complex k-plane where $\Re\varphi(k) \leq -M$, where $M \gg 1$. Likewise a *hill* is a region in the complex k-plane where $\Re\varphi(k) \geq M$, where $M \gg 1$. The asymptotic centers of the valleys (resp. hills) can be identified by rays along which $\Re\varphi(k)$ decreases (resp. increases) most rapidly as $k \to \infty$. If $k = Re^{i\theta}$ we have that

$$\Re\varphi(Re^{i\theta}) = -bR\sin(\theta) + \rho R^r \sin(r\theta).$$

Hence the asymptotic center of the valleys are along the rays $\theta = V_j = \frac{1}{r}(-\frac{\pi}{2} + 2\pi j)$, where $j \in \mathbb{Z}$. Likewise the asymptotic center of the hills are along the rays

$\theta = H_j = \frac{1}{r}(\frac{\pi}{2} + 2\pi j)$, where $j \in \mathbb{Z}$. We will refer to hills and valleys by the arguments of their asymptotic centers.

The initial contour of integration (i.e. $\theta = 0$) is between a hill (at $\theta = H_0 = \frac{\pi}{2r}$) and a valley (at $\theta = V_0 = -\frac{\pi}{2r}$). The ray of integration cannot be rotated in the direction of the hill, but it can be rotated toward the valley. But it cannot be rotated beyond that valley up onto the next hill, i.e. beyond $\theta = -\frac{\pi}{r}$. Thus the deformed path of integration, whatever else it does, must eventually go to infinity in the valley V_0. We call this valley the *final* valley.

A steepest descent contour (SDC) starting at a (simple) saddle point k_j leaves tangent to a certain line through the saddle point and proceeds downhill as fast as possible, i.e. always travelling in the opposite direction as the gradient vector of $\Re\varphi(k)$. In order to determine this line we must expand $\varphi(k)$ in the neighborhood of the saddle point. A short calculation will verify that

$$\varphi(k_j) = (-1)^\alpha i(r-1)\rho B^{\frac{r}{r-1}} e^{iS_j^\alpha},$$

$$-\varphi''(k_j) = ir(r-1)\rho B^{\frac{r-2}{r-1}} e^{i(r-2)S_j^\alpha}.$$

Suppose k is on the ray $k - k_j = Re^{i\omega}$. Then

$$t\varphi(k) = t\varphi(k_j) + \frac{t}{2}\varphi''(k_j)R^2 e^{i2\omega} + O(R^3)$$

$$= t\varphi(k_j) - \frac{|t|}{2}|\varphi''(k_j)|R^2 \exp\left[i\arg(t) + i\omega_0 + i2\omega\right]$$

$$+ O(R^3),$$

where $0 < R < 1$ and $\omega_0 = \arg(-\varphi''(k_j)) = \frac{\pi}{2} + (r-2)S_j^\alpha + 2\pi\mathbb{Z}$. Hence we must choose $\arg(t) + \omega_0 + 2\omega = 0$ if the ray is to be tangent to a SDC. This equation determines ω up to an integral multiple of π:

$$\omega = -\frac{\arg(t)}{2} - \frac{(2\alpha+1)}{4}\pi + \frac{1}{2}S_j^\alpha + \pi\mathbb{Z}.$$

Following Olver, we will first make a choice of ω according to the above. Since square roots of the quantity $-\varphi''(k_j)t$ appear in the terms of the expansion arising from this saddle point, it will be convenient to require that $\arg(t) + \omega_0$ be determined by the choice of ω by the relation $\arg(t) + \omega_0 + 2\omega = 0$. The value of the square root will then be determined by that choice of argument. So if ω changes by π, such as happens when one departs the saddle point in the opposite direction as before, the value of $[-\varphi''(k_j)t]^{1/2}$ changes by a factor of -1. Obviously, it is important that this choice of branch be used consistently.

Steepest descent contours turn out to be along level curves of $\Im\varphi(k)$ (see [BH]). We note that

$$\Im\varphi(Re^{i\theta}) = bR\cos(\theta) - \rho R^r \cos(r\theta).$$

Thus the SDC through the saddle point k_j is part of the polar locus

$$bR\cos(\theta) - \rho R^r \cos(r\theta) = (-1)^\alpha(r-1)\rho B^{\frac{r}{r-1}} \cos(S_j^\alpha).$$

The SDC through $k = 0$ is part of the polar locus

$$bR\cos(\theta) - \rho R^r \cos(r\theta) = 0.$$

It will turn out that there is a new contour of integration, formed out of SDCs, which is homotopic (endpoints held fixed) to the original contour of integration on the Riemann sphere. This polar representation will be useful in the proof of this fact, because each SDC will lie in a certain sector whose endpoints correspond to valleys. (This is more or less true, except for the SDC through $k = 0$, where one endpoint does not correspond to a valley. Also we will see it is possible for a SDC to "overshoot" its sector a little bit, but as $k \to \infty$ it happens that $\arg(k)$ tends to the argument of the center of the valley.) A study of the necessary contours and the proof that the contour can be changed as we indicated above is given in §1.2.

But for our analysis of the error terms we will need another parameterization of the SDCs, one which allows us to prove estimates which hold independently of b. Let $l = k_j$. Since $\varphi'(l) = ib - ir\rho l^{r-1} = 0$, we have $b = r\rho l^{r-1}$. Therefore $\varphi(k) = i\rho(rl^{r-1}k - k^r)$ and $\varphi(l) = i\rho(r-1)l^r$. Let C_j denote the SDC passing through l. If k is on C_j then $\varphi(k)$ differs from $\varphi(l)$ by a negative real number, say $-w^2$, where $w \in \mathbb{R}$. So $\varphi(k) = \varphi(l) - w^2$. Note that $\varphi''(k) = -i\rho r(r-1)l^{r-2}$. Following Olver, we define $p(k) = -\varphi(k)$ and $p_0 = -\varphi''(l)/2 = i\rho r(r-1)l^{r-2}/2$. Notice that $\omega_0 = \arg(p_0)$. Rewriting the relation $\varphi(k) = \varphi(l) - w^2$ in terms of these quantities we get

$$\left(1 + \frac{k}{l} - 1\right)^r - 1 - r\left(\frac{k}{l} - 1\right) = \frac{r(r-1)}{2}\left(\frac{w}{l\sqrt{p_0}}\right)^2.$$

Thus it is natural to define $\tilde{k} = \frac{k}{l} - 1$ and $\tilde{w} = \frac{w}{l\sqrt{p_0}}$. In these variables the relation becomes

$$(1 + \tilde{k})^r - 1 - r\tilde{k} = \binom{r}{2}\tilde{w}^2.$$

If w is a small positive number, then there are clearly two small complex solutions $\tilde{k}$ to this equation, corresponding to two different points k on C_j, one on either side of the saddle point l. We can use the above relation to define $\tilde{k}$ implicitly as a function of $\tilde{w}$, at least for $|\tilde{w}|$ sufficiently small. There are actually two such implicit functions; we will consider only the one where $\tilde{k} = \tilde{w} + O(\tilde{w}^2)$ for $|\tilde{w}|$ sufficiently small. We will suppose this function is continued analytically along the ray $\tilde{w} = \frac{w}{l\sqrt{p_0}}$, where $w \geq 0$. This will always be possible. Here again we see that the choice of the square root of p_0 determines which direction along C_j one departs from l as w increases from 0 to ∞. Estimates of this implicit function, which is clearly independent of l, are what makes our proof work; these estimates can be found in §1.7.

Let $\tilde{C}$ denote the SDC starting at $k = 0$. Another way of parameterizing this contour is in terms of the nonnegative real variable v, where $\varphi(k) = -v$. So

$ikb - i\rho k^r = -v$. We can rewrite this in the following way:

$$\left(\frac{b}{iv}\right) k - 1 = \frac{\rho}{b} \left(\frac{iv}{b}\right)^{r-1} \left[1 + \left(\frac{b}{iv}\right) k - 1\right]^r.$$

Thus it is natural to define the variables $\tilde{k} = \left(\frac{b}{iv}\right) k - 1$ and $\tilde{v} = \frac{\rho}{b} \left(\frac{iv}{b}\right)^{r-1}$. No confusion will arise between the contours $\tilde{C}$ and C_j, so we will use the same variable $\tilde{k}$ for two different things in these two contexts. In these new variables we have

$$\frac{\tilde{k}}{(1+\tilde{k})^r} = \tilde{v}.$$

For small $\tilde{v}$ we can uniquely determine $\tilde{k}$ as an analytic function of $\tilde{v}$. We then wish to analytically continue this function along the ray $\tilde{v} = \frac{\rho}{b} \left(\frac{iv}{b}\right)^{r-1}$, $v \geq 0$. For most values of r this will be possible, but for certain exceptional values of r this ray will run into a branch point at a finite value of v. This corresponds to the situation where $\tilde{C}$ runs into a saddle point instead of simply passing out to infinity in some valley. This (important!) exceptional case must be dealt with in a slightly more complicated manner. In the non-exceptional case the crucial fact we must prove is that the $\tilde{k}$ image of the above ray in the $\tilde{v}$ plane remains bounded. This allows us to prove global estimates of k as a function of v which are independent of b. All this is done in §1.4.

The relevant saddle points and the new contour of integration can be described as follows. When $b < 0$ there are J relevant saddle points, where $J = \left[\!\left[\frac{r+1}{4}\right]\!\right]$. These are $k_{-j}(b)$, $j = 1, \ldots, J$. When $b > 0$ there are $K + 1$ relevant saddle points, where $K = \left[\!\left[\frac{r-1}{4}\right]\!\right]$. These are $k_j(b)$, $j = 0, 1, \ldots, K$. First one follows $\tilde{C}$ until one ends up at infinity in a valley. (In the special circumstance where $\tilde{C}$ hits a saddle point, after arriving at the saddle point one must follow the SDC proceeding from this saddle point in the general direction of the final valley, which will also end up at infinity in some valley.) Then it may be necessary to follow successively some SDCs, called C_j for certain values of j, through saddle points which allow us to pass from one valley to the next valley closer to the final valley. After a finite number of such SDCs, one will end up at infinity in the final valley. All this takes place on the main sheet (provided $r \geq 3/2$) of the Riemann surface (domain) of the function $\varphi(k)$ (see below). A more precise description, and some pictures, can be found in §1.2.

If $f(z)$ is an analytic function defined on some open disc in the complex z-plane, we denote by $S\{f(z)\}$ the Riemann surface which is the natural domain of the global analytic function associated to f; see Ahlfors [Ahl]. So, for example, as a point set $S\{\ln z\} = \{(z,w) \in \mathbb{C}^2 \mid z = e^w\}$. The covering projection $\pi_1 \colon S\{\ln z\} \to \mathbb{C} \backslash \{0\} \colon (z,w) \mapsto z$ renders $S\{\ln z\}$ as an infinite sheeted covering space. In particular, the map π_1 allows us to give the set $S\{\ln z\}$ the structure of a one dimensional complex manifold. The global natural logarithm function is $\ln \colon S\{\ln z\} \to \mathbb{C} \colon (z,w) \mapsto w$.

Explicit Coefficients in Laplace Expansions. We will finish our discussion of generalities in this section by giving some explicit formulae for the coefficients in the asymptotic expansions of Laplace contour integrals. We will follow very closely the discussion in Olver's book [O1], pages 121–125. Olver uses t for the variable of integration, ranging over a contour connecting the initial point a to the final point b, and z for the variable which tends to infinity. Our only change from Olver's notation will be to use k for the variable of integration, ranging over a contour connecting the initial point l to the final point L, and t for the large parameter. Thus we seek an asymptotic expansion of the integral

$$\int_l^L e^{-p(k)t} q(k)\, dk$$

as $t \to \infty$. $p(k)$ and $q(k)$ are assumed to have known convergent expansions

$$p(k) - p(l) = (k-l)^\mu \sum_{s=0}^\infty p_s(k-l)^s, \qquad q(k) = (k-l)^{\lambda-1}\sum_{s=0}^\infty q_s(k-l)^s,$$

where $p_0 \neq 0$, $\mu > 0$ and $\Re\lambda > 0$. We make all the assumptions that Olver did on pages 121–122; basically, the order of the "saddle point" at $k = l$ is μ, and λ controls the degree of vanishing (or singularity) of $q(k)$ at $k = l$. One defines

$$w = [p(k) - p(l)]^{1/\mu} = p_0^{1/\mu}(k-l)\left[\sum_{s=0}^\infty \frac{p_s}{p_0}(k-l)^s\right]^{1/\mu},$$

where the $1/\mu$ power of the sum is the principal value for $k-l$ small. Thus $k-l$ can be expressed as a convergent power series in w, at least for $|w|$ sufficiently small:

$$k - l = \sum_{s=1}^\infty c_s w^s.$$

Our first order of business is to obtain explicit formulae for the coefficients c_s, $s = 1, 2, \ldots$.

Lemma 1.1.1. *For all $s \geq 1$ we have*

$$c_s = \frac{1}{s p_0^{s/\mu}} \sum_{m=0}^{s-1}\left[\prod_{h=1}^m\left(-\frac{s}{\mu} - h + 1\right)\right]\sum_{(\alpha_1,\ldots,\alpha_s)}\prod_{h=1}^{s-1}\frac{1}{\alpha_h!}\left(\frac{p_h}{p_0}\right)^{\alpha_h},$$

where the (finite) sum is over all s-tuples $(\alpha_1,\ldots,\alpha_s)$ of nonnegative integers satisfying the following restrictions

$$\sum_{h=1}^s \alpha_h = m \qquad\qquad \sum_{h=1}^s h\alpha_h = s-1.$$

Proof. By Lagrange's formula for reversion of series (see page 14 of [AS]) we have

$$k - l = \sum_{s=1}^{\infty} \frac{w^s}{s!} \left(\frac{d}{dk} \right)^{s-1} \left\{ \frac{k-l}{[p(k) - p(l)]^{1/\mu}} \right\}^s \Bigg|_{k=l}.$$

To compute this derivative we use Faà di Bruno's formula (see page 823 of [AS])

$$\frac{1}{n!} \left(\frac{d}{dk} \right)^n f(g(k)) = \sum_{m=0}^{n} f^{(m)}(g(k)) \sum_{(\alpha_1,\ldots,\alpha_{n+1})} \prod_{h=1}^{n} \frac{1}{\alpha_h!} \left(\frac{g^{(h)}(k)}{h!} \right)^{\alpha_h},$$

where the sum is over all $n + 1$-tuples $(\alpha_1, \ldots, \alpha_{n+1})$ of nonnegative integers satisfying the conditions

$$\sum_{h=1}^{n+1} \alpha_h = m \qquad \sum_{h=1}^{n+1} h\alpha_h = n.$$

This formula is valid for all $n \geq 0$. Every $n + 1$-tuple satisfying the above conditions has $\alpha_{n+1} = 0$; however we include it to make clear that when $n = 0$ the sum is not empty—it has one term which equals 1 (an empty product). We will apply Faà di Bruno's formula with $n = s - 1$, $f(u) = u^{-s/\mu}$ and $g(k) = \sum_{j=0}^{\infty} \frac{p_j}{p_0} (k - l)^j$. Computing we get

$$f^{(m)}(u) = u^{-m-s/\mu} \prod_{h=1}^{m} \left(-\frac{s}{\mu} - h + 1 \right),$$

$$\frac{g^{(h)}(k)}{h!} \Bigg|_{k=l} = \frac{p_h}{p_0}, \qquad h \geq 0.$$

Also we have

$$\left\{ \frac{k-l}{[p(k) - p(l)]^{1/\mu}} \right\}^s = \left\{ \frac{p(k) - p(l)}{(k-l)^\mu} \right\}^{-s/\mu} = p_0^{-s/\mu} f(g(k)).$$

Thus we finally have

$$c_s = \frac{1}{s!} \left(\frac{d}{dk} \right)^{s-1} \left\{ \frac{k-l}{[p(k) - p(l)]^{1/\mu}} \right\}^s \Bigg|_{k=l}$$

$$= \frac{1}{s p_0^{s/\mu}} \sum_{m=0}^{s-1} g(l)^{-m-s/\mu} \left[\prod_{h=1}^{m} \left(-\frac{s}{\mu} - h + 1 \right) \right] \sum_{(\alpha_1,\ldots,\alpha_s)} \prod_{h=1}^{s-1} \frac{1}{\alpha_h!} \left(\frac{p_h}{p_0} \right)^{\alpha_h},$$

which yields the desired result. □

Notice that if we define $\tilde{c}_s = p_0^{s/\mu} c_s$ for all $s \geq 1$, then $\tilde{c}_s$ does not depend on the choice of the phase of p_0.

Since $q(k)$ can be expanded in powers of $k - l$ and $k - l$ is a power series in w, we can expand $q(k)$ in powers of w

$$q(k) = w^{\lambda-1} \sum_{s=0}^{\infty} b_s w^s.$$

Our next lemma gives explicit formulae for the coefficients b_s, $s \geq 0$.

Lemma 1.1.2. *For all $s \geq 0$ we have*

$$b_s = \frac{1}{p_0^{(s+\lambda-1)/\mu}} \sum_{m=0}^{s} q_{s-m} \sum_{n=0}^{m} \left[\prod_{h=1}^{n} (s - m + \lambda - h) \right] \sum_{(\alpha_1, \ldots, \alpha_{m+1})} \prod_{j=1}^{m} \frac{\tilde{c}_{j+1}^{\alpha_j}}{\alpha_j!},$$

where the sum is over all $(m+1)$-tuples $(\alpha_1, \ldots, \alpha_{m+1})$ of nonnegative integers satisfying the conditions

$$\sum_{j=1}^{m+1} \alpha_j = n \qquad \sum_{j=1}^{m+1} j\alpha_j = m.$$

In the special case where $\lambda = 1$ the above expression for b_s can be rewritten in a more compact form

$$b_s = \frac{1}{p_0^{s/\mu}} \sum_{m=0}^{s} q_m m! \sum_{(\beta_1, \ldots, \beta_{s+1})} \prod_{j=1}^{s} \frac{\tilde{c}_j^{\beta_j}}{\beta_j!},$$

where the sum is over all $(s+1)$-tuples $(\beta_1, \ldots, \beta_{s+1})$ of nonnegative integers satisfying the conditions

$$\sum_{j=1}^{s+1} \beta_j = m \qquad \sum_{j=1}^{s+1} j\beta_j = s.$$

Proof. Define $\tilde{w} = w/p_0^{1/\mu}$. Then $k - l = \sum_{j=1}^{\infty} \tilde{c}_j \tilde{w}^j = \tilde{w} \sum_{j=0}^{\infty} \tilde{c}_{j+1} \tilde{w}^j$. So

$$q(k) = \sum_{s=0}^{\infty} q_s (k - l)^{s+\lambda-1} = \sum_{s=0}^{\infty} q_s \tilde{w}^{s+\lambda-1} \left(\sum_{j=0}^{\infty} \tilde{c}_{j+1} \tilde{w}^j \right)^{s+\lambda-1}.$$

The $s + \lambda - 1$-power of the analytic function $g(\tilde{w}) = \sum_{j=0}^{\infty} \tilde{c}_{j+1} \tilde{w}^j$ can be expanded using Taylor's theorem

$$g(\tilde{w})^{s+\lambda-1} = \sum_{m=0}^{\infty} \frac{1}{m!} \left(\frac{d}{d\tilde{w}} \right)^m \left[g(\tilde{w})^{s+\lambda-1} \right] \Big|_{\tilde{w}=0} \tilde{w}^m.$$

We compute the derivative using Faà di Bruno's formula:

$$\frac{1}{m!} \left(\frac{d}{d\tilde{w}} \right)^m \left[g(\tilde{w})^{s+\lambda-1} \right]$$

$$= \sum_{n=0}^{m} \left(\frac{d}{du} \right)^n (u^{s+\lambda-1}) \Big|_{u=g(\tilde{w})} \sum_{(\alpha_1, \ldots, \alpha_{m+1})} \prod_{j=1}^{m} \frac{1}{\alpha_j!} \left(\frac{g^{(j)}(\tilde{w})}{j!} \right)^{\alpha_j},$$

where $(\alpha_1, \ldots, \alpha_{m+1})$ must satisfy the conditions stated. We must evaluate this at $\tilde{w} = 0$. Clearly we have $g^{(j)}(0)/j! = \tilde{c}_{j+1}$ for all $j \geq 0$ and

$$\left(\frac{d}{du}\right)^n (u^{s+\lambda-1})\bigg|_{u=g(0)} = \prod_{h=1}^{n} (s + \lambda - h).$$

Thus

$$\frac{1}{m!}\left(\frac{d}{d\tilde{w}}\right)^m [g(\tilde{w})^{s+\lambda-1}]\bigg|_{\tilde{w}=0} = \sum_{n=0}^{m}\left[\prod_{h=1}^{n}(s+\lambda-h)\right] \sum_{(\alpha_1,\ldots,\alpha_{m+1})} \prod_{j=1}^{m} \frac{\tilde{c}_{j+1}^{\alpha_j}}{\alpha_j!}.$$

So

$$q(k) = \sum_{s=0}^{\infty} q_s \tilde{w}^{s+\lambda-1} \sum_{m=0}^{\infty} \tilde{w}^m \sum_{n=0}^{m}\left[\prod_{h=1}^{n}(s+\lambda-h)\right] \sum_{(\alpha_1,\ldots,\alpha_{m+1})} \prod_{j=1}^{m} \frac{\tilde{c}_{j+1}^{\alpha_j}}{\alpha_j!}$$

$$= \tilde{w}^{\lambda-1} \sum_{s=0}^{\infty} \tilde{w}^s \sum_{m=0}^{s} q_{s-m} \sum_{n=0}^{m}\left[\prod_{h=1}^{n}(s-m+\lambda-h)\right] \sum_{(\alpha_1,\ldots,\alpha_{m+1})} \prod_{j=1}^{m} \frac{\tilde{c}_{j+1}^{\alpha_j}}{\alpha_j!}.$$

The result for general λ follows immediately from this. When $\lambda = 1$, one can compute $(k - l)^s$ using the multinomial theorem, instead of having to separate out $\tilde{w}$ and use Taylor's Theorem and Faà di Bruno's formula. One can also show that the two expressions we have given in the case of $\lambda = 1$ are equivalent by noting that $\alpha_j = \beta_{j+1}$ for $1 \leq j \leq m$. (One needs to replace m by $s - m$ in the second expression before attempting a comparison.) $\square$

Notice again that if we define $\tilde{b}_s = p_0^{(s+\lambda-1)/\mu} b_s$ for all $s \geq 0$ then $\tilde{b}_s$ does not depend on a choice of the phase of p_0.

The final step is to introduce the variable $v = w^\mu$ and change variables in the integral to be expanded

$$\int_l^L e^{-p(k)t} q(k)\, dk = e^{-p(l)t} \int_0^V e^{-vt} q(k) \frac{dk}{dw}\frac{dw}{dv}\, dv.$$

Define $f(v) = q(k)\frac{dk}{dw}\frac{dw}{dv} = \frac{1}{\mu} q(k)\frac{dk}{dw} v^{\frac{1}{\mu}-1}$. The expansion is obtained from Watson's lemma as soon as one has $f(v)$ expanded in powers of v

$$f(v) = v^{\frac{\lambda}{\mu}-1} \sum_{s=0}^{\infty} a_s v^{s/\mu}.$$

It is comparatively simple now to get formulae for a_s, $s \geq 0$.

Lemma 1.1.3. *For all $s \geq 0$ we have*

$$a_s = \frac{1}{\mu} \sum_{n=0}^{s} (s - n + 1) b_n c_{s-n+1}.$$

Proof. Differentiate the series for $k - l$ term-by-term with respect to w and multiply by the series for $q(k)$ to obtain the result. $\square$

Notice that

$$a_s = \frac{1}{\mu} \sum_{n=0}^{s} (s-n+1) \frac{\tilde{b}_n}{p_0^{(n+\lambda-1)/\mu}} \frac{\tilde{c}_{s-n+1}}{p_0^{(s-n+1)/\mu}} = \frac{1}{\mu p_0^{(s+\lambda)/\mu}} \sum_{n=0}^{s} (s-n+1) \tilde{b}_n \tilde{c}_{s-n+1}.$$

So we define $\tilde{a}_s = p_0^{(s+\lambda)/\mu} a_s$ for all $s \geq 0$. This quantity does not depend on the choice of argument for p_0.

The final result, as Olver states and proves, is the asymptotic expansion

$$\int_l^L e^{-p(k)t} q(k)\, dk \sim e^{-p(l)t} \sum_{s=0}^{\infty} \Gamma\left(\frac{s+\lambda}{\mu}\right) \frac{a_s}{t^{(s+\lambda)/\mu}}$$

$$= e^{-p(l)t} \sum_{s=0}^{\infty} \Gamma\left(\frac{s+\lambda}{\mu}\right) \frac{\tilde{a}_s}{(p_0 t)^{(s+\lambda)/\mu}}.$$

From our point of view, the last expression is preferable, since the effect of a choice of the phase of $p_0 t$ is immediately evident.

1.2 Saddle points and contours

The goal of this section is to describe in detail the new contour of integration which is homotopic (on the Riemann sphere) to the original contour $[0, \infty)$. A consequence of our detailed description is a proof of the fact that the new contour can always be constructed so as to consist entirely of steepest descent contours (SDCs) passing through certain saddle points and so as to be homotopic to the original contour. The details of the contour depend strongly on whether $b < 0$ (case A) or $b > 0$ (case B), although there are many parallels and similarities between the two cases. We will treat the two cases together, giving complete proofs for the case $b < 0$; the proofs for $b > 0$ are extremely similar and are left to the reader. An attempt to present unified statements or proofs would make everything obscure.

Recall $J = \lceil \frac{r+1}{4} \rceil$ and $K = \lceil \frac{r-1}{4} \rceil$. So $4J - 1 \leq r < 4J + 3$ and $4K + 1 \leq r < 4K + 5$. J (resp. $K + 1$) will turn out to be the number of saddle points relevant to the long-time asymptotics when $b < 0$ (resp. $b > 0$). Recall we always assume $\rho > 0$ and $r > 1$; so $J, K \geq 0$. First we will discuss the SDC emanating from $k = 0$, which we have named $\tilde{C}$.

Lemma 1.2.1A. *Suppose $b < 0$. Then $\tilde{C}$ is described as follows.*

(1) $r = 4J - 1$. *The ray $k = -iR$, $0 \le R \le B^{\frac{1}{r-1}}$, is the SDC starting at $k = 0$ and ending at the saddle point k_{-J}.*

(2) $4J - 1 < r < 4J + 1$. *The SDC starting at $k = 0$ (when $\theta = -\pi/2$) and ending at infinity (as $\theta \to V^+_{-J}$) has the polar equation*

$$R = B^{\frac{1}{r-1}} \left[\frac{-r\cos(\theta)}{\cos(r\theta)} \right]^{\frac{1}{r-1}}, \qquad V_{-J} < \theta \le -\frac{\pi}{2}.$$

(3) $r = 4J + 1$. *The ray $k = -iR$, $0 \le R < \infty$, is the SDC starting at $k = 0$ and ending at infinity in the valley V_{-J}.*

(4) $4J + 1 < r < 4J + 3$. *The SDC starting at $k = 0$ (when $\theta = -\pi/2$) and ending at infinity (as $\theta \to V^-_{-J}$) has the polar equation*

$$R = B^{\frac{1}{r-1}} \left[\frac{-r\cos(\theta)}{\cos(r\theta)} \right]^{\frac{1}{r-1}}, \qquad -\frac{\pi}{2} \le \theta < V_{-J}.$$

Proof. (1) If $\theta = -\pi/2$ then $\cos(\theta) = 0$ and $\cos(r\theta) = \cos((4J-1)(-\pi/2)) = 0$. So this ray is a level curve for $\Im\varphi(k)$. The saddle point $k_{-J} = B^{\frac{1}{r-1}} e^{i\pi \frac{(-2J+1)}{4J-2}}$ is on this ray. Also

$$\varphi(k_{-J}) = -(r-1)\rho B^{\frac{r}{r-1}} < 0.$$

Since $\Re\varphi(k)$ decreases as we move along the ray from $k = 0$ to the saddle point, we see that this segment of the ray is a SDC.

(2) First we must check that the quotient $b\cos(\theta)/(\rho\cos(r\theta))$ is nonnegative when θ is in the indicated interval. Since $1 < r < 4J + 1$ we must have $J \ge 1$. Therefore

$$\frac{4J+1}{r} < \frac{4J+1}{4J-1} \le \frac{5}{3},$$

so $-5\pi/6 < -\pi(4J+1)/(2r) = V_{-J} < \theta \le -\pi/2$. Therefore $b\cos(\theta) \ge 0$. Also since $r > 4J - 1$ we have $-\pi r/2 < -\pi(4J-1)/2 = -2J\pi + \pi/2$. Therefore

$$-2J\pi - \frac{\pi}{2} = -\frac{\pi(4J+1)}{2} < r\theta \le -\frac{\pi}{2}r < -2J\pi + \frac{\pi}{2},$$

and hence $\rho\cos(r\theta) > 0$. Thus the expression for R makes sense. Clearly $R = 0$ when $\theta = -\pi/2$ and $R \to \infty$ as $\theta \to V^+_{-J}$. This curve is a level curve of $\Im\varphi(k)$. It tends to infinity in the valley V_{-J}. Thus it is a SDC.

(3) As in (1) it is easy to check that the entire ray is a level curve for $\Im\varphi(k)$. It tends to infinity in the valley V_{-J}, where $V_{-J} = -\frac{\pi}{2}$. So it is a SDC.

(4) Clearly $b\cos(\theta) \le 0$. Since $r < 4J + 3$ we have $-2J\pi - 3\pi/2 = -\pi(4J+3)/2 < -\pi r/2$. Thus

$$-2J\pi - \frac{3\pi}{2} < -\frac{\pi r}{2} \le r\theta < -\frac{\pi}{2}(4J+1) = -2J\pi - \frac{\pi}{2},$$

and thus $\rho\cos(r\theta) < 0$. Thus the expression for R makes sense. The rest of the argument proceeds as in (2). $\square$

Lemma 1.2.1B. *Suppose $b > 0$. Then $\tilde{C}$ is described as follows.*

(1) $r = 4K + 1$. *The ray $k = iR$, $0 \leq R \leq B^{\frac{1}{r-1}}$, is the SDC starting at $k = 0$ and ending at the saddle point k_K.*

(2) $4K + 1 < r < 4K + 3$. *The SDC starting at $k = 0$ (when $\theta = \pi/2$) and ending at infinity (as $\theta \to V_{K+1}^-$) has the polar equation*

$$R = B^{\frac{1}{r-1}} \left[\frac{r \cos(\theta)}{\cos(r\theta)} \right]^{\frac{1}{r-1}}, \qquad \frac{\pi}{2} \leq \theta < V_{K+1}.$$

(3) $r = 4K + 3$. *The ray $k = iR$, $0 \leq R < \infty$, is the SDC starting at $k = 0$ and ending at infinity in the valley V_{K+1}.*

(4) $4K + 3 < r < 4K + 5$. *The SDC starting at $k = 0$ (when $\theta = \pi/2$) and ending at infinity (as $\theta \to V_{K+1}^+$) has the polar equation*

$$R = B^{\frac{1}{r-1}} \left[\frac{r \cos(\theta)}{\cos(r\theta)} \right]^{\frac{1}{r-1}}, \qquad V_{K+1} < \theta \leq \frac{\pi}{2}.$$

The reader should consult the end of this section where some figures showing examples of the SDCs are given. The contour $\tilde{C}$ is labeled "Ct" in those figures. The centers of the hills and valleys are drawn as dashed lines, and labeled in an obvious manner.

In the special cases $r = 4J - 1$ or $r = 4K + 1$, $\tilde{C}$ hits a saddle point, either k_{-J} or k_K. If one were to continue on the ray one would find oneself going up the center of the hill $H_{-J} = -\frac{\pi}{2}$ or $H_K = \frac{\pi}{2}$. Instead we must take one branch of the SDC through the saddle point. The behavior of this SDC, given the name C_{-J}^+ or C_K^+, is described in the next lemma.

Lemma 1.2.2A. *Suppose $b < 0$ and $r = 4J - 1$, where $J \geq 1$. Then the SDC C_{-J}^+, starting at the saddle point k_{-J} and departing tangent to the ray $k - k_{-J} = Re^{i\omega}$ with $\omega = \frac{1}{2}S_{-J}^1 + \frac{\pi}{4} = 0$ (this takes us in the general direction of the final valley V_0) has the polar equation*

$$R = B^{\frac{1}{r-1}} \left[\frac{-r \cos(\theta)}{\cos(r\theta)} \right]^{\frac{1}{r-1}}, \qquad -\frac{\pi}{2} < \theta < V_{-J+1}.$$

We have $R \to B^{1/(r-1)}$ as $\theta \to -\pi/2^+$ and $R \to \infty$ as $\theta \to V_{-J+1}^-$.

Proof. Clearly $b \cos(\theta) \leq 0$. But since $r = 4J - 1$ we have

$$-2J\pi + \frac{\pi}{2} = -\frac{\pi}{2}(4J - 1) < r\theta < -\frac{\pi}{2}(4J - 3) = -2J\pi + \frac{3\pi}{2},$$

and thus $\rho \cos(r\theta) < 0$. Thus the expression for R makes sense. The limit as $\theta \to -\pi/2^+$ follows from L'Hospital's rule, and the other limit is obvious. This is a level curve of $\Im\varphi(k)$ which tends to infinity in the valley V_{-J+1}. Thus it is a SDC. To show that it leaves the saddle point tangent to the indicated ray one expands

$$R\left(\alpha - \frac{\pi}{2}\right) = B^{\frac{1}{4J-2}} \left[\frac{r \sin(\alpha)}{\sin(r\alpha)} \right]^{\frac{1}{r-1}} = B^{\frac{1}{4J-2}} \left[1 + \frac{2J}{3}\alpha^2 + O(\alpha^3) \right]. \qquad \square$$

Lemma 1.2.2B. *Suppose $b > 0$ and $r = 4K + 1$, where $K \geq 1$. Then the SDC C_K^+, starting at the saddle point k_K and departing tangent to the ray $k - k_K = Re^{i\omega}$ with $\omega = \frac{1}{2}S_K^0 - \frac{\pi}{4} = 0$ (this takes us in the general direction of the final valley V_0) has the polar equation*

$$R = B^{\frac{1}{r-1}} \left[\frac{r\cos(\theta)}{\cos(r\theta)} \right]^{\frac{1}{r-1}}, \qquad V_K < \theta < \frac{\pi}{2}.$$

We have $R \to B^{1/(r-1)}$ as $\theta \to \pi/2^-$ and $R \to \infty$ as $\theta \to V_K^+$.

Examples of these SDCs when $r = 3, b < 0$, or $r = 5, b > 0$, can be seen illustrated in the figures at the end of this section. C_{-1}^+ is labeled "Cm1+", and C_1^+ is labeled "C1+" in those figures.

Thus we always eventually end up at infinity in a valley with asymptotic center

$$\tilde{V} \stackrel{\text{def}}{=} \begin{cases} \begin{cases} V_{-J+1} & r = 4J - 1, \\ V_{-J} & 4J - 1 < r < 4J + 3, \end{cases} & b < 0, \\ \begin{cases} V_K & r = 4K + 1, \\ V_{K+1} & 4K + 1 < r < 4K + 5, \end{cases} & b > 0. \end{cases}$$

If $1 < r \leq 3$ (when $b < 0$) then this is the final valley, and therefore the entire new contour of integration has been found already. When $r > 3$ (when $b < 0$) or $r > 1$ (when $b > 0$), we must pass from the valley we are in to the final valley. So we must study the terrain which intervenes between these two valleys. As we have already seen, the asymptotic centers of valleys and hills alternate as we procede clockwise around $k = 0$, i.e. $\cdots > V_2 > H_1 > V_1 > H_0 > V_0 > H_{-1} > V_{-1} > H_{-2} > \ldots$ etc.. The asymptotic centers of an adjacent valley and hill are separated by π/r radians, hence two successive valleys are separated by $2\pi/r$ radians. The most natural and in some sense canonical way to ascend out of any valley is to follow a path of steepest ascent. However, unless such a path hits a saddle point it will continue uphill forever, tending to infinity along the asymptotic center of some hill. The integral along such a contour is obviously divergent on one end. So we want to insure that a saddle point is positioned appropriately so that our path can pass through it and then continue downhill into the valley on the other side. It is frequently possible (on certain portions of $S\{\ln k\}$, the Riemann surface for $\ln(k)$) that two successive valleys need not be connected by steepest ascent/descent curves in such a simple way. But our next lemma will show that the saddle points are positioned (in regard to angle) appropriately in the particular area we must traverse.

Lemma 1.2.3.

A) *Suppose $b < 0$. In the sectoral region (with vertex at $k = 0$) $H_{-J-1} \leq \theta \leq H_0$ on $S\{\ln k\}$ there are exactly J saddle points $k_{-1}, k_{-2}, \ldots, k_{-J}$. For $J \geq 1$ these can be more precisely located: for $j = 1, 2, \ldots, J$ we have $H_{-j} \leq S_{-j}^1 < V_{-j+1}$. In the first inequality, equality holds if and only if $r = 4J - 1$ and $j = J$.*

B) *Suppose $b > 0$. In the sectoral region (with vertex at $k = 0$) $H_{-1} \leq \theta \leq H_{K+1}$ on $S\{\ln k\}$ there are exactly $K + 1$ saddle points $k_0, k_1, \ldots, k_K$. These can be more precisely located: for $j = 0, 1, \ldots, K$ we have $V_j < S_j^0 \leq H_j$. In the second inequality, equality holds if and only if $r = 4K+1$ and $j = K$.*

Remarks.

(1) The sectoral region in the preceding lemma always contains $-\pi/2 \leq \theta \leq 0$ (when $b < 0$) or $0 \leq \theta \leq \pi/2$ (when $b > 0$) as a subsector. This subsector always contains all the relevant saddle points.

(2) The Riemann surface $S\{\ln k\}$ is an infinite sheeted covering space of the base space $\mathbb{C}^* = \mathbb{C} \setminus \{0\}$. It is important to realize that the saddle points live in this covering space, and not in the base space. For example, when r is irrational, the same sector in the base space contains the images of infinitely many saddle points under the covering projection.

Proof of A). Because the saddle points are indexed by increasing argument, the first assertion will follow if we show that $S_{-J-1}^1 < H_{-J-1}$ and $H_0 < S_0^0$. The first inequality $-\frac{\pi}{r-1}(2J + 1) < -\frac{\pi}{r}(2J + \frac{3}{2})$ follows since $r < 4J + 3$. The second, $\frac{\pi}{2r} < \frac{\pi}{r-1}$, is obviously true.

Now suppose $1 \leq j \leq J$. The first inequality, $H_{-j} \leq S_{-j}^1$, i.e. $\frac{\pi}{r}(\frac{1}{2} - 2j) \leq \frac{\pi}{r-1}(-2j + 1)$ is equivalent to $4j - 1 \leq r$. Clearly this holds. Furthermore, equality holds if and only if $r = 4J - 1$ and $j = J$. The second inequality, $S_{-j}^1 < V_{-j+1}$, i.e. $\frac{\pi}{r-1}(-2j+1) < \frac{\pi}{r}(\frac{3}{2} - 2j)$, is equivalent to $-4j + 3 < r$, which is always true since $j \geq 1$. $\square$

So, for example, when $3 \leq r < 5$, i.e. $J = 1$ and $K = 0$, we have

$$V_{-1} < H_{-1} \leq S_{-1}^1 < V_0 < S_0^0 < H_0 < V_1,$$

where equality holds in the second inequality if and only if $r = 3$. When $r = 3$ and $b < 0$, $\tilde{C}$ starts at $k = 0$ and hits k_{-1}; then C_{-1}^+ passes to infinity in V_0 (the final valley). When $3 < r < 5$ and $b < 0$, $\tilde{C}$ tends to infinity in V_{-1}; another path (namely C_{-1}) out of V_{-1}, through k_{-1}, and out into V_0 is necessary. When $3 \leq r < 5$ and $b > 0$, $\tilde{C}$ starts at $k = 0$ and passes to infinity in V_1; another path (namely C_0) is needed to get from V_1, through k_0, and out into V_0.

Our next task is to describe these additional paths in general. They go by the names C_j, $-J \leq j \leq K$. This path C_j, approximately speaking, lies in the sector $V_j \leq \theta \leq V_{j+1}$, and passes through the saddle point $k_j(b)$. It breaks naturally (from the standpoint of simplicity of description) into two segments: the segment where $\theta < H_j$ and the segment where $\theta > H_j$. These segments will be treated separately in the following four lemmas.

Lemma 1.2.4A. *Suppose $b < 0$ and $1 \leq j \leq J$ (if $r = 4J - 1$ suppose $1 \leq j \leq J - 1$). There are two forms that the segment $\theta < H_{-j}$ can take, depending on whether $V_{-j} \leq -\pi/2$ or $V_{-j} > -\pi/2$. If $V_{-j} \leq -\pi/2$ then the entire segment is contained in the sector $V_{-j} < \theta < H_{-j}$, and is described in* *more detail in (2) below. If $V_{-j} > -\pi/2$ then the segment "overshoots" the sector $V_{-j} < \theta < H_{-j}$, i.e. there is a small positive number ϵ_j such that the segment is contained in the slightly larger sector $V_{-j} - \epsilon_j \leq \theta < H_{-j}$. The "overshoot", i.e. that part in the sector $V_{-j} - \epsilon_j \leq \theta \leq V_{-j}$, is described in (1) below. The main part contained in the sector $V_{-j} \leq \theta < H_{-j}$ is described in (2) below. We now give the sectoral description of this segment.*

(1) *Suppose $V_{-j} > -\pi/2$ and define $\tilde{\epsilon}_j = \frac{1}{r}\sin^{-1}\left[\frac{\cos^r(V_{-j})}{\cos^{r-1}(S^1_{-j})}\right]$. Then $V_{-j} - \tilde{\epsilon}_j > -\pi/2$ and there exists a unique number $\epsilon_j \in (0, \tilde{\epsilon}_j)$ such that*

$$\epsilon_j = \frac{1}{r}\sin^{-1}\left[\frac{\cos^r(V_{-j} - \epsilon_j)}{\cos^{r-1}(S^1_{-j})}\right].$$

If $V_{-j} - \epsilon_j \leq \theta < V_{-j}$ define $R_c(\theta) = \left[\frac{B\cos(\theta)}{-\cos(r\theta)}\right]^{\frac{1}{r-1}}$; this is always well-defined and positive; $R_c(\theta) \to \infty$ as $\theta \to V^-_{-j}$. Then for θ in the interval $V_{-j} - \epsilon_j < \theta < V_{-j}$ there exists exactly one solution $R = R_1(\theta)$ of the equation

(*) $$BR\cos(\theta) + \frac{1}{r}R^r\cos(r\theta) = \frac{r-1}{r}B^{\frac{r}{r-1}}\cos(S^1_{-j})$$

in the interval $0 < R < R_c(\theta)$, and there exists exactly one solution $R = R_2(\theta)$ of the equation (*) in the interval $R_c(\theta) < R < \infty$. These solutions are the only two positive solutions of (*) and are continuously differentiable functions of θ and have the properties:

$$\lim_{\theta \to (V_{-j}-\epsilon_j)^+} R_1(\theta) = \lim_{\theta \to (V_{-j}-\epsilon_j)^+} R_2(\theta) = R_c(V_{-j}-\epsilon_j) = B^{\frac{1}{r-1}}\frac{\cos(S^1_{-j})}{\cos(V_{-j}-\epsilon_j)},$$

$$\lim_{\theta \to V^-_{-j}} R_1(\theta) = \frac{r-1}{r}B^{\frac{1}{r-1}}\frac{\cos(S^1_{-j})}{\cos(V_{-j})}, \qquad \lim_{\theta \to V^-_{-j}} R_2(\theta) = \infty.$$

When $\theta = V_{-j} - \epsilon_j$ there is exactly one positive solution of (*), namely $R = R_c(V_{-j} - \epsilon_j)$. In a sufficiently small neighborhood of the solution

$$(R, \theta) = (R_c(V_{-j}-\epsilon_j), V_{-j}-\epsilon_j)$$

of (*), the set of all solutions (R, θ) of (*) can be described as the graph of a continuously differentiable function $\theta(R)$, defined for all R sufficiently near $R_c(V_{-j} - \epsilon_j)$, and which has a single isolated minimum value at $R = R_c(V_{-j} - \epsilon_j)$.

(2) *If $V_{-j} < \theta < H_{-j}$ then there exists exactly one positive solution $R = R(\theta)$ of equation (*). $R(\theta)$ is a continuously differentiable function of θ and has the properties*

$$\lim_{\theta \to V^+_{-j}} R(\theta) = \begin{cases} \infty & \text{if } V_{-j} \leq -\pi/2, \\ \frac{r-1}{r}B^{\frac{1}{r-1}}\frac{\cos(S^1_{-j})}{\cos(V_{-j})} & \text{if } V_{-j} > -\pi/2, \end{cases}$$

$$\lim_{\theta \to H^-_{-j}} R(\theta) = \frac{r-1}{r}B^{\frac{1}{r-1}}\frac{\cos(S^1_{-j})}{\cos(H_{-j})}.$$

Proof. (1) Define

$$f(R; \theta, B, r, j) = BR\cos(\theta) + \frac{1}{r}R^r\cos(r\theta) - \frac{r-1}{r}B^{\frac{r}{r-1}}\cos(S^1_{-j}).$$

$f'(R) = B\cos(\theta) + R^{r-1}\cos(r\theta)$, so $f'(R) = 0$ if and only if $R^{r-1} = \frac{B\cos(\theta)}{-\cos(r\theta)}$.
Assume $\max\{H_{-j-1}, -\frac{\pi}{2}\} < \theta < V_{-j}$. Then $B\cos(\theta) > 0$. Also since $\frac{\pi}{r}(-\frac{3}{2} - 2j) < \theta < \frac{\pi}{r}(-\frac{1}{2} - 2j)$ we have $-\cos(r\theta) > 0$. Thus $R_c(\theta)$ is well-defined and
positive. So, for θ in this range, $f(R)$ is increasing for $R \in (0, R_c(\theta))$ (since
$f'(0) > 0$), and decreasing for $R \in (R_c(\theta), \infty)$. By assumption and by Lemma
1.2.3 we have $-\frac{\pi}{2} < V_{-j} < H_{-j} < S^1_{-j}$. Thus $f(0) = -\frac{r-1}{r}B^{\frac{r}{r-1}}\cos(S^1_{-j}) < 0$. Also $\lim_{R\to\infty} f(R) = -\infty$. Hence the equation $f(R) = 0$ will have 2, 1,
or 0 positive solutions according to whether $f(R_c(\theta)) > 0$, $f(R_c(\theta)) = 0$, or
$f(R_c(\theta)) < 0$. A short computation shows that

$$f(R_c(\theta)) = \frac{r-1}{r}B^{\frac{r}{r-1}}\left\{\left[\frac{\cos^r(\theta)}{-\cos(r\theta)}\right]^{\frac{1}{r-1}} - \cos(S^1_{-j})\right\}.$$

For $\theta < V_{-j}$ and sufficiently near to V_{-j} we have $f(R_c(\theta)) > 0$. We must show
that this is the case for all $\theta \in (V_{-j} - \epsilon_j, V_{-j})$ and that $f(R_c(\theta)) = 0$ when
$\theta = V_{-j} - \epsilon_j$. The equation to be satisfied by ϵ_j is therefore

$$\left[\frac{\cos^r(V_{-j} - \epsilon_j)}{-\cos(rV_{-j} - r\epsilon_j)}\right]^{\frac{1}{r-1}} = \cos(S^1_{-j}).$$

Since $-\cos(rV_{-j} - r\epsilon_j) = -\cos(-\frac{\pi}{2} - 2j\pi - r\epsilon_j) = \sin(r\epsilon_j)$, we can rearrange
the equation for ϵ_j to be in the form

$$\frac{\cos^r(V_{-j} - \epsilon_j)}{\sin(r\epsilon_j)} = \cos^{r-1}(S^1_{-j}).$$

From the definition of $\tilde{\epsilon}_j$ we get the estimates $\tilde{\epsilon}_j \leq \frac{\pi}{2r} < \frac{\pi}{2(r-1)}(1 + \frac{5}{r}) \leq \frac{1}{r-1}(\frac{\pi}{2} - V_{-j})$ (since $V_{-j} \leq V_{-1} = -\frac{5\pi}{2r}$) and $\tilde{\epsilon}_j \leq \frac{\pi}{2r}(V_{-j} + \frac{\pi}{2}) < V_{-j} + \frac{\pi}{2}$. The
last inequality follows since $V_{-j} > -\pi/2$ is only possible if $r > 5$. Thus the
function

$$g(\epsilon) = \frac{\cos^r(V_{-j} - \epsilon)}{\sin(r\epsilon)}$$

is well-defined for $\epsilon \in (0, \tilde{\epsilon}_j)$. Furthermore our estimates on $\tilde{\epsilon}_j$ imply

$$g'(\epsilon) = -r\frac{\cos^{r-1}(V_{-j} - \epsilon)\cos(V_{-j} + (r-1)\epsilon)}{\sin^2(r\epsilon)} < 0.$$

Thus $g(\epsilon)$ is decreasing on the interval $(0, \tilde{\epsilon}_j)$. Clearly

$$g(\tilde{\epsilon}_j) < \frac{\cos^r(V_{-j})}{\sin(r\tilde{\epsilon}_j)} = \cos^{r-1}(S^1_{-j})$$

and $\lim_{\epsilon \to 0+} g(\epsilon) = \infty$. So there exists a unique $\epsilon_j \in (0, \tilde{\epsilon}_j)$ such that $g(\epsilon_j) = \cos^{r-1}(S^1_{-j})$. Clearly $g(\epsilon) > \cos^{r-1}(S^1_{-j})$ for all $\epsilon \in (0, \epsilon_j)$. This implies that $f(R_c(\theta)) > 0$ for all $\theta \in (V_{-j} - \epsilon_j, V_{-j})$, as we wanted to show. Thus the solutions $R_1(\theta)$ and $R_2(\theta)$ exist by the intermediate value theorem and are unique by monotonicity. These are continuously differentiable functions of θ on this interval by the implicit function theorem, which applies since $f'(R_i(\theta)) \neq 0$. When $\theta = V_{-j}$ the equation (*) has the exact solution $R_1 = \frac{r-1}{r} B^{\frac{1}{r-1}} \frac{\cos(S^1_{-j})}{\cos(V_{-j})}$. Using the implicit function theorem we can obtain R as a continuously differentiable function of θ defined in a neighborhood of $\theta = V_{-j}$ ($f'(R_1) = B\cos(V_{-j}) \neq 0$). By uniqueness, this function must agree with $R_1(\theta)$ for all $\theta < V_{-j}$ in that neighborhood. Thus $\lim_{\theta \to V_{-j}^-} R_1(\theta) = R_1$. Since $R_2(\theta) > R_c(\theta)$ and $R_c(\theta) \to \infty$ as $\theta \to V_{-j}^-$, we have $\lim_{\theta \to V_{-j}^-} R_2(\theta) = \infty$.

When $\theta = V_{-j} - \epsilon_j$ the equation (*) has a unique positive solution $R = R_c(V_{-j} - \epsilon_j)$. Since, as one can check,

$$\frac{\partial f}{\partial \theta}(R_c(V_{-j} - \epsilon_j); V_{-j} - \epsilon_j) = B^{\frac{r}{r-1}} \frac{\cos(S^1_{-j}) \cos(V_{-j} + (r-1)\epsilon_j)}{\sin(r\epsilon_j) \cos(V_{-j} - \epsilon_j)} \neq 0,$$

we can obtain θ as a continuously differentiable function of R defined in a neighborhood of $|R - R_c(V_{-j} - \epsilon_j)| < \delta_0$ using the implicit function theorem. $\theta(R) \geq V_{-j} - \epsilon_j$ for all R in this neighborhood, since the equation $f(R; \theta) = 0$ has no solutions for $\theta \in (V_{-j} - \tilde{\epsilon}_j, V_{-j} - \epsilon_j)$, since $f(R_c(\theta)) < 0$ for such θ. Also $\theta(R) > V_{-j} - \epsilon_j$ for all $0 < |R - R_c(V_{-j} - \epsilon_j)| < \delta_0$. Now let $0 < \delta < \delta_0$ be given. We want to find $\epsilon_0 > 0$ such that if $V_{-j} - \epsilon_j < \theta < V_{-j} - \epsilon_j + \epsilon_0$ then both $R_1(\theta)$ and $R_2(\theta)$ are within δ of $R_c(V_{-j} - \epsilon_j)$. Since $\theta(R)$ is a continuous function on the interval $|R - R_c(V_{-j} - \epsilon_j)| \leq \delta$ we can define

$$\theta_1 = \sup_{-\delta \leq R - R_c(V_{-j} - \epsilon_j) \leq 0} \theta(R), \qquad \theta_2 = \sup_{0 \leq R - R_c(V_{-j} - \epsilon_j) \leq \delta} \theta(R).$$

Define $\epsilon_0 = \min\{\theta_1, \theta_2\} - (V_{-j} - \epsilon_j)$. If $V_{-j} - \epsilon_j < \theta < V_{-j} - \epsilon_j + \epsilon_0$ then by the Intermediate Value Theorem there exist numbers $0 < R_1 < R_c(V_{-j} - \epsilon_j) < R_2$, both R_1, R_2 are within δ of $R_c(V_{-j} - \epsilon_j)$, such that $\theta(R_1) = \theta(R_2) = \theta$. Hence both (R_1, θ), and (R_2, θ) are solutions of (*), and $R_1 < R_2$. By uniqueness we must have $R_1 = R_1(\theta)$ and $R_2 = R_2(\theta)$. Thus we have shown that both $R_1(\theta)$ and $R_2(\theta)$ tend to $R_c(V_{-j} - \epsilon_j)$ as $\theta \to (V_{-j} - \epsilon_j)^+$.

(2) Now suppose $V_{-j} < \theta < H_{-j}$, and let $f(R; \theta)$ and $R_c(\theta)$ be defined as in (1). Since $\pi(-\frac{1}{2} - 2j) < r\theta < \pi(\frac{1}{2} - 2j)$ we have $\cos(r\theta) > 0$. Thus $\lim_{R \to \infty} f(R) = \infty$. Also $R_c(\theta)$ is well-defined if $\theta \leq -\pi/2$ and is not a positive number if $\theta > -\pi/2$. Since $f'(0) = B\cos(\theta)$, this means that $f(R; \theta)$ is monotone increasing for $R \in (0, \infty)$ if $\theta > -\pi/2$. The same thing is true for $\theta = -\pi/2$, since $R_c(-\pi/2) = 0$. For $\theta < -\pi/2$, $f(R; b)$ is decreasing on $(0, R_c(\theta))$ and increasing on $(R_c(\theta), \infty)$. As in (1), $f(0) < 0$. Thus in all cases the intermediate value theorem implies that there exists exactly one positive solution $R(\theta)$ of equation (*). When $\theta \leq -\pi/2$ we know that $R_c(\theta) < R(\theta)$. By the implicit

function theorem $R(\theta)$ is a continuously differentiable function of θ. To evaluate the limit of $R(\theta)$ as $\theta \to V_{-j}^{+}$, we must consider three cases:

a) $V_{-j} < -\pi/2$. Since $R_c(\theta) < R(\theta)$ and $R_c(\theta) \to \infty$ as $\theta \to V_{-j}^{+}$, we have that $R(\theta) \to \infty$ as well.

b) $V_{-j} = -\pi/2$. Since both $B\cos(\theta)$ and $\frac{1}{r}\cos(r\theta)$ tend to zero (through positive values) as $\theta \to V_{-j}^{+}$, we must have that $R(\theta) \to \infty$.

c) $V_{-j} > -\pi/2$. We note that equation (*) has the exact solution $\theta = V_{-j}$, $R = \frac{r-1}{r}B^{\frac{1}{r-1}}\frac{\cos(S_{-j}^{1})}{\cos(V_{-j})}$. So by the implicit function theorem, there is a continuously differentiable function $\tilde{R}(\theta)$, defined for θ in a neighborhood of V_{-j}, such that $f(\tilde{R}(\theta);\theta) = 0$ for such θ. By uniqueness, we must have $\tilde{R}(\theta) = R(\theta)$ for $\theta > V_{-j}$ and sufficiently near V_{-j}. The desired limit now follows from the continuity of $\tilde{R}(\theta)$.

An argument very similar to c) above proves the limit as $\theta \to H_{-j}^{-}$. □

Lemma 1.2.4B. *Suppose $b > 0$ and $0 \le j \le K$ (if $r = 4K + 1$ suppose $0 \le j \le K - 1$). There are two forms that the segment $\theta > H_j$ can take, depending on whether $V_{j+1} \ge \pi/2$ or $V_{j+1} < \pi/2$. If $V_{j+1} \ge \pi/2$ then the entire segment is contained in the sector $H_j < \theta < V_{j+1}$, and is described in more detail in (2) below. If $V_{j+1} < \pi/2$ then the segment "overshoots" the sector $H_j < \theta < V_{j+1}$, i.e. there is a small positive number ϵ_j such that the segment is contained in the slightly larger sector $H_j < \theta \le V_{j+1} + \epsilon_j$. The "overshoot", i.e. that part in the sector $V_{j+1} \le \theta \le V_{j+1} + \epsilon_j$, is described in (1) below. The main part contained in the sector $H_j < \theta \le V_{j+1}$ is described in (2) below. We now give the sectoral description of this segment.*

(1) *Suppose $V_{j+1} < \pi/2$ and define $\tilde{\epsilon}_j = \frac{1}{r}\sin^{-1}\left[\frac{\cos^r(V_{j+1})}{\cos^{r-1}(S_j^0)}\right]$. Then $V_{j+1} + \tilde{\epsilon}_j < \pi/2$ and there exists a unique number $\epsilon_j \in (0, \tilde{\epsilon}_j)$ such that*

$$\epsilon_j = \frac{1}{r}\sin^{-1}\left[\frac{\cos^r(V_{j+1} + \epsilon_j)}{\cos^{r-1}(S_j^0)}\right].$$

If $V_{j+1} < \theta \le V_{j+1} + \epsilon_j$ define $R_c(\theta) = \left[\frac{B\cos(\theta)}{\cos(r\theta)}\right]^{\frac{1}{r-1}}$; this is always well-defined and positive; $R_c(\theta) \to \infty$ as $\theta \to V_{j+1}^{+}$. Then for θ in the interval $V_{j+1} < \theta < V_{j+1} + \epsilon_j$ there exists exactly one solution $R = R_1(\theta)$ of the equation

(**)
$$BR\cos(\theta) - \frac{1}{r}R^r\cos(r\theta) = \frac{r-1}{r}B^{\frac{r}{r-1}}\cos(S_j^0)$$

in the interval $0 < R < R_c(\theta)$, and there exists exactly one solution $R = R_2(\theta)$ of the equation (**) in the interval $R_c(\theta) < R < \infty$. These solutions are the only two positive solutions of (**) and are continuously *differentiable functions of θ and have the properties:*

$$\lim_{\theta \to (V_{j+1}+\epsilon_j)^-} R_1(\theta) = \lim_{\theta \to (V_{j+1}+\epsilon_j)^-} R_2(\theta) = R_c(V_{j+1}+\epsilon_j) = \frac{B^{\frac{1}{r-1}}\cos(S_j^0)}{\cos(V_{j+1}+\epsilon_j)},$$

$$\lim_{\theta \to V_{j+1}^+} R_1(\theta) = \frac{r-1}{r}B^{\frac{1}{r-1}}\frac{\cos(S_j^0)}{\cos(V_{j+1})}, \qquad \lim_{\theta \to V_{j+1}^+} R_2(\theta) = \infty.$$

When $\theta = V_{j+1}+\epsilon_j$ there is exactly one positive solution of $(**)$, namely $R = R_c(V_{j+1}+\epsilon_j)$. In a sufficiently small neighborhood of the solution

$$(R,\theta) = (R_c(V_{j+1}+\epsilon_j), V_{j+1}+\epsilon_j)$$

of $(**)$, the set of all solutions (R,θ) of $(**)$ can be described as the graph of a continuously differentiable function $\theta(R)$, defined for all R sufficiently near $R_c(V_{j+1}+\epsilon_j)$, and which has a single isolated maximum value at $R = R_c(V_{j+1}+\epsilon_j)$.

(2) *If $H_j < \theta < V_{j+1}$ then there exists exactly one positive solution $R = R(\theta)$ of equation $(**)$. $R(\theta)$ is a continuously differentiable function of θ and has the properties*

$$\lim_{\theta \to V_{j+1}^-} R(\theta) = \begin{cases} \infty & \text{if } V_{j+1} \geq \pi/2, \\ \dfrac{r-1}{r}B^{\frac{1}{r-1}}\dfrac{\cos(S_j^0)}{\cos(V_{j+1})} & \text{if } V_{j+1} < \pi/2, \end{cases}$$

$$\lim_{\theta \to H_j^+} R(\theta) = \frac{r-1}{r}B^{\frac{1}{r-1}}\frac{\cos(S_j^0)}{\cos(H_j)}.$$

The "overshoot" described in part (1) of Lemma 1.2.4B is illustrated in the figures at the end of this section. It is present in the case $r = 4, b > 0$, and is most clearly evident in the case $r = 5, b > 0$. It is also present for $b < 0$ if $r > 5$, but we have not given a figure for such a case. The curves $R = R_c(\theta)$ are also indicated in these figures by dashed lines with the label "Rc".

Lemma 1.2.5A. *Suppose $b < 0$ and $1 \leq j \leq J$ (if $r = 4J - 1$ suppose $1 \leq j \leq J - 1$). Define $R_c(\theta) = \left[\dfrac{B\cos(\theta)}{-\cos(r\theta)}\right]^{\frac{1}{r-1}}$ for all $H_{-j} < \theta < V_{-j+1}$. The segment $(H_{-j} < \theta < V_{-j+1})$ of the SDC C_{-j} through the saddle point $k_{-j}(b)$ can be described as follows.*

(1) *For θ satisfying $H_{-j} < \theta < S_{-j}^1$ there is exactly one solution $R = R(\theta)$ of $(*)$ (see Lemma 1.2.4A) in the interval $(0, R_c(\theta))$. It is a continuously differentiable function of θ and*

$$\lim_{\theta \to H_{-j}^+} R(\theta) = \frac{r-1}{r}B^{\frac{1}{r-1}}\frac{\cos(S_{-j}^1)}{\cos(H_{-j})},$$

$$\lim_{\theta \to (S_{-j}^1)^-} R(\theta) = B^{\frac{1}{r-1}}.$$

(2) *For θ satisfying $S^1_{-j} < \theta < V_{-j+1}$ there is exactly one solution $R = R(\theta)$ of (*) in the interval $(R_c(\theta), \infty)$. It is a continuously differentiable function of θ and*

$$\lim_{\theta \to (S^1_{-j})^+} R(\theta) = B^{\frac{1}{r-1}},$$

$$\lim_{\theta \to V^-_{-j+1}} R(\theta) = \infty.$$

(3) *For θ in a sufficiently small neighborhood of S^1_{-j} define $\alpha = \theta - S^1_{-j}$. Then for $|\alpha|$ sufficiently small the functions $R(\theta)$ defined in (1) and (2) above satisfy*

$$R(S^1_{-j} + \alpha) = B^{\frac{1}{r-1}}\left[1 + \tan(\tfrac{1}{2}S^1_{-j} + \tfrac{\pi}{4})\alpha + O(\alpha^2)\right].$$

In particular, if Arg z denotes the argument, with values in $(-\pi, \pi]$, of a nonzero complex number z, then

$$\lim_{\theta \to (S^1_{-j})^-} \mathrm{Arg}(R(\theta)e^{i\theta} - k_{-j}) = \tfrac{1}{2}S^1_{-j} - \tfrac{3}{4}\pi,$$

$$\lim_{\theta \to (S^1_{-j})^+} \mathrm{Arg}(R(\theta)e^{i\theta} - k_{-j}) = \tfrac{1}{2}S^1_{-j} + \tfrac{1}{4}\pi.$$

Remark. In parts (1) and (2) of the above Lemma we must specify an interval strictly smaller than $(0, \infty)$ in which the desired solution lies. This is because there is another solution to (*) giving the steepest *ascent* contour emanating from the saddle point k_{-j}.

Proof. (1) and (2). Define $f(R; \theta)$ as in the proof of Lemma 1.2.4. If $H_{-j} < \theta < V_{-j+1}$ then we have

$$\frac{\pi}{r}(\tfrac{1}{2} - 2j) < \theta < \frac{\pi}{r}(\tfrac{3}{2} - 2j),$$

so that $-\cos(r\theta) > 0$. Thus $\lim_{R \to \infty} f(R) = -\infty$. Note that we always have $r > 4j - 1$. This is equivalent to $H_{-j} > -\pi/2$. Thus $\cos(\theta) > 0$, and hence $R_c(\theta)$ is well-defined. As in the previous Lemma $f(0) < 0$. $f(R_c(\theta))$ is given by the same expression as in the proof of the previous Lemma. Define for $H_{-j} < \theta < V_{-j+1}$

$$g(\theta) = -\frac{\cos^r(\theta)}{\cos(r\theta)}.$$

Computing the derivative we get

$$g'(\theta) = -\frac{r\cos^{r-1}(\theta)\sin[(r-1)\theta]}{\cos^2(r\theta)}.$$

So $g'(\theta) = 0$ when $(r-1)\theta \in \pi\mathbb{Z}$. Since $V_{-j+1} - H_{-j} = \frac{\pi}{r} < \frac{\pi}{r-1}$, the interval (H_{-j}, V_{-j+1}) can contain at most one zero of $g'(\theta)$. $\theta = S^1_{-j}$ is therefore the only such zero in that interval. If $H_{-j} < \theta < S^1_{-j}$ then $-2j\pi < (r-1)\theta < \pi(-2j+1)$, so $\sin[(r-1)\theta] > 0$ and $g(\theta)$ is decreasing. If $S^1_{-j} < \theta < V_{-j+1}$ then $\pi(-2j+1) < \theta < \pi(-2j+2)$, so $\sin[(r-1)\theta] < 0$ and $g(\theta)$ is increasing. So on the interval (H_{-j}, V_{-j+1}) the function $g(\theta)$ assumes its minimum value at the unique point $\theta = S^1_{-j}$. A short computation reveals that $g(S^1_{-j}) = \cos^{r-1}(S^1_{-j})$. Thus $f(R_c(\theta)) > 0$ for all $H_{-j} < \theta < V_{-j+1}$ except $\theta = S^1_{-j}$, where $f(R_c(\theta)) = 0$. Thus $f(R; \theta) = 0$ has two positive solutions, one in $(0, R_c(\theta))$ and the other in $(R_c(\theta), \infty)$, for all $\theta \in (H_{-j}, V_{-j+1})$ except $\theta = S^1_{-j}$, where it has only one solution, namely $R = R_c(S^1_{-j}) = B^{1/(r-1)}$.

In part (1) of the present Lemma, where $H_{-j} < \theta < S^1_{-j}$ define $R(\theta)$ to be the solution in $(0, R_c(\theta))$. This is appropriate, since $R_c(\theta) \to \infty$ as $\theta \to H^+_{-j}$, and thus the solution in $(R_c(\theta), \infty)$ could never match up to the solution branch described in part (2) of Lemma 1.2.4. The fact that the branch we have chosen does match up follows from the implicit function theorem. This proves the limit assertion as $\theta \to H^+_{-j}$. The other limit assertion follows from part (3), proved below.

In part (2) of the present Lemma, where $S^1_{-j} < \theta < V_{-j+1}$ define $R(\theta)$ to be the solution in $(R_c(\theta), \infty)$. This is appropriate, since $R_c(\theta) \to \infty$ as $\theta \to V^-_{-j+1}$, which forces $R(\theta) \to \infty$ in that same limit, as desired. The other limit assertion follows from part (3), proved below.

(3) In order to study the behavior of $R(\theta)$ as $\theta \to S^1_{-j}$, we need to give a finer analysis of the solutions of the equation $f(R; \theta) = 0$ when θ is in a small neighborhood of S^1_{-j}. For this purpose, we introduce new variables: $R = R_c(\theta)(1 + \tilde{R})$ and $\theta = S^1_{-j} + \alpha$. For convenience let $C = \tan(S^1_{-j})$. Then some manipulations yield the equation

$$\frac{2f(R)}{(r-1)B^{\frac{r}{r-1}}g(\theta)^{\frac{1}{r-1}}}$$
$$= \frac{2}{r}\left\{1 - \left[\frac{\cos(r\alpha) - C\sin(r\alpha)}{(\cos(\alpha) - C\sin(\alpha))^r}\right]^{\frac{1}{r-1}}\right\} - \frac{2[(1+\tilde{R})^r - 1 - r\tilde{R}]}{r(r-1)}$$
$$= f_1(\alpha) - f_2(\tilde{R}).$$

We are interested in the pairs $(\tilde{R}, \alpha)$ where this expression vanishes and α is small. We will find two values of $\tilde{R}$ for a given fixed value of α except when $\alpha = 0$, and the values of $\tilde{R}$ will be small when α is small. Since we already know that these represent the only two solutions, it is clear that only the case where $\tilde{R}$ is small is interesting or important. Expanding $f_1(\alpha)$ as $\alpha \to 0$ and $f_2(\tilde{R})$ as $\tilde{R} \to 0$ we get:

$$f_1(\alpha) = (1 + C^2)\alpha^2 + \frac{2}{3}(r+1)C(1+C^2)\alpha^3 + O(\alpha^4),$$
$$f_2(\tilde{R}) = \tilde{R}^2 + \frac{r-2}{3}\tilde{R}^3 + O(\tilde{R}^4).$$

For $\tilde{R}$ sufficiently small, we can find two inverse functions $\tilde{R}(f_2)$ to $f_2(\tilde{R})$, one with positive values and one with negative values. These two inverse functions can be expanded at $f_2 = 0$ in powers of $\sqrt{f_2}$:

$$\tilde{R} = \pm\sqrt{f_2} - \frac{r-2}{6}f_2 + \mathrm{O}(f_2^{3/2}).$$

Plugging $f_2 = f_1(\alpha)$ into these two inverse functions $\tilde{R}(f_2)$ gives two different expressions for $\tilde{R}$ as a function of α. Taking a power series expansion of these as $\alpha \to 0$ we get

$$\tilde{R} = \pm\sqrt{1 + C^2}\alpha + \mathrm{O}(\alpha^2).$$

When $\theta < S^1_{-j}$, i.e. $\alpha < 0$, we want the solution $R(\theta) < R_c(\theta)$, i.e. $\tilde{R}(\alpha) < 0$. When $\theta > S^1_{-j}$, i.e. $\alpha > 0$, we want the solution $R(\theta) > R_c(\theta)$, i.e. $\tilde{R}(\alpha) > 0$. Thus the solution we want is obtained by always choosing the $+$ sign in the above. The $-$ sign gives the curve of steepest ascent leaving the saddle point k_{-j}. Expanding $R_c(\theta)$ as a power series in α we get

$$R_c(S^1_{-j} + \alpha) = B^{1/(r-1)}[1 + C\alpha + \mathrm{O}(\alpha^2)].$$

Thus we have

$$R(S^1_{-j} + \alpha) = B^{1/(r-1)}[1 + (C + \sqrt{1 + C^2})\alpha + \mathrm{O}(\alpha^2)]$$

as an expansion valid as $\alpha \to 0$. The limit assertions as $\theta \to (S^1_{-j})^\pm$ stated in (1) and (2) are now proved. Furthermore, since

$$C + \sqrt{1 + C^2} = \tan(S^1_{-j}) + \sec(S^1_{-j}) = \tan(\frac{1}{2}S^1_{-j} + \frac{\pi}{4})$$

we obtain the expansion asserted in (3). Using this expansion, we can obtain the following

$$R(S^1_{-j} + \alpha)e^{i(S^1_{-j} + \alpha)} - k_{-j} = B^{1/(r-1)}e^{iS^1_{-j}}[|\alpha|\sec(D)e^{i(-D + \mathrm{sgn}(\alpha)\frac{\pi}{2})} + \mathrm{O}(\alpha^2)],$$

where $D = \frac{1}{2}S^1_{-j} + \frac{\pi}{4}$. The $\mathrm{O}(\alpha^2)$ term cannot affect the limiting value of the argument of this quantity, so the limits as $\alpha \to 0^\pm$ asserted in (3) now follow immediately. $\square$

Lemma 1.2.5B. *Suppose $b > 0$ and $0 \le j \le K$ (if $r = 4K + 1$ suppose $0 \le j \le K - 1$). Define $R_c(\theta) = \left[\frac{B\cos(\theta)}{\cos(r\theta)}\right]^{\frac{1}{r-1}}$ for all $V_j < \theta < H_j$. The segment $(V_j < \theta < H_j)$ of the SDC C_j through the saddle point $k_j(b)$ can be described as follows.*

(1) *For θ satisfying $S^0_j < \theta < H_j$ there is exactly one solution $R = R(\theta)$ of (**) (see Lemma 1.2.4B) in the interval $(0, R_c(\theta))$. It is a continuously differentiable function of θ and*

$$\lim_{\theta \to H^-_j} R(\theta) = \frac{r-1}{r}B^{\frac{1}{r-1}}\frac{\cos(S^0_j)}{\cos(H_j)},$$

$$\lim_{\theta \to (S^0_j)^+} R(\theta) = B^{\frac{1}{r-1}}.$$

(2) *For θ satisfying $V_j < \theta < S_j^0$ there is exactly one solution $R = R(\theta)$ of (**) in the interval $(R_c(\theta), \infty)$. It is a continuously differentiable function of θ and*

$$\lim_{\theta \to (S_j^0)^-} R(\theta) = B^{\frac{1}{r-1}},$$

$$\lim_{\theta \to V_j^+} R(\theta) = \infty.$$

(3) *For θ in a sufficiently small neighborhood of S_j^0 define $\alpha = \theta - S_j^0$. Then for $|\alpha|$ sufficiently small the functions $R(\theta)$ defined in (1) and (2) above satisfy*

$$R(S_j^0 + \alpha) = B^{\frac{1}{r-1}} \left[1 + \tan(\frac{1}{2} S_j^0 - \frac{\pi}{4})\alpha + O(\alpha^2) \right].$$

In particular, if $\operatorname{Arg} z$ denotes the argument, with values in $(-\pi, \pi]$, of a nonzero complex number z, then

$$\lim_{\theta \to (S_j^0)^+} \operatorname{Arg}[R(\theta)e^{i\theta} - k_j(b)] = \frac{1}{2}S_j^0 + \frac{3}{4}\pi,$$

$$\lim_{\theta \to (S_j^0)^-} \operatorname{Arg}[R(\theta)e^{i\theta} - k_j(b)] = \frac{1}{2}S_j^0 - \frac{1}{4}\pi.$$

As a matter of terminology, we will denote by C_j^+ the branch of C_j starting at $k_j(b)$ and tending to infinity in the valley V_{j+1} (when $-J \leq j \leq -1$) or V_j (when $0 \leq j \leq K$). This branch never overshoots the sector $S_j^1 < \theta < V_{j+1}$ (when $-J \leq j \leq -1$) or $V_j < \theta < S_j^0$ (when $0 \leq j \leq K$). The entire remainder of C_j will be denoted C_j^-; it starts at ∞ and ends at the saddle point. As we have seen, C_j^- has been rather complicated to describe in polar coordinates (c.f. Lemma 1.2.4 and Lemma 1.2.5(1)). However, it will be helpful to have this precise information about the sector containing this branch, namely $V_{-j} - \epsilon_j \leq \theta \leq S_{-j}^1$ (when $b < 0$) and $S_j^0 \leq \theta \leq V_{j+1} + \epsilon_j$ (when $b > 0$). (In the above ϵ_j has a different definition, depending on the sign of b, and in some cases it may be zero and the sector half-open.)

Finally, we justify the change of contour of integration. First of all. we remark that the original integral defining $I(x, t)$ need not converge absolutely. Let Γ_θ denote the ray with argument θ starting at $k = 0$ and tending to infinity. We define $I(x, t)$ to be the integral over Γ_θ of the same integrand, where $V_0 \leq \theta < 0$. Since $V(k)$ is the Fourier transform of a function with compact support, it has bounds of the type

$$|V(k)| \leq C(1 + |k|)^N e^{a|\Im k|},$$

where $k \in \mathbb{C}, C > 0, N \in \mathbb{R}, a > 0$. Such bounds are clearly sufficient to show that the integral over Γ_θ does not depend on the value of $\theta \in [V_0, 0)$, so $I(x, t)$ is well-defined.

It remains to show that the integral over Γ_{V_0} is equal to the integral over

$$\Gamma = \begin{cases} \begin{cases} \tilde{C} \cup C_{-J}^+ \cup \bigcup_{j=1}^{J-1} C_{-j} & r = 4J - 1, \\ \tilde{C} \cup \bigcup_{j=1}^{J} C_{-j} & 4J - 1 < r < 4J + 3, \end{cases} & b < 0, \\ \begin{cases} \tilde{C} \cup C_K^+ \cup \bigcup_{j=0}^{K-1} C_j & r = 4K + 1, \\ \tilde{C} \cup \bigcup_{j=0}^{K} C_j & 4K + 1 < r < 4K + 5, \end{cases} & b > 0. \end{cases}$$

Consider a circle centered at $k = 0$ and of radius $M > 1$. Let $\Gamma(M)$ be the path Γ whenever it lies inside this circle, and the portions of the circle necessary to make a continuous path from $k = 0$ to the point Me^{iV_0}. Clearly, the integral over $\Gamma(M)$ equals the integral from 0 to Me^{iV_0} along Γ_{V_0}. Now let $M \to \infty$. We must show that the contributions from the little segments along the circle vanish in the limit. We will explain how to prove this in the case $b < 0$, $4J - 1 < r < 4J + 3$, $1 \le j \le J - 1$, and $V_{-j} > -\pi/2$ (and hence $r > 7$), for the segment between C_{-j-1} and C_{-j}. (These two contours cannot intersect since they are level curves for the imaginary part of $\varphi(k)$ for two different levels.) The other cases are exactly similar. Suppose $V_{-j} - \epsilon_j < \theta < V_{-j}$. Let $R_1(\theta)$ and $R_0(\theta)$ be polar representations of C_{-j-1} and C_{-j} respectively. We have $R_1(\theta) > R_0(\theta)$ for all such θ (otherwise the contours would have to intersect). Let $M = R_1(\theta)$; as $\theta \to V_{-j}^-$ we have $M \to \infty$. The key observation is that C_{-j} crosses the circle of radius $R_1(\theta)$ between the angle θ and the angle V_{-j}. This is because C_{-j} "overshoots" its sector and approaches V_{-j} from the same side that C_{-j-1} does. Thus the arclength of this segment of the circle is bounded by $(V_{-j} - \theta)R_1(\theta)$. Now we can estimate the integral over this segment in an elementary manner to be less than something like

$$Ce^{-bR_1(\theta)\sin(V_{-j})t + \rho R_1(\theta)^r \sin(r\theta)t}(1 + R_1(\theta))^{N+\gamma+1}e^{a|\sin(\theta)|R_1(\theta)}(V_{-j} - \theta).$$

Clearly this vanishes as $\theta \to V_{-j}^-$, since the strong exponential decay from the valley overwhelms even the exponential growth coming from $V(k)$.

On the following pages some figures illustrating the contour Γ for different selected values of r are given. The number B in the captions stands for the quantity $\frac{b}{r\rho}$ (in section 1.1 B was defined as the absolute value of this quantity) which is normalized to have either of the values 1 or -1.

r=1.5, B=1

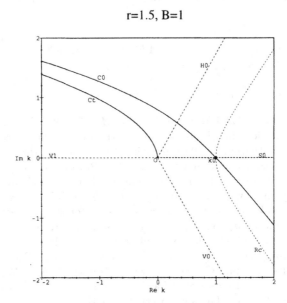

r=1.5, B=-1

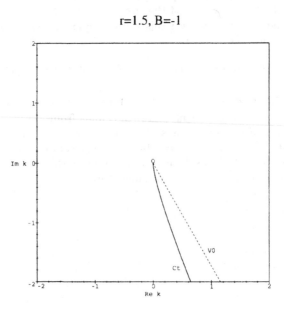

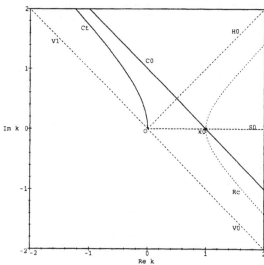

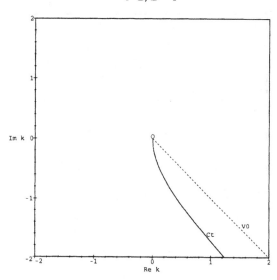

r=3, B=1

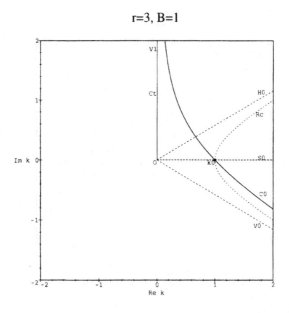

r=3, B=-1

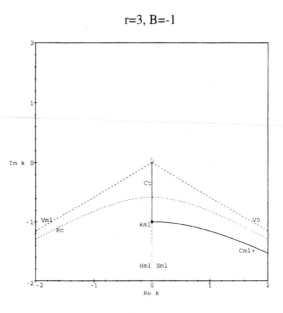

r=4, B=1

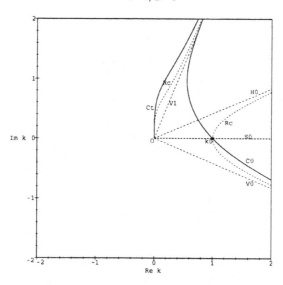

r=4, B=-1

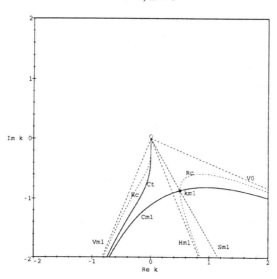

r=5, B=1

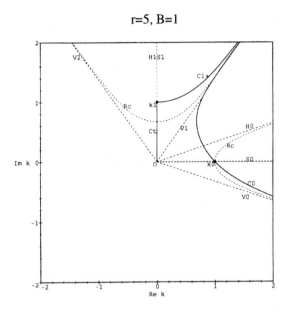

r=5, B=-1

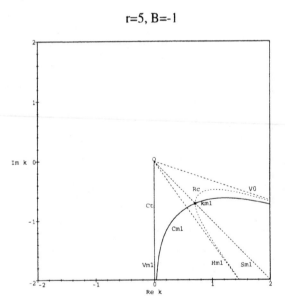

1.3 Expansion and Error terms, for the contour $\tilde{C}$

In this section we will compute explicitly all the terms of an asymptotic expansion as $t \to \infty$ of the integral

$$I(x, t; r, \rho, \gamma, V, \tilde{C}) \overset{\text{def}}{=} \int_{\tilde{C}} e^{i(kb - \rho k^r)t} k^\gamma V(k)\, dk,$$

where the contour $\tilde{C} = \tilde{C}(b)$ depends on $b = x/t$. We will also derive expressions for the error terms, which will be estimated in the next section. Despite the generality of the explicit formulae derived in §1.1 for the terms of the asymptotic expansion of integrals like the above, we cannot use them here (except when $r \in \mathbb{Z}$) because (when $r \notin \mathbb{Z}$) the series for $p(k) = -ikb + i\rho k^r$ does not ascend in integral powers of k at $k = 0$. Even when $r \geq 3$ is an integer, it is more cumbersome to apply those formulae than to use the general procedure we will develop here. The key to our approach is the observation already made in §1.1 that k as a function of $v = p(k)$ can be expressed in terms of an analytic function $\tilde{v} = \tilde{k}(1 + \tilde{k})^{-r}$, where $k = \frac{iv}{b}(1 + \tilde{k})$ and $\tilde{v} = \frac{\rho}{b}\left(\frac{iv}{b}\right)^{r-1}$. However, what we need in order to derive the terms of the asymptotic expansion of the above Laplace contour integral is an expansion of $f(v) = k^\gamma V(k)\frac{dk}{dv}$ in powers of v. If we expand $V(k)$ by its Taylor series at $k = 0$ we get

$$f(v) = \sum_{j=0}^\infty \frac{V^{(j)}(0)}{j!} k^{\gamma+j} \frac{dk}{dv}.$$

So it remains for each $j \geq 0$ to expand $k^{\gamma+j}\frac{dk}{dv}$ in powers of v. Since $k = \frac{iv}{b}(1 + \tilde{k})$, we have $k^{\gamma+j} = \left(\frac{iv}{b}\right)^{\gamma+j}(1 + \tilde{k})^{\gamma+j}$, where for $v > 0$ and $\tilde{k}$ sufficiently small we use the principal branch to define fractional powers. We also get

$$\frac{dk}{dv} = \frac{i}{b}(1 + \tilde{k}) + \frac{iv}{b}\frac{d\tilde{k}}{d\tilde{v}}\frac{d\tilde{v}}{dv}.$$

But from the definition of $\tilde{v}$ we get

$$\frac{iv}{b}\frac{d\tilde{v}}{dv} = \frac{iv}{b}\frac{\rho}{b}(r - 1)\left(\frac{iv}{b}\right)^{r-2}\frac{i}{b} = \frac{i}{b}(r - 1)\tilde{v}.$$

Therefore

$$\frac{dk}{dv} = \frac{i}{b}\left[1 + \tilde{k} + (r - 1)\tilde{v}\frac{d\tilde{k}}{d\tilde{v}}\right].$$

Thus we have

$$k^{\gamma+j}\frac{dk}{dv} = \frac{i}{b}\left(\frac{iv}{b}\right)^{\gamma+j}\left[(1 + \tilde{k})^{1+\gamma+j} + \frac{(r - 1)\tilde{v}}{1 + \gamma + j}\frac{d}{d\tilde{v}}(1 + \tilde{k})^{1+\gamma+j}\right].$$

Our first lemma computes an expansion of $(1 + \tilde{k})^{1+\gamma+j}$.

Lemma 1.3.1. *Suppose $a \in \mathbb{R}$ and $\tilde{v} = \tilde{k}(1 + \tilde{k})^{-r}$. Then for $|\tilde{v}|$ sufficiently small we have the expansion*

$$(1 + \tilde{k})^a = 1 + a \sum_{n=1}^{\infty} \frac{\tilde{v}^n}{n!} \prod_{h=1}^{n-1} (nr + a - h).$$

Proof. We apply the generalized Lagrange formula (see page 14 of [AS]):

$$\psi(\tilde{k}) = \psi(0) + \sum_{n=1}^{\infty} \frac{\tilde{v}^n}{n!} \left\{ \left(\frac{d}{d\tilde{k}} \right)^{n-1} \left[\psi'(\tilde{k}) \left(\frac{\tilde{k}}{g(\tilde{k})} \right)^n \right] \right\} \Bigg|_{\tilde{k}=0},$$

where $\psi(\tilde{k}) = (1 + \tilde{k})^a$ and $g(\tilde{k}) = \tilde{k}(1 + \tilde{k})^{-r}$. Thus

$$\psi'(\tilde{k}) \left(\frac{\tilde{k}}{g(\tilde{k})} \right)^n = a(1 + \tilde{k})^{nr+a-1}.$$

The $(n - 1)$st derivative of this quantity with respect to $\tilde{k}$ is

$$a(1 + \tilde{k})^{nr+a-n} \prod_{h=1}^{n-1} (nr + a - h).$$

Setting $\tilde{k} = 0$ finishes the proof. $\square$

This series can be differentiated term-by-term to get

$$\frac{(r-1)\tilde{v}}{1 + \gamma + j} \frac{d}{d\tilde{v}} (1 + \tilde{k})^{1+\gamma+j} = (r-1)\tilde{v} \sum_{n=1}^{\infty} \frac{\tilde{v}^{n-1}}{(n-1)!} \prod_{h=1}^{n-1} (nr + 1 + \gamma + j - h)$$

$$= \sum_{n=1}^{\infty} \frac{\tilde{v}^n}{n!} (nr - n) \prod_{h=1}^{n-1} (nr + 1 + \gamma + j - h).$$

Thus we get the following relatively simple closed form expansion

$$k^{\gamma+j} \frac{dk}{dv}$$

$$= \frac{i}{b} \left(\frac{iv}{b} \right)^{\gamma+j} \left[1 + \sum_{n=1}^{\infty} \frac{\tilde{v}^n}{n!} (1 + \gamma + j + nr - n) \prod_{h=1}^{n-1} (nr + 1 + \gamma + j - h) \right]$$

$$= \frac{i}{b} \left(\frac{iv}{b} \right)^{\gamma+j} \sum_{n=0}^{\infty} \frac{\tilde{v}^n}{n!} \prod_{h=1}^{n} (nr + 1 + \gamma + j - h)$$

$$= \frac{i}{b} \sum_{n=0}^{\infty} \binom{\gamma + j + nr}{n} \left(\frac{\rho}{b} \right)^n \left(\frac{iv}{b} \right)^{\gamma+j+n(r-1)}.$$

Thus we have proved the following.

Lemma 1.3.2. *If $v > 0$ is sufficiently small then we have the following convergent expansion*

$$f(v) = \frac{i}{b} \sum_{j=0}^{\infty} \sum_{n=0}^{\infty} \frac{V^{(j)}(0)}{j!} \binom{\gamma+j+nr}{n} \left(\frac{\rho}{b}\right)^n \left(\frac{iv}{b}\right)^{\gamma+j+n(r-1)}.$$

Clearly when r is an integer, this expansion ascends in integral powers of v; in general, however, this will not be true. The full asymptotic expansion of our Laplace contour integral is obtained by computing $\int_0^{\infty} e^{-vt} f(v)\, dv$ by substituting for $f(v)$ its series expansion, and formally integrating term-by-term. This yields

$$\int_{\tilde{C}} e^{i(kb-\rho k^r)t} k^{\gamma} V(k)\, dk$$

$$\sim \sum_{j=0}^{\infty} \sum_{n=0}^{\infty} \frac{V^{(j)}(0)}{j!n!} \Gamma(1+\gamma+j+nr) \left(\frac{\rho}{b}\right)^n \left(\frac{i}{bt}\right)^{1+\gamma+j+n(r-1)}.$$

An alternate derivation of this expansion is to expand $e^{-i\rho k^r t} k^{\gamma} V(k)$ in powers of k and integrate term by term in the sense of Hardy, c.f. [O2]. This alternate derivation is essentially a formal application of the methodology of Watson's lemma to extract an asymptotic expansion of the integral as $\pm x \to \infty$, where t is treated as a parameter. The contour $\tilde{C}$ is not appropriate if t is allowed to become small.

Asymptotic Character of the Expansion. When b is a nonzero constant, these terms are ordered as $t \to \infty$ according to the value of $j+n(r-1)$. However, since we must consider such expansions in the outer region $R_{\pm}$, we need to check if (and how) these terms are ordered in this considerably larger asymptotic region. For this purpose, we consider two such terms, indexed by (j_1, n_1) and (j_2, n_2), where we desire the second term to be asymptotically smaller in $R_{\pm}$ than the first as $t \to \infty$. So, ignoring constants which do not depend on x or t, we consider the absolute value of the quotient of the second by the first:

$$|b|^{n_1-n_2} |bt|^{j_1-j_2+(n_1-n_2)(r-1)}$$

We want this quotient to tend to zero as $t \to \infty$. In the outer regions $R_{\pm}$ the natural variable is $\zeta = xt^{-\beta}$, where $\frac{1}{r} < \beta < 1$. In $R_{\pm}$ we have $|\zeta| \geq \epsilon > 0$. Expressing the quotient of our two terms in this variable, we get

$$|\zeta|^{j_1+rn_1-(j_2+rn_2)} (t^{\beta})^{j_1+(r-1/\beta)n_1-[j_2+(r-1/\beta)n_2]}.$$

Since in $R_{\pm}$ it is possible for $|\zeta| \to \infty$, in order for these two terms to be strictly ordered as $t \to \infty$ we need

$$j_1 + rn_1 \leq j_2 + rn_2, \qquad j_1 + (r-1/\beta)n_1 < j_2 + (r-1/\beta)n_2.$$

If strict inequality holds in the first of these two inequalities, then the two terms are also ordered in an additional sense, namely, for each fixed $t \geq 1$ the second term decays faster than the first term as $|\zeta| \to \infty$ (or $|x| \to \infty$). Thus it will be possible for us to give a single expansion (the one already written down above) of our Laplace contour integral which is valid (in a certain sense to be explained below) as $|x| + t \to \infty$ in the regions $R_\pm$. When $|\zeta| \geq \epsilon$ is a constant, the terms are ordered in terms of increasing rate of decay as $t \to \infty$ according to the value of $j + (r - 1/\beta)n$. (Notice how non-uniformity manifests itself, namely if $\beta = 1/r$ then increasing n leaving j fixed does not increase the decay rate of the term, hence one must sum infinitely many terms in order to gain an asymptotic improvement.) But if $t \geq 1$ is a constant, the terms are ordered in terms of increasing rate of decay as $|\zeta| \to \infty$ according to the value of $j + rn$. Thus we cannot expect to be able show that the error committed by approximating the integral by some collection of terms decays as $|x| + t \to \infty$ $((x,t) \in R_\pm)$ more rapidly than the most rapidly decaying term in the collection, since the identity of that term changes with β.

However, we do not really need to show this; rather it suffices to show that by choosing the collection appropriately we can make the corresponding error as rapidly decaying as we like in both senses. For example, suppose for each integer $N \geq 0$ we define the collection

$$T_N = \{(j,n) \mid j \geq 0, n \geq 0, j + (r - 1/\beta)n < N\}.$$

Define the error associated to this collection as

$$E_N(\zeta, t) = \int_{\tilde{C}} e^{i(kb - \rho k^r)t} k^\gamma V(k) \, dk$$
$$- \sum_{(j,n) \in T_N} \frac{V^{(j)}(0)}{j! n!} \Gamma(1 + \gamma + j + nr) \left(\frac{\rho}{b}\right)^n \left(\frac{i}{bt}\right)^{1 + \gamma + j + n(r-1)}.$$

We will prove (see the next section) that for each such N we have

$$\sup_{t \geq 1} \sup_{\pm\zeta \geq \epsilon} |\zeta t^\beta|^{1 + \gamma + N} |E_N(\zeta, t)| < \infty.$$

It follows from this same result with N replaced by $N + 1$ that this is the strongest possible weighted estimate of the error $E_N(\zeta, t)$ which holds true, since $(N, 0) \in T_{N+1}$ and that term has exactly the same behavior as that which will be proved for $E_N(\zeta, t)$. Also, if ζ is fixed ($\pm\zeta \geq \epsilon$), then this result implies that $E_N(\zeta, t)$ decays as $t \to \infty$ faster than each one of the terms coming from the collection T_N; thus each one of those terms must have been included in order to make such a result true. However, if $t \geq 1$ is fixed, it is usually true that $E_N(\zeta, t)$ decays more slowly as $\pm\zeta \to \infty$ than certain of the terms coming from the collection T_N; consequently, it would seem that such terms do not give any useful information, and should not be included in the collection. However, such terms are seen to be relevant when one considers the analogous result for another sufficiently large value of N. For such larger values of N the error $E_N(\zeta, t)$ will decay as $\pm\zeta \to \infty$ faster than the terms which were once thought to be irrelevant, which will also occur in the larger new collection T_N. Thus those terms are in fact essential in the limit $\pm\zeta \to \infty$. The problem was that the original collection did not contain all the terms necessary to completely capture the asymptotic behavior of the integral as $\pm\zeta \to \infty$ to that order. By enlarging N, and hence the collection T_N, we obtain all the terms necessary to capture exactly the asymptotic behavior to that order. Since we will prove our result for each N, all of the terms we have written down are essential.

Error Terms. In preparation for our proof of the above error estimate, we will now express $E_N(\zeta, t)$ in terms of certain Taylor remainders. First of all, we expand $V(k)$ in a Taylor series at $k = 0$

$$V(k) = \sum_{j=0}^{N-1} \frac{V^{(j)}(0)}{j!} k^j + R_N(k).$$

Thus we have

$$\int_{\tilde{C}} e^{i(kb-\rho k^r)t} k^\gamma V(k) \, dk$$

$$= \sum_{j=0}^{N-1} \frac{V^{(j)}(0)}{j!} \int_{\tilde{C}} e^{i(kb-\rho k^r)t} k^{\gamma+j} \, dk + \int_{\tilde{C}} e^{i(kb-\rho k^r)t} k^\gamma R_N(k) \, dk$$

$$= \sum_{j=0}^{N-1} \frac{V^{(j)}(0)}{j!} \int_0^\infty e^{-vt} k^{\gamma+j} \frac{dk}{dv} \, dv + \int_0^\infty e^{-vt} k^\gamma R_N(k) \frac{dk}{dv} \, dv.$$

If $r = 4J - 1$ (when $b < 0$) or $r = 4K + 1$ (when $b > 0$) then the upper limit in the v integrals is not ∞, but rather the particular finite value $p(k_j(b))$, where $j = -J$ when $b < 0$ and $j = K$ when $b > 0$. This will give rise to one additional error term in these cases (see below).

Near the beginning of this section we gave a formula for $k^{\gamma+j} \frac{dk}{dv}$, which expressed it as the product of $\frac{i}{b} \left(\frac{iv}{b}\right)^{\gamma+j}$ and a certain holomorphic function of $\tilde{v}$

$$(1+\tilde{k})^{1+\gamma+j} + \frac{(r-1)\tilde{v}}{1+\gamma+j} \frac{d}{d\tilde{v}}(1+\tilde{k})^{1+\gamma+j} = (1+\tilde{k})^{1+\gamma+j} + (r-1)(1+\tilde{k})^{\gamma+j}\tilde{v}\frac{d\tilde{k}}{d\tilde{v}}.$$

The Taylor expansion of this function at $\tilde{v} = 0$ gave rise to the terms of our expansion. Let us derive a simplified formula for this function. From the equation $\tilde{v} = \tilde{k}(1+\tilde{k})^{-r}$ we compute $\frac{d\tilde{v}}{d\tilde{k}} = [1 - (r-1)\tilde{k}](1+\tilde{k})^{-r-1}$, and therefore

$$\tilde{v}\frac{d\tilde{k}}{d\tilde{v}} = \frac{\tilde{k}(1+\tilde{k})}{1 - (r-1)\tilde{k}}.$$

Thus the function in question can be written

$$(1+\tilde{k})^{1+\gamma+j} + (r-1)(1+\tilde{k})^{\gamma+j}\tilde{v}\frac{d\tilde{k}}{d\tilde{v}} = (1+\tilde{k})^{1+\gamma+j}\left[1 + \frac{(r-1)\tilde{k}}{1-(r-1)\tilde{k}}\right]$$

$$= \frac{(1+\tilde{k})^{1+\gamma+j}}{1-(r-1)\tilde{k}}.$$

Thus we have

$$k^{\gamma+j}\frac{dk}{dv} = \frac{i}{b}\left(\frac{iv}{b}\right)^{\gamma+j}\frac{(1+\tilde{k})^{1+\gamma+j}}{1-(r-1)\tilde{k}}.$$

Suppose $0 \le j \le N-1$, and let n_j denote the smallest integer such that $j + n_j(r-1/\beta) \ge N$. As we saw earlier, we have the following explicit Taylor expansion:

$$\frac{(1+\tilde{k})^{1+\gamma+j}}{1-(r-1)\tilde{k}} = \sum_{n=0}^{n_j-1}\binom{\gamma+j+nr}{n}\tilde{v}^n + R_N^j(\tilde{v}).$$

Using this expansion in our above formula for our Laplace contour integral we get the following

$$\int_{\tilde{C}} e^{i(kb-\rho k^r)t} k^\gamma V(k)\, dk$$

$$= \sum_{j=0}^{N-1}\frac{V^{(j)}(0)}{j!}\sum_{n=0}^{n_j-1}\binom{\gamma+j+nr}{n}\frac{i}{b}\int_0^\infty e^{-vt}\left(\frac{iv}{b}\right)^{\gamma+j}\tilde{v}^n\, dv$$

$$+ \sum_{j=0}^{N-1}\frac{V^{(j)}(0)}{j!}\frac{i}{b}\int_0^\infty e^{-vt}\left(\frac{iv}{b}\right)^{\gamma+j}R_N^j(\tilde{v})\, dv$$

$$+ \frac{i}{b}\int_0^\infty e^{-vt}\left(\frac{iv}{b}\right)^\gamma\frac{(1+\tilde{k})^{1+\gamma}}{1-(r-1)\tilde{k}}R_N(k)\, dv.$$

The double sum on the right is the sum over $(j, n) \in T_N$ of the terms of the expansion. The other two summands on the right are the error terms. Thus when $\tilde{C}$ does not hit a saddle point we have

$$E_N(\zeta, t) = \sum_{j=0}^{N-1}\frac{V^{(j)}(0)}{j!}\left(\frac{i}{b}\right)^{1+\gamma+j}\int_0^\infty e^{-vt}v^{\gamma+j}R_N^j(\tilde{v})\, dv$$

$$+ \left(\frac{i}{b}\right)^{1+\gamma}\int_0^\infty e^{-vt}v^\gamma\frac{(1+\tilde{k})^{1+\gamma}}{1-(r-1)\tilde{k}}R_N(k)\, dv.$$

As we have noted, there is one further error term in the case where $\tilde{C}$ hits a saddle point. In that case all the integrals over $(0, \infty)$ must be replaced by integrals over $(0, v_0)$, where $v_0 = (r-1)\rho B^{r/(r-1)}$. But

$$
\int_0^{v_0} e^{-vt} v^{\gamma+j+n(r-1)}\, dv = \int_0^{\infty} e^{-vt} v^{\gamma+j+n(r-1)}\, dv - \int_{v_0}^{\infty} e^{-vt} v^{\gamma+j+n(r-1)}\, dv
$$

$$
= \frac{\Gamma(1+\gamma+j+n(r-1))}{t^{1+\gamma+j+n(r-1)}} - \frac{\Gamma(1+\gamma+j+n(r-1), v_0 t)}{t^{1+\gamma+j+n(r-1)}},
$$

where the last term involves the incomplete Gamma function. The second term decays exponentially in t, and so is asymptotically insignificant when compared to the first term. However, sometimes, as remarked in Olver [O1], it might be advantageous to retain this exponentially small term in order to increase numerical accuracy of the expansion when t is not that large. In the following section we will treat this exponentially small term as an error term, but sometimes in remarks we will treat it as a term to be retained. Thus when $\tilde{C}$ does hit a saddle point we have

$$
E_N(\zeta, t) = -\sum_{(j,n)\in T_N} \frac{V^{(j)}(0)}{j!} \binom{\gamma+j+nr}{n} \cdot
$$

$$
\cdot \left(\frac{i}{bt}\right)^{1+\gamma+j+n(r-1)} \left(\frac{\rho}{b}\right)^n \Gamma(1+\gamma+j+n(r-1), v_0 t)
$$

$$
+ \sum_{j=0}^{N-1} \frac{V^{(j)}(0)}{j!} \left(\frac{i}{b}\right)^{1+\gamma+j} \int_0^{v_0} e^{-vt} v^{\gamma+j} R_N^j(\tilde{v})\, dv
$$

$$
+ \left(\frac{i}{b}\right)^{1+\gamma} \int_0^{v_0} e^{-vt} v^\gamma \frac{(1+\tilde{k})^{1+\gamma}}{1-(r-1)\tilde{k}} R_N(k)\, dv.
$$

1.4 Error Control, for the contour $\tilde{C}$

In this section we prove certain estimates on the error $E_N(\zeta, t)$, which was defined in the previous section. The treatment is the simplest when $\tilde{C}$ does not hit a saddle point, so we will concentrate on that case first, returning to the other case later in this section. Thus for the present we are assuming $r \neq 4J-1$ when $b < 0$ and $r \neq 4K+1$ when $b > 0$. Recall we have parameterized the contour $\tilde{C}$ in terms of the variable $v \geq 0$ by defining $\tilde{v} = \frac{\rho}{b}\left(\frac{iv}{b}\right)^{r-1}$, $\tilde{k}$ implicitly from the equation $\tilde{k} = \tilde{v}(1+\tilde{k})^r$, and finally $k = \frac{iv}{b}(1+\tilde{k})$, which lies on $\tilde{C}$. k depends on both v and b. In order to obtain estimates of $E_N(\zeta, t)$ holding independently of b, we will need to obtain estimates of k in terms of v only. This will be done by studying the analytic continuation of $\tilde{k}$ as a function of $\tilde{v}$ along the ray $\tilde{v} = \frac{\rho}{b}\left(\frac{iv}{b}\right)^{r-1}$, $v > 0$. Our first lemma shows that the $\tilde{k}$-image of this ray in the $\tilde{v}$-plane is bounded.

Lemma 1.4.1. *Suppose $r \neq 4J-1$ when $b < 0$ and $r \neq 4K+1$ when $b > 0$. For $\tilde{v}$ sufficiently small, $\tilde{k}$ is an analytic function of $\tilde{v}$ determined implicitly from the relation $\tilde{k} = \tilde{v}(1+\tilde{k})^r$. This function continues analytically along the ray*

$\tilde{v} = \frac{\ell}{b}\left(\frac{iv}{b}\right)^{r-1}$, $v > 0$. *Furthermore, as $\tilde{v}$ ranges over this ray, $\tilde{k}$ remains in a bounded set, the size of which depends only on r, and $|1 - (r-1)\tilde{k}|$ remains bounded away from zero.*

Proof. The ray has argument $-\pi/2 \pm \pi/2 + (\pm\pi/2)(r-1) = (\pm r - 1)\pi/2$ according as $\pm b > 0$. So

$$b < 0 \text{ and } 4J - 1 < r < 4J + 3 \text{ implies } -2\pi - 2J\pi < -\frac{\pi}{2}(r+1) < -2J\pi,$$

$$b > 0 \text{ and } 4K + 1 < r < 4J + 5 \text{ implies } \qquad 2K\pi < \frac{\pi}{2}(r-1) < 2K\pi + 2\pi.$$

The case $r = 4J + 1$ (when $b < 0$) or $r = 4K + 3$ (when $b > 0$) is especially simple since in those cases the ray in question is the negative real $\tilde{v}$-axis. Since the function $\tilde{k} \mapsto \tilde{k}(1+\tilde{k})^{-r}$ maps the interval $(-1, 0]$ monotonically (increasing) onto $(-\infty, 0]$, the inverse function maps the ray $\tilde{v} \in (-\infty, 0]$ monotonically (increasing) onto $\tilde{k} \in (-1, 0]$. Thus in this case the function $\tilde{v} \mapsto \tilde{k}$ has an analytic continuation along the negative real axis, and the $\tilde{k}$-image of this ray is the bounded set $(-1, 0]$.

When $4J + 1 < r < 4J + 3$ ($b < 0$) and when $4K + 1 < r < 4K + 3$ ($b > 0$) the $\tilde{v}$ ray is in the upper-half $\tilde{v}$-plane. For the other cases the ray is in the lower-half plane. We will give the proof only for the upper-half plane, since the other case is very similar. From now on, nothing in the proof depends on the sign of b (or on $|b|$), but only on $0 < \phi = \arg(\tilde{v}) < \pi$, where $\tilde{v}$ is on the ray. ϕ depends on r and the sign of b; it is given by the formula

$$\phi = \begin{cases} \frac{\pi}{2}(4J + 3 - r) & b < 0, \\ \frac{\pi}{2}(r - 4K - 1) & b > 0. \end{cases}$$

We will show that the image of the ray under the map $\tilde{v} \mapsto \tilde{k}$ is a continuous path connecting $\tilde{k} = 0$ to $\tilde{k} = -1$ lying in the upper-half $\tilde{k}$-plane. In fact, this curve has an expression in terms of polar coordinates (R, θ), where $\tilde{k} = -1 + Re^{i\theta}$, $R > 0$ and $0 < \theta < \theta_{\max} = \frac{\pi - \phi}{r}$. If we can prove all this, then the statement of this Lemma follows.

Consider $\tilde{v} = \tilde{k}(1+\tilde{k})^{-r}$, for $\tilde{k}$ in the wedge $\tilde{k} = -1 + Re^{i\theta}$, where $R > 0$ and $0 < \theta < \theta_{\max}$. We will compute $\arg(\tilde{v})$ in terms of the polar coordinates (R, θ):

$$\tilde{v} = R^{-r}e^{-ir\theta}[e^{i\pi} + Re^{i\theta}] = R^{-r}e^{i(\pi - r\theta)}[1 + Re^{i(\theta - \pi)}],$$

and hence

$$\arg(\tilde{v}) = \pi - r\theta + \arg(1 + R\cos(\theta - \pi) + iR\sin(\theta - \pi))$$
$$= \pi - r\theta + \arg(1 - R\cos(\theta) - iR\sin(\theta))$$
$$= \pi - r\theta - \cos^{-1}\left(\frac{1 - R\cos(\theta)}{\sqrt{1 - 2R\cos(\theta) + R^2}}\right).$$

Thus the polar equation of the curve we are looking for is

$$\phi = g(R; \theta) = \pi - r\theta - \cos^{-1}\left(\frac{1 - R\cos(\theta)}{\sqrt{1 - 2R\cos(\theta) + R^2}}\right).$$

$g(R; \theta)$ is a monotonically decreasing function of $R > 0$ (for $\theta \in (0, \theta_{\max})$ fixed). Furthermore, $g(0; \theta) = \pi - r\theta > \pi - r\theta_{\max} = \phi$ and $\lim_{R \to \infty} g(R, \theta) = \pi - r\theta - \cos^{-1}(-\cos(\theta)) = -(r-1)\theta < 0 < \phi$. So for each $\theta \in (0, \theta_{\max})$ there is a unique $R(\theta) > 0$ where $g(R(\theta); \theta) = \phi$. The implicit function theorem implies that $R(\theta)$ is a continuous function of $\theta \in (0, \theta_{\max})$.

Since (by Lemma 1.3.1 with $a = 1$) $\tilde{k} = \tilde{v} + O(\tilde{v}^2)$ for $\tilde{v}$ sufficiently small, the curve (the $\tilde{k}$-image of the ray $\arg(\tilde{v}) = \phi$) departs $\tilde{k} = 0$ tangent to the ray $\arg(\tilde{k}) = \phi$ as $\tilde{v}$ departs from 0, and hence lies in the wedge. By uniqueness, we see that $\lim_{\theta \to 0^+} R(\theta) = 1$.

On the other hand, for $0 \leq R \ll 1$ fixed, we can consider $g(R; \theta) = \phi$ as an equation to determine $\theta \in (0, \theta_{\max}]$. For such R, $g(R; \theta)$ is a decreasing function of θ. $g(R; \theta_{\max}) < \pi - r\theta_{\max} = \phi$. $g(R; 0) = \pi - \cos^{-1}(1) = \pi > \phi$. Hence there is a unique $\theta(R) \in (0, \theta_{\max})$ such that $g(R; \theta) = \phi$. $\theta(R)$ is a continuous function of R. Furthermore, since $\phi = \pi - r\theta_{\max}$ we have

$$\theta_{\max} - \theta(R) = \frac{1}{r}\cos^{-1}\left(\frac{1 - R\cos(\theta(R))}{\sqrt{1 - 2R\cos(\theta(R)) + R^2}}\right) \to 0^+$$

as $R \to 0^+$. We claim this implies $\lim_{\theta \to \theta_{\max}^-} R(\theta) = 0$. To see this, let $0 < \epsilon \ll 1$ be given, and let θ_0 be the minimum value of the continuous function $\theta(R)$ on the interval $[0, \epsilon]$. We know $0 < \theta_0 < \theta_{\max}$. Then for all $\theta_1 \in (\theta_0, \theta_{\max})$ we have by the Intermediate Value Theorem that there exists an $R \in (0, \epsilon)$ such that $\theta(R) = \theta_1$. But by uniqueness this implies that $R(\theta_1) = R$. Thus our claim is true.

So $R(\theta)$ can be extended to be a continuous function on $[0, \theta_{\max}]$. The graph of this polar equation is the curve we are looking for. It cannot pass through the point $\tilde{k} = 1/(r-1)$. Hence the derivative of the function $\tilde{k} \mapsto \tilde{k}(1 + \tilde{k})^{-r}$ never vanishes on the curve. The $\tilde{v}$-image of the point $\tilde{k} = -1 + R(\theta)e^{i\theta}$ as tends to infinity as $\theta \to \theta_{\max}^-$. Thus the interval $[0, \theta_{\max})$ is mapped surjectively onto the ray via the map $\theta \mapsto \tilde{k} = -1 + R(\theta)e^{i\theta} \mapsto \tilde{v}$. Now we can prove that the function $\tilde{v} \mapsto \tilde{k}$ has an analytic continuation along the ray $\arg(\tilde{v}) = \phi$. There is a maximal half-open subinterval (containing $\tilde{v} = 0$) of the ray along which an analytic continuation can be performed. If this subinterval is not the entire ray, then the endpoint of the subinterval is a finite point on the ray. By the surjectivity we mentioned above, there is a number $\theta \in [0, \theta_{\max})$ such that $\tilde{k} = -1 + R(\theta)e^{i\theta}$ is the preimage of the endpoint under the map $\tilde{k} \mapsto \tilde{v}$. Since the derivative does not vanish at this point, we have by the inverse function theorem that the map $\tilde{k} \mapsto \tilde{v}$ can be inverted holomorphically in a neighborhood of the endpoint. This is a contradiction of the maximality of the subinterval.

Finally, by the uniqueness of the solutions of the equation $g(R; \theta) = \phi$, it follows that the image of the ray under this analytic continuation is contained in the curve we have described in polar coordinates. Hence that image is bounded in the $\tilde{k}$-plane. $\square$

This lemma allows us to use an alternate parameterization of the contour $\tilde{C}$, namely $k(v)$ ranges over $\tilde{C}$ as v ranges over $[0, \infty)$. Now a bound on $k(v)$ in terms of v holding uniformly in b can be proved.

Lemma 1.4.2. *Under the assumptions of Lemma 1.4.1, Let $C = \sup |\tilde{k}/\rho|^{\frac{1}{r}}$, where $\tilde{k}$ ranges over the image of the ray $\tilde{v} = \frac{\rho}{b} \left(\frac{iv}{b} \right)^{r-1}$, $v > 0$, under the analytic continuation (described in the previous lemma) of the local inverse mapping $\tilde{v} \mapsto \tilde{k}$. Then for all $v \geq 0$ we have*

$$|k(v)| \leq C v^{1/r}.$$

Proof. For $\tilde{v}$ on this ray, and $\tilde{k}(\tilde{v})$ denoting the value of the analytic continuation at $\tilde{v}$, we have $\tilde{v} = \tilde{k}(\tilde{v})[1 + \tilde{k}(\tilde{v})]^{-r}$. Thus $|1 + \tilde{k}(\tilde{v})| = |\tilde{k}(\tilde{v})/\tilde{v}|^{1/r}$. Therefore

$$|k(v)| = \frac{v}{|b|}|1 + \tilde{k}(\tilde{v})| = \frac{v}{|b|} \frac{|\tilde{k}(\tilde{v})|^{1/r}}{\left[\frac{\rho}{|b|} \frac{v^{r-1}}{|b|^{r-1}} \right]^{1/r}} = \left| \frac{\tilde{k}(\tilde{v})}{\rho} \right|^{1/r} v^{1/r} \leq C v^{1/r}. \quad \square$$

This estimate faithfully reflects the behavior of of $k(v)$ for large v, but not generally for small v. However, it is sufficient to establish all our decay estimates, as the following lemmas show. The first concerns that part of $E_N(\zeta, t)$ arising from the Taylor remainder of $V(k)$.

Lemma 1.4.3. *Under the assumptions of Lemma 1.4.1, suppose in addition that $V(k)$ is an entire function satisfying the estimate*

$$|V(k)| \leq C \frac{e^{a \Im k}}{(1 + |k|)^{\lambda}},$$

uniformly for k in the smallest closed sector containing the contour $\tilde{C}$, where $C > 0, a \in \mathbb{R}, \lambda \in \mathbb{R}$. Let $N \geq 0$ be an integer, and $R_N(k; 0)$ the Taylor remainder of $V(k)$ at $k = 0$ defined by

$$R_N(k; 0) = V(k) - \sum_{j=0}^{N-1} \frac{V^{(j)}(0)}{j!} k^j$$

for all $k \in \mathbb{C}$. Then

$$\sup_{t \geq 1} \sup_{\pm \zeta \geq \epsilon} |\zeta t^{\beta}|^{1+\gamma+N} \left| \left(\frac{i}{b} \right)^{1+\gamma} \int_0^{\infty} e^{-vt} v^{\gamma} \frac{(1 + \tilde{k})^{1+\gamma}}{1 - (r-1)\tilde{k}} R_N(k; 0) \, dv \right| < \infty.$$

Proof. First we note that we have the estimate

$$\left| \frac{R_N(k;0)}{k^N} \right| \leq C' \frac{e^{|ak|}}{(1+|k|)^{\lambda'}},$$

where $\lambda' = \min\{\lambda + N, 0\}$. To obtain this, we used the fact that $|R_N(k)/k^N|$ is bounded for $|k| \leq 1$, and for $|k| \geq 1$ we used the definition of $R_N(k;0)$ and the estimates on $V(k)$ given in the statement of the lemma. If we now use Lemma 1.4.2 we get

$$\left| \frac{R_N(k;0)}{k^N} \right| \leq C'(1 + Cv^{1/r})^{-\lambda'} e^{|a|Cv^{1/r}}.$$

Note that $k^N = \left(\frac{iv}{b}\right)^N (1 + \tilde{k})^N$. From Lemma 1.4.1 and Lemma 1.4.2 we see that

$$\frac{(1+\tilde{k})^{1+\gamma+N}}{1-(r-1)\tilde{k}}$$

is bounded by an absolute constant. Since $\zeta t^\beta = bt$ we have

$$|\zeta t^\beta|^{1+\gamma+N} \left| \left(\frac{i}{b}\right)^{1+\gamma} \int_0^\infty e^{-vt} v^\gamma \frac{(1+\tilde{k})^{1+\gamma}}{1-(r-1)\tilde{k}} R_N(k;0)\, dv \right|$$

$$\leq C'' |bt|^{1+\gamma+N} \frac{1}{|b|^{1+\gamma+N}} \int_0^\infty e^{-vt} v^{\gamma+N} (1+v)^{-\lambda'/r} e^{|a|Cv^{1/r}}\, dv$$

$$\leq C'' \int_0^\infty e^{-y} y^{\gamma+N} (1+y/t)^{-\lambda'/r} e^{|a|C(y/t)^{1/r}}\, dy$$

$$\leq C'' \int_0^\infty e^{-y} y^{\gamma+N} (1+y)^{-\lambda'/r} e^{|a|Cy^{1/r}}\, dy < \infty,$$

since $r > 1$ and $t \geq 1$. $\square$

Our next lemma concerns that part of $E_N(\zeta, t)$ coming from the Taylor remainder $R_N^j(\tilde{v})$.

Lemma 1.4.4. *Under the assumptions of Lemma 1.4.1, suppose in addition that $\frac{1}{r} < \beta < 1$, $N \geq 1$ and $0 \leq j \leq N-1$ are integers and n_j is the smallest integer such that $j + n_j(r - 1/\beta) \geq N$. Define*

$$R_N^j(\tilde{v}) = \frac{(1+\tilde{k})^{1+\gamma+j}}{1-(r-1)\tilde{k}} - \sum_{n=0}^{n_j-1} \binom{\gamma+j+nr}{n} \tilde{v}^n.$$

for all $\tilde{v} = \frac{\rho}{b}\left(\frac{iv}{b}\right)^{r-1}$, $v \geq 0$. Then we have that

$$\sup_{t \geq 1} \sup_{\pm \zeta \geq \epsilon} \frac{|\zeta t^\beta|^{1+\gamma+j+n_j r}}{t^{n_j}} \left| \left(\frac{i}{b}\right)^{1+\gamma+j} \int_0^\infty e^{-vt} v^{\gamma+j} R_N^j(\tilde{v})\, dv \right| < \infty.$$

Remark. This part of the error can decay faster than the part dealt with in Lemma 1.4.3. We have stated the optimal result here. To see that the above result implies that this part of the error decays like $O(|\zeta t^\beta|^{-(1+\gamma+N)})$, observe that since $\pm\zeta \geq \epsilon > 0$ we have

$$|\zeta t^\beta|^{1+\gamma+N} \leq C_\epsilon |\zeta t^\beta|^{1+\gamma+j+n_j(r-1/\beta)} \leq C'_\epsilon |\zeta|^{1+\gamma+j+n_jr}(t^\beta)^{1+\gamma+j+n_j(r-1/\beta)}$$
$$= C'_\epsilon \frac{|\zeta t^\beta|^{1+\gamma+j+n_jr}}{t^{n_j}}.$$

Proof. We begin with the observation that $R^j_N(\tilde{v})/\tilde{v}^{n_j}$ is bounded by a constant independent of $\tilde{v}$ (on the ray). For $|\tilde{v}| \leq 1$ this follows from the fact that $R^j_N(\tilde{v})$ is a Taylor remainder of the correct order. For $|\tilde{v}| \geq 1$ it follows from Lemma 1.4.1. Estimating as in the previous lemma we get

$$\frac{|\zeta t^\beta|^{1+\gamma+j+n_jr}}{t^{n_j}} \left| \left(\frac{i}{b}\right)^{1+\gamma+j} \int_0^\infty e^{-vt} v^{\gamma+j} R^j_N(\tilde{v})\, dv \right|$$
$$\leq C \frac{|bt|^{1+\gamma+j+n_jr}}{t^{n_j}} \frac{1}{|b|^{1+\gamma+j+n_jr}} \int_0^\infty e^{-vt} v^{\gamma+j+n_j(r-1)}\, dv$$
$$= C \int_0^\infty e^{-y} y^{\gamma+j+n_j(r-1)}\, dy < \infty. \qquad \Box$$

The results of the previous two lemmata can be assembled to prove the desired result in the case where $\tilde{C}$ does not hit a saddle point. In the remainder of this section we must deal with the important special case where $\tilde{C}$ does hit a saddle point, i.e. $r = 4J - 1$ when $b < 0$ and $r = 4K + 1$ when $b > 0$. The formulae at the beginning of the proof of Lemma 1.4.1 show that in this case the ray containing $\tilde{v}$ is contained in the positive real axis. This ray collides with a branch point, $\tilde{v}_0 = (r - 1)^{r-1}/r^r$, of the inverse map $\tilde{v} \mapsto \tilde{k}$. So as v ranges over the interval $[0, v_0]$ (where $v_0 = (r - 1)\rho B^{r/(r-1)}$), $\tilde{v}$ ranges over the interval $[0, \tilde{v}_0]$, and $\tilde{k}$ ranges over $[0, 1/(r - 1)]$. In particular, the $\tilde{k}$-image, i.e. $[0, 1/(r - 1)]$, of the $\tilde{v}$-interval $[0, \tilde{v}_0]$ is a bounded set whose size depends only on r. Thus the conclusion of Lemma 1.4.1 holds in this case as well. Because of this boundedness, the estimate $|k(v)| \leq Cv^{1/r}$ holds for all $v \in [0, v_0]$, where C depends only on r and ρ. The proof is the same as that of Lemma 1.4.2. However, since $\tilde{k} \to 1/(r - 1)$ as $\tilde{v} \to \tilde{v}_0$, we need to get some control over the quantity $1/[1 - (r - 1)\tilde{k}]$. This is provided in the next lemma.

Lemma 1.4.5. *Suppose* $r = 4J - 1$ *when* $b < 0$ *and* $r = 4K + 1$ *when* $b > 0$, *and that* $\tilde{v} \in [0, \tilde{v}_0)$ *and* $\tilde{k} \in [0, 1/(r - 1))$ *is a function of* $\tilde{v}$ *such that* $\tilde{v} = \tilde{k}(1 + \tilde{k})^{-r}$. *Then*

$$\frac{1}{1 - (r - 1)\tilde{k}} \leq \frac{1}{\sqrt{1 - \tilde{v}/\tilde{v}_0}}.$$

Proof. Consider the function $(1 + x)(1 - x/r)^r$ defined for $x \in [0, 1]$. Since $r > 1$ this function is non-increasing on $[0, 1]$. Hence $(1+x)(1-x/r)^r \leq 1$ on that interval. Rearranging we get $1/(1-x/r)^r \geq 1+x$; multiplying by the nonnegative quantity $1 - x$ we get $(1 - x)/(1 - x/r)^r \geq 1 - x^2$. Since $\tilde{k} \in [0, 1/(r - 1))$, we have $(r - 1)\tilde{k} \in [0, 1)$ and $1 - (r - 1)\tilde{k} \in (0, 1]$. Substituting $x = 1 - (r - 1)\tilde{k}$ into our inequality, we get

$$1 - [1 - (r - 1)\tilde{k}]^2 \leq \frac{(r - 1)\tilde{k}}{\{1 - \frac{1}{r}[1 - (r - 1)\tilde{k}]\}^r} = \frac{r^r}{(r - 1)^{r-1}}\frac{\tilde{k}}{(1 + \tilde{k})^r} = \frac{\tilde{v}}{\tilde{v}_0}.$$

The desired estimate follows by rearranging this. $\quad\square$

In the following three lemmas we must deal with the three types of error terms which are present in this case. The first type arises from the Taylor remainder $R_N(k)$ for the function $V(k)$.

Lemma 1.4.6. *Under the assumptions of Lemma 1.4.5, make the additional assumptions on $V(k)$, N, and $R_N(k; 0)$ as in Lemma 1.4.3. Then*

$$\sup_{t \geq 1} \sup_{\pm \zeta \geq \epsilon} |\zeta t^\beta|^{1+\gamma+N} \left| \left(\frac{i}{b}\right)^{1+\gamma} \int_0^{v_0} e^{-vt} v^\gamma \frac{(1 + \tilde{k})^{1+\gamma}}{1 - (r - 1)\tilde{k}} R_N(k; 0) \, dv \right| < \infty.$$

Proof. Since $\tilde{k}$ remains bounded, the estimate in Lemma 1.4.2 remains true, holding now for $v \in [0, v_0]$. Hence we can estimate just as in Lemma 1.4.3 except we must also use Lemma 1.4.5. So the quantity to be estimated is bounded by

$$C'' t^{1+\gamma+N} \int_0^{v_0} e^{-vt} \frac{v^{\gamma+N}(1 + v)^{-\lambda'/r}}{\sqrt{1 - \tilde{v}/\tilde{v}_0}} e^{|a|Cv^{1/r}} \, dv.$$

Now we split the integral in two according as $\tilde{v} \in [0, \frac{3}{4}\tilde{v}_0]$ or as $\tilde{v} \in (\frac{3}{4}\tilde{v}_0, \tilde{v}_0]$. On the first subinterval, $(1-\tilde{v}/\tilde{v}_0)^{-1/2} \leq 2$, and hence one can extend the interval of integration to $[0, \infty)$ and estimate as in Lemma 1.4.3. On the second subinterval, namely $v \in (v_1, v_0]$, where $v_1 = (\frac{3}{4})^{\frac{1}{r-1}} v_0$, we can remove from the integral the supremum of everything except the square root, obtaining as a bound

$$C'' t \left[\sup_{v_1 < v \leq v_0} e^{-vt}(vt)^{\gamma+N}(1 + v)^{-\lambda'/r} e^{|a|Cv^{1/r}} \right] \int_{v_1}^{v_0} \frac{dv}{\sqrt{1 - \tilde{v}/\tilde{v}_0}}.$$

Notice that

$$\frac{\tilde{v}}{\tilde{v}_0} = \frac{\frac{\rho}{b}\left(\frac{iv}{b}\right)^{r-1}}{\frac{\rho}{b}\left(\frac{iv_0}{b}\right)^{r-1}} = \left(\frac{v}{v_0}\right)^{r-1}.$$

Therefore, when we change variables in the integral we get

$$\int_{v_1}^{v_0} \frac{dv}{\sqrt{1 - \tilde{v}/\tilde{v}_0}} = v_0 \int_{(\frac{3}{4})^{\frac{1}{r-1}}}^1 \frac{dv'}{\sqrt{1 - (v')^{r-1}}}.$$

Thus we must bound (up to a constant multiple)

$$v_0 t \sup_{v_1 < v \le v_0} e^{-vt}(vt)^{\gamma+N}(1+vt)^{-\lambda'/r}e^{|a|C(vt)^{1/r}}$$

$$\le v_0 t \sup_{y > (\frac{3}{4})^{\frac{1}{r-1}} v_0 t} e^{-y} y^{\gamma+N}(1+y)^{M'/r}e^{|a|Cy^{1/r}}.$$

Since the function under the supremum is eventually exponentially decreasing, it is clear that the above quantity is bounded independently of $t \ge 1$ and $\pm\zeta \ge \epsilon$. $\square$

The next lemma handles the contribution to the error arising from the Taylor remainder $R_N^j(\tilde{v})$.

Lemma 1.4.7. *Under the assumptions of Lemma 1.4.5, assume in addition the assumptions of Lemma 1.4.4 concerning N, j, n_j, and $R_N^j(\tilde{v})$. Then*

$$\sup_{t\ge 1} \sup_{\pm\zeta\ge\epsilon} \frac{|\zeta t^\beta|^{1+\gamma+j+n_j r}}{t^{n_j}} \left|\left(\frac{i}{b}\right)^{1+\gamma+j} \int_0^{v_0} e^{-vt} v^{\gamma+j} R_N^j(\tilde{v})\, dv\right| < \infty.$$

Proof. As in the proof of Lemma 1.4.4 we see we must estimate

$$\int_0^{v_0} e^{-vt}(vt)^{\gamma+j+n_j(r-1)}\left|\frac{R_N^j(\tilde{v})}{\tilde{v}^{n_j}}\right| t\, dv = \int_0^{v_1} + \int_{v_1}^{v_0},$$

where $v_1 = (\frac{3}{4})^{\frac{1}{r-1}} v_0$. In the first integral, $\tilde{v}$ ranges over the interval $[0, \frac{3}{4}\tilde{v}_0]$, where $\tilde{v}_0 = (r-1)^{r-1}/r^r$. Over this interval $R_N^j(\tilde{v})/\tilde{v}^{n_j}$ is bounded by an absolute constant. Hence the first integral can be bounded as in Lemma 1.4.4. The second integral corresponds to $\tilde{v} \in (\frac{3}{4}\tilde{v}_0, \tilde{v}_0)$, and on this interval we have, using the definition and Lemma 1.4.5, the estimate

$$\left|\frac{R_N^j(\tilde{v})}{\tilde{v}^{n_j}}\right| \le \frac{C}{\sqrt{1-\tilde{v}/\tilde{v}_0}} = \frac{C}{\sqrt{1-(v/v_0)^{r-1}}}.$$

Thus we can bound this second integral in a similar way to what was done in the proof of Lemma 1.4.6. $\square$

Finally, we must treat the portion of the error term arising from the complementary incomplete Gamma function.

Lemma 1.4.8. *Under the assumptions of Lemma 1.4.7, suppose further that $n_j \ge 1$ and $0 \le n \le n_j - 1$ is an integer. Then*

$$\sup_{t\ge 1}\sup_{\pm\zeta\ge\epsilon} |\zeta t^\beta|^{1+\gamma+N}\left|\left(\frac{i}{bt}\right)^{1+\gamma+j+n(r-1)}\left(\frac{\rho}{b}\right)^n \Gamma(1+\gamma+j+n(r-1), v_0 t)\right|$$

$$< \infty.$$

Proof. We will use the estimate $|\Gamma(a,x)| \leq Cx^{a-1}e^{-x}$, which holds when $a > 0$ and $x \geq M > 0$. So up to a multiplicative constant we have

$$|\zeta t^\beta|^{N-j-nr} t^n (v_0 t)^{\gamma+j+n(r-1)} e^{-v_0 t} = t^{\frac{N-j}{r}} |\zeta t^{\beta-\frac{1}{r}}|^{N-j-nr} (v_0 t)^{\gamma+j+n(r-1)} e^{-v_0 t}$$

But

$$v_0 t = (r-1)\rho \left| \frac{b}{r\rho} \right|^{\frac{r}{r-1}} \qquad t = (r-1)\rho \left| \frac{btt^{-1/r}}{r\rho} \right|^{\frac{r}{r-1}} = (r-1)\rho \left| \frac{\zeta t^{\beta-1/r}}{r\rho} \right|^{\frac{r}{r-1}}.$$

So again up to a multiplicative constant we have

$$t^{(N-j)/r} |\zeta t^{\beta-1/r}|^{N+(r\gamma+1)/(r-1)} e^{-c|\zeta t^{\beta-1/r}|^{\frac{r}{r-1}}}$$

Let $y_0 \geq 0$ be the abscissa where the function $y^{N+(r\gamma+1)/(r-1)} e^{-cy^{\frac{r}{r-1}}}$ attains its maximum value. This number does not depend on ζ or t. The function is decreasing on the interval (y_0, ∞). Let $t_0 \geq 1$ be as small as possible such that $\epsilon t_0^{\beta-1/r} \geq y_0$. If $1 \leq t \leq t_0$ then the quantity we must bound is no larger than $t_0^{(N-j)/r}$ times the maximum value of this function. On the other hand, if $t > t_0$ then

$$t^{(N-j)/r} \sup_{\pm\zeta \geq \epsilon} |\zeta t^{\beta-1/r}|^{N+(r\gamma+1)/(r-1)} e^{-c|\zeta t^{\beta-1/r}|^{\frac{r}{r-1}}}$$

$$\leq t^{(N-j)/r} |\epsilon t^{\beta-1/r}|^{N+(r\gamma+1)/(r-1)} e^{-c|\epsilon t^{\beta-1/r}|^{\frac{r}{r-1}}}$$

which is clearly bounded independently of $t > t_0$. □

So we have finally finished the proof of the main result concerning the Laplace integral over the contour $\tilde{C}$. Although this result was stated informally in the previous section, we will now state it formally.

Theorem 1.4.9. *Suppose $r > 1$, $\rho > 0$, $\gamma > -1$, $\frac{1}{r} < \beta < 1$, and $\epsilon > 0$. Suppose $(x,t) \in R_\pm$, and $b = x/t$, $\zeta = xt^{-\beta}$. Let $\tilde{C}$ be the contour, depending on b, described in Lemma 1.2.1. Suppose $V(k)$ is an entire function of k which satisfies the estimate*

$$|V(k)| \leq C\frac{e^{a\Im k}}{(1+|k|)^\lambda},$$

uniformly for k in the smallest closed sector containing the contour $\tilde{C}$, where $C > 0, a \in \mathbb{R}, \lambda \in \mathbb{R}$. Let $N \geq 0$ be an integer, and for all integers $0 \leq j \leq N-1$ let $n_j \geq 0$ be the smallest integer such that $j + (r - 1/\beta)n_j \geq N$. Define

$$E_N(\zeta, t) = \int_{\tilde{C}} e^{i(kb-\rho k^r)t} k^\gamma V(k)\, dk$$

$$- \sum_{j=0}^{N-1} \sum_{n=0}^{n_j-1} \frac{V^{(j)}(0)}{j! n!} \Gamma(1+\gamma+j+nr) \left(\frac{\rho}{b}\right)^n \left(\frac{i}{bt}\right)^{1+\gamma+j+n(r-1)}.$$

Then the following estimate holds.

$$\sup_{t \geq 1} \sup_{\pm \zeta \geq \epsilon} |\zeta t^{\beta}|^{1+\gamma+N} |E_N(\zeta, t)| < \infty.$$

Remark. If in the case when $\tilde{C}(b)$ hits a saddle point we wish to include the exponentially small terms instead of considering them to be part of the error term, then in the above expansion we should replace the factor $\frac{1}{n!}\Gamma(1+\gamma+j+nr)$ by

$$\frac{\Gamma(1+\gamma+j+nr)}{n!} - \binom{\gamma+j+nr}{n} \Gamma\left(1+\gamma+j+n(r-1), (r-1)\rho \left|\frac{b}{r\rho}\right|^{\frac{r}{r-1}} t\right).$$

1.5 Asymptotic behavior of $V^{(n)}(k)$ as $k \to \infty$

In order to analyze the relative sizes of the terms of the asymptotic series we will derive in the next section it will be essential to understand exactly how

$$V^{(n)}(k) = \hat{v}^{(n)}(k) = \int_{-\infty}^{\infty} e^{-iky}(-iy)^n v(y)\, dy$$

behaves as $k \to \infty$. For this purpose we will need to understand this behavior as k tends to infinity along a ray which makes an angle θ with the positive real axis. Later, in order to control the error terms, we will need to impose estimates on $V^{(n)}(k)$ which hold uniformly for k in sectors adjoining $k = \infty$. Our error estimates will involve weights which reflect the behavior of $V(k)$ on the ray containing a saddle point. If the sectoral estimates we assume on $V(k)$ do not agree with the actual behavior of $V(k)$ on that ray (which is contained in the sector), then the error estimates we prove will fail to be sharp. To insure that the error estimates we prove are optimal, we must demonstrate for a reasonably broad class of functions $v(y)$ that the estimates we will assume on sectors are sharp, i.e agree with the leading order behavior of $V(k)$, and therefore no weaker than the sharpest estimates which hold on the ray containing the saddle point. Thus in this section we will prove detailed estimates of $V^{(n)}(k)$ on both rays and sectors, under clearly stated conditions on $v(y)$.

The behavior along rays $\arg l = \theta$ with $\theta \neq 0$ and $\theta = 0$ are similar in form but require different types of assumptions on the function $v(y)$. Of course there are many possibilities, but we will discuss a situation which we feel is sufficiently broad so as to give a good idea of the behavior we desire to understand. We suppose that $v(y)$ is defined for all $y \in \mathbb{R}$ and vanishes outside a compact interval. Generally speaking, each "singularity" of $v(y)$ contributes a term to the asymptotic behavior of $V(k)$. The nature of each term depends on the type and placement of the singularity. So not all singularities are equally important. We are only interested in the leading-order behavior, so many singularities can be effectively ignored. We will assume it is possible to divide the interval of support of v into finitely many intervals whose endpoints are the "important singularities". In the next two lemmas we will examine the case of a single subinterval.

The general case of multiple subintervals can be understood by simply adding the contributions of each subinterval and noting the relative sizes of the terms and errors.

The first lemma concerns the case $\theta \neq 0$. Together with it we can treat the case of sectors not containing the ray $\theta = 0$ (and such that $\Im k$ has a definite sign throughout the sector). In this case, the behavior of $y^n v(y)$ at one of the two endpoints of the interval of support completely determines the leading-order asymptotic behavior of $V^{(n)}(l)$ as $l \to \infty$. Which endpoint depends on the sign of θ.

Lemma 1.5.1 ($\theta \neq 0$). *Suppose* $V(l) = \int_{y_-}^{y_+} e^{-ily} v(y) \, dy$, *where* $y_- < y_+$ *and* l *tends to* ∞ *along the ray* $\arg(l) = \theta \in [-\pi/2, \pi/2] \setminus \{0\}$. *In the following whenever the symbols* $\pm$ *or* $\mp$ *are encountered, choose the upper sign if* $\theta > 0$ *and the lower sign if* $\theta < 0$. *Suppose that*

$$\phi(x) = v(\mp y + y_\pm) - a y^{\lambda-1} = O(y^{\epsilon+\lambda-1}), \qquad \text{for } y \in (0, c],$$
$$v(\pm y + y_\mp) \text{ is integrable} \qquad \text{for } y \in (0, c],$$

where $a \neq 0$, $\lambda > 0$, $c = (y_+ - y_-)/2$, $0 < \epsilon \leq 1$. *Let* $n \geq 0$ *be an integer. We have two cases.*

(1) $y_\pm \neq 0$. *Then we have the asymptotic result*

$$V^{(n)}(l) = \frac{e^{-ily_\pm}(-iy_\pm)^n \Gamma(\lambda) a}{[\mp il]^\lambda} + O\left(\frac{e^{y_\pm \Im l}}{|l|^{\epsilon+\lambda}}\right), \qquad \text{as } l \to \infty.$$

(2) $y_\pm = 0$. *Then we have the asymptotic result*

$$V^{(n)}(l) = \frac{\Gamma(\lambda+n) a}{[\mp il]^\lambda (-l)^n} + O\left(\frac{1}{|l|^{\epsilon+\lambda+n}}\right), \qquad \text{as } l \to \infty.$$

In both cases the stated error estimates also hold uniformly for $|l|$ *sufficiently large and* $L_0 \leq \pm \arg l \leq L_1$ *for any* $0 < L_0 < |\theta| < L_1 < \pi$.

Remark. It is inessential that we have divided the interval of integration into two *equal* subintervals and imposed some control on $v(y)$ only on the subinterval containing the dominant endpoint. We could impose control on any subinterval of positive length containing the dominant endpoint and obtain the same conclusion with a similar proof.

Proof. We split the interval of integration in half, and change variables in each of the resulting integrals to obtain

$$V^{(n)}(l) = (-i)^n e^{-ily_-} \int_0^c e^{-ily}(+y + y_-)^n v(+y + y_-) \, dy$$
$$+ (-i)^n e^{-ily_+} \int_0^c e^{ily}(-y + y_+)^n v(-y + y_+) \, dy.$$

One of these two integrals contributes the dominant term, depending on the sign of θ. If $\theta < 0$ (i.e. $\Im(l) < 0$) the first term dominates; if $\theta > 0$ (i.e. $\Im(l) > 0$) the second term dominates. To see this, rewrite the subordinate term in the following manner

$$(-i)^n e^{-ily_\mp} \int_0^c e^{\mp ily}(\pm y + y_\mp)^n v(\pm y + y_\mp)\, dy$$

$$= (-i)^n e^{-ily_\pm} \int_0^c e^{\pm il(y_+ - y - -y)}(\pm y + y_\mp)^n v(\pm y + y_\mp)\, dy.$$

Because of the estimate $|e^{\pm il(y_+ - y - -y)}| \le e^{-|\Im(l)|(2c-y)} \le e^{-|\Im(l)|c}$ we can bound the subordinate term by

$$e^{y_\pm \Im(l) - |\Im(l)|c} \int_0^c |\pm y + y_\mp|^n |v(\pm y + y_\mp)|\, dy.$$

If $L_0 \le \pm \arg l \le L_1$, where $0 < L_0 < L_1 < \pi$, then $|l| \to \infty$ implies $|\Im l| \to \infty$. So the subordinate term is uniformly exponentially smaller than the error terms we have stated for the dominant term.

So it remains to extract the behavior of the integral in the dominant term

(a) $$\int_0^c e^{\pm ily}(\mp y + y_\pm)^n v(\mp y + y_\pm)\, dy.$$

Notice that $\Re(\pm ily) = -|\Im(l)|y$. Here we apply the proof of Theorem 3.2, page 113, of [O1]. By our assumptions we have

(b) $$(\mp y + y_\pm)^n v(\mp y + y_\pm) = y_\pm^n a y^{\lambda - 1}$$
$$+ a y^{\lambda - 1}[(\mp y + y_\pm)^n - y_\pm^n]$$
$$+ (\mp y + y_\pm)^n \phi(y),$$

for all $y \in [0, c]$.

(1) The first term on the right-hand-side of (b) will make the dominant contribution when $n = 0$ or when $y_\pm \ne 0$. The second and third terms are $O(y^\lambda)$ and $O(y^{\epsilon + \lambda - 1})$ respectively. Inserting (b) into (a) and using these estimates we get

$$y_\pm^n a \int_0^\infty e^{\pm ily} y^{\lambda - 1}\, dy - y_\pm^n a \int_c^\infty e^{\pm ily} y^{\lambda - 1}\, dy$$
$$+ O\left(\int_0^c e^{-|\Im(l)|y} y^\lambda\, dy \right) + O\left(\int_0^c e^{-|\Im(l)|y} y^{\epsilon + \lambda - 1}\, dy \right)$$

These estimates hold uniformly in the sector $L_0 \le \pm \arg l \le L_1$. Changing variables in all these integrals we obtain

$$\frac{y_\pm^n a \Gamma(\lambda)}{[\mp il]^\lambda} - \frac{y_\pm^n a \Gamma(\lambda, \mp cil)}{[\mp il]^\lambda} + O\left(\frac{\Gamma(\lambda + 1)}{|\Im(l)|^{\lambda + 1}} \right) + O\left(\frac{\Gamma(\epsilon + \lambda)}{|\Im(l)|^{\epsilon + \lambda}} \right)$$

The term involving the incomplete Gamma function is $O(e^{-c|\Im l|}|l|^{-1})$ since we have the estimate $|\Gamma(a, x)| \leq (1 + \epsilon)|x|^{a-1}e^{-\Re x}$, which holds for $|x|$ sufficiently large and $|\arg(x)| < 3\pi/2$, and $x = \mp cil$ tends to infinity in the sector $-\pi/2 + L_0 \leq \pm \arg(x) \leq L_1 - \pi/2$. Note that $e^{-\Re x} = e^{-c|\Im l|}$. Thus the second and third terms are absorbed into the fourth, and this uniformly for l in the sector $L_0 \leq \pm \arg l \leq L_1$. Inserting this expression in for the integral (a) in the dominant term yields the result we stated.

(2) The first term on the right-hand-side of (b) vanishes however if $n \geq 1$ and $y_\pm = 0$. In that case the second and third terms become $a(\mp 1)^n y^{\lambda+n-1}$ and $(\mp 1)^n y^n \phi(y)$ respectively. Arguing just as in (1) yields the desired result. $\square$

When $\theta = 0$ the situation is more complex. As one can see from the proof of the previous lemma, the dominant term and the sub-dominant term become comparable when $\Im l = 0$. Thus contributions to the leading-order behavior might be spread throughout the interval of support, instead of arising only from one endpoint. Also notice that if $y_\pm$ is an endpoint of the support of v then $\lambda = \lambda_\pm$ is a measure of how smooth $v(y)$ is at the point $y = y_\pm$. "Important singularities" of v will now be locations where v fails to be smooth to a certain degree. Let m be the largest integer such that $v \in C^m(\mathbb{R})$. If v is not even continuous, we set $m = -1$. The situation which we will consider is where v fails to be C^{m+1} at finitely many points y_0, where it can be expanded in the following manner

$$v(y) = [a_0 + a_1 \operatorname{sgn}(y - y_0)]|y - y_0|^{\lambda-1} + \Phi(y),$$

where $0 \leq m + 1 < \lambda \leq m + 2$, and $\Phi^{(m+1)}$ is absolutely continuous in a neighborhood of $y = y_0$. When the non-degeneracy condition

$$a_0 \cos(\tfrac{\pi}{2}\lambda) - ia_1 \sin(\tfrac{\pi}{2}\lambda) \neq 0$$

holds (insuring that there is a genuine singularity at $y = y_0$ for the purposes of taking the limit $l \to \infty$), the number $\lambda - 1$ is called the *degree of smoothness*. Our results are the most interesting when the above non-degeneracy condition is satisfied, but our proofs in this section never use this assumption. The non-degeneracy condition for the purposes of taking the limit $l \to -\infty$ is the same as the above with $-i$ replaced by i. Thus only when λ is an integer is it possible to fail to have a genuine singularity for the purpose of taking both of the limits $l \to \infty$ and $l \to -\infty$. However in such a circumstance we would have that v is C^{m+1} near y_0, which is contrary to our definition of m.

There is no reason why one cannot have two or more terms in the expansion of $v(y)$ near $y = y_0$ with different values of λ, but having the same m. Our results could easily be modified to allow for this. Usually, the leading-order behavior of $V^{(n)}(l)$ as $l \to \infty$ will arise from the point where $\lambda - 1$ is the smallest (an exception to this can happen when $n \geq 1$ and the point is $y = 0$). We will derive the expansion of the integral over a subinterval both of whose endpoints are singularities like the point $y = y_0$ above.

Lemma 1.5.2 $(\theta = 0)$. *Suppose* $V(l) = \int_{y_-}^{y_+} e^{-ily} v(y)\, dy$, *where* $y_- < y_+$. *Suppose* $m \geq -1$ *is an integer,* $v \in C^m[y_-, y_+]$, $\bar{y} = (y_- + y_+)/2$, *and suppose that we define*

$$\phi_-(y) = v(y) - a_- |y - y_-|^{-1+\lambda_-}, \qquad y \in [y_-, \bar{y}],$$
$$\phi_+(y) = v(y) - a_+ |y - y_+|^{-1+\lambda_+}, \qquad y \in [\bar{y}, y_+],$$

where $a_-, a_+ \in \mathbb{C}$ *and* $\lambda_-, \lambda_+ > 0$. *Define* m_-, m_+ *to be integers such that*

$$0 \leq m_- + 1 < \lambda_- \leq m_- + 2,$$
$$0 \leq m_+ + 1 < \lambda_+ \leq m_+ + 2.$$

Assume $m = \min\{m_-, m_+\}$. *Suppose* $0 < L < \pi$, *and that* l *is in the sector* $0 \leq \pm \arg l \leq L$. *This means that we assume that* $\arg l$ *has a definite sign, and that we choose either the upper or lower sign in order to make the inequality* $\pm \arg l \geq 0$ *true. In the following whenever the symbols* $\pm$ *or* $\mp$ *are encountered, choose the upper sign if* $\arg l \geq 0$ *and the lower sign if* $\arg l \leq 0$. *Suppose*

$$\{(y_-, \lambda_-, m_-, a_-, \phi_-, -1), (y_+, \lambda_+, m_+, a_+, \phi_+, +1)\}$$
$$= \{(y_1, \lambda_1, m_1, a_1, \phi_1, \sigma_1), (y_2, \lambda_2, m_2, a_2, \phi_2, \sigma_2)\}$$

where if both y_- *and* y_+ *are nonzero we set* $y_1 = y_\pm$; *if one of* y_- *or* y_+ *vanishes, suppose* $y_1 = 0$. *Assume* $n \geq 0$ *is an integer. We have two cases:*

(1) $y_- \neq 0$ *and* $y_+ \neq 0$. *Assume* $v \in C^{m+1}(y_-, y_+)$, $\phi_-^{(m+1)}$ *is absolutely continuous for* $y \in [y_-, \bar{y}]$, *and* $\phi_+^{(m+1)}$ *is absolutely continuous for* $y \in [\bar{y}, y_+]$. *Then as* $l \to \infty$ *we have*

$$V^{(n)}(l) = -\sum_{j=0}^{m+1} \sum_{h=1}^{2} \sigma_h \frac{[(-iy)^n \phi_h(y)]^{(j)} e^{-ily}}{(il)^{j+1}} \bigg|_{y=y_h}$$
$$+ \frac{e^{-ily_-}(-iy_-)^n \Gamma(\lambda_-) a_-}{(il)^{\lambda_-}} + \frac{e^{-ily_+}(-iy_+)^n \Gamma(\lambda_+) a_+}{(-il)^{\lambda_+}} + o\left(\frac{e^{y_\pm \Im l}}{|l|^{m+2}}\right).$$

(2) *One of* y_- *or* y_+ *vanishes. Define*

$$m' = \min\{m_1 + n, m_2\}, \qquad n' = \max\{0, m' - m_1\} = \max\{0, \min\{n, m_2 - m_1\}\},$$

and assume that $v \in C^{m'+1}(y_-, y_+)$, $\left[y^{n'} \phi_1(y)\right]^{(m'+1)}$ *is absolutely continuous for* y *between* y_1 *and* $\bar{y}$, *and* $\phi_2^{(m'+1)}$ *is absolutely continuous for* y *between* y_2 *and* $\bar{y}$. *Then as* $l \to \infty$ *we have*

$$V^{(n)}(l) = -\sum_{j=0}^{m'+1} \sum_{h=1}^{2} \sigma_h \frac{[(-iy)^n \phi_h(y)]^{(j)} e^{-ily}}{(il)^{j+1}} \bigg|_{y=y_h}$$
$$+ \frac{\Gamma(n + \lambda_1) a_1}{(-\sigma_1 il)^{\lambda_1}(-l)^n} + \frac{e^{-ily_2}(-iy_2)^n \Gamma(\lambda_2) a_2}{(-\sigma_2 il)^{\lambda_2}} + o\left(\frac{e^{y_\pm \Im l}}{|l|^{m'+2}}\right).$$

In both cases the error estimates hold uniformly for $0 \leq \pm \arg l \leq L$.

Remarks. **(1)** If a term in the above expansions is asymptotically smaller than the corresponding error term then it is to be ignored.

(2) The terms of these expansions contained in the summations (the ordinary contributions of the endpoints y_-, y_+) will ultimately vanish when this lemma is applied in the manner we intend. This is because, as in the discussion prior to the statement of the lemma, we can choose ϕ_h to be the restriction of a function, denoted by Φ, defined on an open interval containing y_h, such that $\Phi^{(m+1)}$ is absolutely continuous on that interval. If y_h is a common endpoint of two adjacent intervals, then the function Φ will be a combination of ϕ_+ for the left interval and ϕ_- for the right interval. Hence the sum of the contributions from the endpoint y_h to the left interval and the right interval will be zero. If y_h is an endpoint of the interval of support of v then the regularity of Φ implies $\Phi(y_h) = \Phi'(y_h) = \cdots = \Phi^{(m+1)}(y_h) = 0$, and this implies that these endpoints contribute nothing. Thus at least one of the terms involving λ will be the leading-order term(s).

(3) The statement of part (2) of the lemma is made more complicated by our desire to understand the behavior of $V^{(n)}(l)$ for all $n \geq 0$, expressing the result only in terms of some assumed expansion for $v(y)$, as opposed to expressing the result in terms of an assumed expansion of $y^n v(y)$. We also try to make the minimal regularity assumptions necessary to obtain the result. If $y_1 = 0$ and $\phi_1^{(m_1+1)}$ is absolutely continuous for y between $y_1 = 0$ and $\bar{y}$, then the expansion

$$v(y) = a_1 |y|^{-1+\lambda_1} + \phi_1(y), \qquad -\sigma_1 y > 0,$$

implies that

$$y^n v(y) = a_1(-\sigma_1)^n |y|^{-1+\lambda_1+n} + y^n \phi_1(y), \qquad -\sigma_1 y > 0.$$

But unless $[y^n \phi_1(y)]^{(m_1+n+1)}$ is absolutely continuous for y between 0 and $\bar{y}$ there might be other singularities of ϕ_1 in that interval which would contribute terms of equal or greater importance to the leading-order behavior of $V^{(n)}(l)$ than the one arising from the endpoint y_1. Our results apply to the case where one can insure this additional regularity of ϕ_1 by the way one divides the interval of support of v into subintervals. For example, if $\epsilon > 0$ then $\phi(y) = y^{m+1+\epsilon}$ satisfies the condition that $\phi^{(m+1)}$ is absolutely continuous for $y \in [0, 1]$. Notice that this example also satisfies the conditions $[y^n \phi(y)]^{(m+n+1)}$ is absolutely continuous on $[0, 1]$ for all $n \geq 0$.

(4) Adopt the notation of the previous remark. If $\lambda_1 < m + 2$ then the constant a_1 is determined by our conditions on ϕ_1. However, if $\lambda_1 = m + 2$ it is not, unless we impose the additional constraint that $\phi_1^{(m+1)}(0) = 0$. This extra condition is unnecessary in practice, since we can expect the terms involving ϕ_1 to either vanish or cancel as we noted in remark 2.

(5) The statement of part (2) of the lemma is complicated because it contains several cases in one statement. Term number 1 is the term involving λ_1; likewise term number 2 is the one involving λ_2.

Case	m'	n'	keep	drop
(a) $m_2 < m_1$	m_2	0	2	1
(b) $m_1 = m_2 < m_1 + n$	m_2	$0 = m_2 - m_1$	2	1
(c) $m_1 < m_2 < m_1 + n$	m_2	$m_2 - m_1$	2	1
(d) $m_2 = m_1 + n$	$m_2 = m_1 + n$	$m_2 - m_1 = n$	1 & 2	
(e) $m_1 + n < m_2$	$m_1 + n$	n	1	2

Case (b) is a transition or change in the type of regularity assumption we make on ϕ_1. In case (a) we assume $\phi_1^{(m_2)}$ is absolutely continuous; but in case (c) we only assume $[y^{m_2-m_1}\phi_1(y)]^{(m_2)}$ is absolutely continuous, since this is less restrictive than supposing $\phi_1^{(m_2)}$ is absolutely continuous, and it is sufficient for our result. Our regularity assumptions in cases (a), (b), and (c) imply that $[y^n\phi_1(y)]^{(m_2)}$ is absolutely continuous. Case (d) is the main transition, where both of the terms are not absorbed in the error term. In cases (a)-(c) term 1 should be ignored, since it is asymptotically smaller than the error term. In case (e), term 2 should be ignored as being asymtotically insignificant as compared with the error term.

   (6) Consider the consequences of part (2) of the lemma when $\theta = \arg l = 0$. The case $m_2 < m_1$ (case (a) in the previous remark) would probably only arise if $y_1 = 0$ were an endpoint of the interval of support; even then its contribution would be lost in the error term. Suppose $m_1 < m_2$. Then as n increases (starting at 0) the situation goes from case (e) through case (d) and into cases (a)-(c). So the point $y_1 = 0$ contributes the dominant term when $n = 0$ but the point y_2 contributes the dominant term for $n > m_2 - m_1$. Notice that this sort of n-dependent transition does not occur in part (1) of the lemma.

   (7) Consider the consequences of part (1) of the lemma when $\theta = \arg l \geq 0$. Suppose $m_- = m_+$ and $\lambda_- < \lambda_+$. When $\theta > 0$, y_+ is the dominant endpoint, and the term arising from y_- is absorbed into the error term, giving a result which agrees with Lemma 1.5.1(1). On this ray we obtain an estimate of the form

$$|V^{(n)}(l)| \leq \frac{C_n(\theta)e^{y_+ + \Im l}}{|l|^{\lambda_+}},$$

where $C_n(\theta) > 0$ depends on the ray. On the ray $\theta = 0$ we have the estimate $|V^{(n)}(l)| \leq C_n|l|^{-\lambda_-}$ as $l \to \infty$. But since $\lambda_- < \lambda_+$ we see that it is impossible that $C_n(\theta)$ remain bounded as $\theta \to 0^+$. The sharpest estimate of the same form which holds *uniformly* on the entire sector $0 \leq \arg l \leq L$, $|l|$ sufficiently large, is

$$|V^{(n)}(l)| \leq \frac{C_n e^{y_+ + \Im l}}{|l|^{\lambda_-}},$$

where $C_n > 0$ is independent of θ. This estimate is sharp on the ray $\theta = 0$ but not if $\theta > 0$. Fortunately, our application only requires our sectoral estimates to be sharp on a single ray in the sector, namely the ray containing the saddle point, which turns out whenever the sector contains the ray $\theta = 0$ to be the ray $\theta = 0$.

Proof. These arguments are based on ideas in [O2]. Define $c = (y_+ - y_-)/2$. As in the proof of Lemma 1.5.1 we can split the interval of integration and change variables, obtaining the exact same expression that was obtained in that proof. Changing notations from y_-, y_+ to y_1, y_2 we have

(c) $$V^{(n)}(l) = (-i)^n \sum_{h=1}^{2} e^{-ily_h} \int_0^c e^{i\sigma_h ly}(-\sigma_h y + y_h)^n v(-\sigma_h y + y_h) \, dy.$$

By our assumed expansions we have for $y \in [0, c]$ that
(d)
$$(-\sigma_h y + y_h)^n v(-\sigma_h y + y_h) = a_h y_h^n y^{-1+\lambda_h} + a_h \left[(-\sigma_h y + y_h)^n - y_h^n \right] y^{-1+\lambda_h}$$
$$+ (-\sigma_h y + y_h)^n \phi_h(-\sigma_h y + y_h).$$

For convenience, define

$$U_h(y) = (-\sigma_h y + y_h)^n \phi_h(-\sigma_h y + y_h) + a_h \left[(-\sigma_h y + y_h)^n - y_h^n \right] y^{-1+\lambda_h}.$$

(1) The $(m+2)$nd derivative of $U_h(y)$ is integrable on $[0, c]$. Since $y_h \neq 0$, the second term in the definition of $U_h(y)$ is $O(y^{\lambda_h})$, and hence its $(m+2)$nd derivative is integrable on $[0, c]$ and also

$$\left(\frac{d}{dy} \right)^j \left[(-\sigma_h y + y_h)^n - y_h^n \right] y^{-1+\lambda_h} \bigg|_{y=0} = 0, \qquad j = 0, 1, \ldots, m+1.$$

Thus $U_h^{(j)}(0) = (-\sigma_h)^j [y^n \phi_h(y)]^{(j)}(y_h)$ for $j = 0, 1, \ldots, m+1$. The $(m+2)$nd derivative of the first term on the right-hand-side of (d) will not be integrable on $[0, c]$, although the $(m+1)$st derivative definitely is, and we have

$$\left(\frac{d}{dy} \right)^j y^{-1+\lambda_h} \bigg|_{y=0} = 0, \qquad j = 0, 1, \ldots, m.$$

Thus when we insert (d) into the integral in (c) and integrate by parts $(m+1)$ times (in the first term) and $(m+2)$ times (in the second and third terms, i.e. $U_h(y)$) we get

$$\int_0^c e^{i\sigma_h ly}(-\sigma_h y + y_h)^n v(-\sigma_h y + y_h) \, dy$$

$$= -\sum_{j=0}^{m+1} \frac{U_h^{(j)}(y) e^{i\sigma_h ly}}{(-i\sigma_h l)^{j+1}} \bigg|_{y=0}^{y=c} + \frac{1}{(-i\sigma_h l)^{m+2}} \int_0^c e^{i\sigma_h ly} U_h^{(m+2)}(y) \, dy$$

$$- \sum_{j=0}^{m} \frac{c_j y^{\lambda_h - 1 - j} e^{i\sigma_h ly}}{(-i\sigma_h l)^{j+1}} \bigg|_{y=0}^{y=c} + \frac{c_{m+1}}{(-i\sigma_h l)^{m+1}} \int_0^c e^{i\sigma_h ly} y^{\lambda_h - m - 2} \, dy,$$

where $c_j = a_h y_h^n \prod_{k=0}^{j-1}(\lambda_h - 1 - k)$, $j = 0, 1, \ldots$. To evaluate the last integral we use the following identity:

$$\int_0^c e^{-xy} y^{a-1} \, dy = \frac{\Gamma(a)}{x^a} - \frac{\Gamma(a, xc)}{x^a},$$

where $\Gamma(a, x) = \int_x^\infty e^{-t} t^{a-1} \, dt$ is the incomplete Gamma function. This identity is more or less obvious if $|\arg x| < \pi/2$, and can be inferred to hold elsewhere ($x \neq 0$) by analytic continuation. We need to apply it when $x = -i\sigma_h l$ and $a = \lambda_h - m - 1 > 0$. We agree that $\arg x = \arg(-i\sigma_h l) = -\sigma_h \pi/2 + \arg l$. With these conventions we have

$$\int_0^c e^{i\sigma_h l y} y^{\lambda_h - m - 2} \, dy = \frac{\Gamma(\lambda_h - m - 1)}{(-i\sigma_h l)^{\lambda_h - m - 1}} - \frac{\Gamma(\lambda_h - m - 1, -i\sigma_h l c)}{(-i\sigma_h l)^{\lambda_h - m - 1}}.$$

Integrating by parts in the integral defining the incomplete Gamma function $\Gamma(a, x)$ yields the formula $\Gamma(a, x) = x^{a-1} e^{-x} + (a - 1)\Gamma(a - 1, x)$ when $a \neq 1$. On the other hand, when $a = 1$ we have simply that $\Gamma(1, x) = e^{-x}$; thus the formula holds then as well. So

$$\frac{\Gamma(\lambda_h - m - 1, -i\sigma_h l c)}{(-i\sigma_h l)^{\lambda_h - m - 1}}$$
$$= \frac{c^{\lambda_h - m - 2} e^{i\sigma_h l c}}{-i\sigma_h l} + \frac{(\lambda_h - m - 2)\Gamma(\lambda_h - m - 2, -i\sigma_h l c)}{(-i\sigma_h l)^{\lambda_h - m - 1}}.$$

Because of the estimate $|\Gamma(a, x)| \leq (1 + \epsilon)|x|^{a-1} e^{-\Re x}$, which holds for $|x|$ sufficiently large, $|\arg(x)| < 3\pi/2$ (see page 263 of [AS]), we see that the second summand is $O(e^{-c\sigma_h \Im l}|l|^{-2})$ as $l \to \infty$ uniformly in the sector $0 \leq \pm \arg l \leq L$. Thus for all $m + 1 < \lambda_h$ we have

$$\int_0^c e^{i\sigma_h l y} y^{\lambda_h - m - 2} \, dy = \frac{\Gamma(\lambda_h - m - 1)}{(-i\sigma_h l)^{\lambda_h - m - 1}} - \frac{c^{\lambda_h - m - 2} e^{i\sigma_h l c}}{-i\sigma_h l} + O(e^{-c\sigma_h \Im l}|l|^{-2}).$$

From elementary properties of the Gamma function we have

$$\Gamma(\lambda_h - m - 1) \prod_{k=0}^m (\lambda_h - 1 - k) = \Gamma(\lambda_h).$$

Using all this we have

$$\int_0^c e^{i\sigma_h l y}(-\sigma_h y + y_h)^n v(-\sigma_h y + y_h) \, dy$$

$$= -\sum_{j=0}^{m+1} \frac{U_h^{(j)}(c) e^{i\sigma_h l c}}{(-i\sigma_h l)^{j+1}} - \sum_{j=0}^{m+1} \frac{c_j c^{\lambda_h - 1 - j} e^{i\sigma_h l c}}{(-i\sigma_h l)^{j+1}}$$

$$+ \sum_{j=0}^{m+1} \frac{[(-\sigma_h y + y_h)^n v(-\sigma_h y; y_h, m_h)]^{(j)} e^{i\sigma_h l y}}{(-i\sigma_h l)^{j+1}} \bigg|_{y=0}$$

$$+ \frac{a_h y_h^n \Gamma(\lambda_h)}{(-i\sigma_h l)^{\lambda_h}} + O\left(\frac{e^{-c\sigma_h \Im l}}{|l|^{m+3}}\right)$$

$$+ \frac{1}{(-i\sigma_h l)^{m+2}} \int_0^c e^{i\sigma_h l y} U_h^{(m+2)}(y) \, dy$$

for $|l|$ sufficiently large. By (d) we can rewrite the first two terms as

$$-\sum_{j=0}^{m+1} \frac{[(-\sigma_h y + y_h)^n v(-\sigma_h y + y_h)]^{(j)}}{(-i\sigma_h l)^{j+1}}\bigg|_{y=c} e^{i\sigma_h l c}$$

$$= \sigma_h \sum_{j=0}^{m+1} \frac{[y^n v(y)]^{(j)}}{(il)^{j+1}}\bigg|_{y=-\sigma_h c + y_h} e^{i\sigma_h l c}$$

Similarly we can rewrite the third term as

$$-\sigma_h \sum_{j=0}^{m+1} \frac{[y^n \phi_h(y)]^{(j)}}{(il)^{j+1}}\bigg|_{y=y_h}.$$

Inserting all this into (c) we get

$$V^{(n)}(l) = \sum_{h=1}^{2} \sigma_h \sum_{j=0}^{m+1} \frac{[(-iy)^n v(y)]^{(j)} e^{-ily}}{(il)^{j+1}}\bigg|_{y=-\sigma_h c + y_h}$$

$$-\sum_{h=1}^{2} \sigma_h \sum_{j=0}^{m+1} \frac{[(-iy)^n \phi_h(y)]^{(j)} e^{-ily}}{(il)^{j+1}}\bigg|_{y=y_h}$$

$$+\sum_{h=1}^{2} \frac{a_h(-iy_h)^n \Gamma(\lambda_h) e^{-ily_h}}{(-i\sigma_h l)^{\lambda_h}} + O\left(\frac{e^{(-c\sigma_h + y_h)\Im l}}{|l|^{m+3}}\right)$$

$$+\sum_{h=1}^{2} \frac{(-i)^n e^{-ily_h}}{(-i\sigma_h l)^{m+2}} \int_0^c e^{i\sigma_h ly} U_h^{(m+2)}(y)\,dy.$$

Since $-\sigma_h c + y_h = (y_- + y_+)/2 = \bar{y}$ for both $h = 1, 2$, we have that the first (double) sum is zero. We also used this fact when we combined the two big O terms, which are seen to be absorbed into our stated error term (since $\bar{y}\Im l \le y_\pm \Im l$). The second (double) sum is the cancellable contribution from y_- and y_+ that we stated. The third sum are the two main (non-cancelling) terms we stated. It remains to show that the last sum can also be absorbed in the error term.

If in the two integrals comprising the final sum we revert back to the original variable of integration, $y' = -\sigma_h y + y_h$, we get two integrals, one over $y' \in [y_-, \bar{y}]$ and the other over $y' \in [\bar{y}, y_+]$, which we can combine into a single integral over $y' \in [y_-, y_+]$. Multiplying and dividing this integral by $e^{-ily_\pm}$ and introducing yet another variable of integration $y'' = y' - y_\pm$ we obtain (after dropping the primes)

$$\frac{e^{-ily_\pm}}{(il)^{m+2}} \int_{y_- - y_\pm}^{y_+ - y_\pm} e^{-ily} U(y)\,dy,$$

where we know simply that $U(y)$ is integrable on the interval $[y_- - y_\pm, y_+ - y_\pm]$. We must show that this integral is $o(1)$ as $l \to \infty$ uniformly in the sector

$0 \leq \pm \arg l \leq L$. Notice that $|e^{-ily}| \leq 1$ for l in this sector and y in the interval of integration. Thus we can follow the usual proof of the Riemann-Lebesgue Lemma. Let $\epsilon > 0$ be given. Approximate U in the L^1-norm to within $\epsilon/2$ by a C^1 function U_1. Write $U = [U - U_0] + U_0$, split into two integrals, estimating the integral involving $U - U_0$ by the L^1-norm of this difference, which is less than $\epsilon/2$. Integrate by parts in the integral involving U_0 and estimate the result as $O(|l|^{-1})$, which for sufficiently large $|l|$ is less than $\epsilon/2$. Thus the above integral is $o(e^{y \pm \Im l}|l|^{-m-2})$ as $|l| \to \infty$ uniformly in the sector $0 \leq \pm \arg l \leq L$. This completes the proof of (1).

(2) As before we split into two integrals as in (c). Since $y_2 \neq 0$ the expansion (d) is relevant, and we can proceed exactly as in part (1) for the case $h = 2$ provided m is replaced by m'. The outcome is

$$
(-i)^n e^{-ily_2} \int_0^c e^{i\sigma_2 ly}(-\sigma_2 y + y_2)^n v(-\sigma_2 y + y_2)\, dy
$$

$$
= \sigma_2 \sum_{j=0}^{m'+1} \frac{[(-iy)^n v(y)]^{(j)} e^{-ily}}{(il)^{j+1}} \Bigg|_{y=\bar{y}}
$$

$$
- \sigma_2 \sum_{j=0}^{m'+1} \frac{[(-iy)^n \phi_2(y)]^{(j)} e^{-ily}}{(il)^{j+1}} \Bigg|_{y=y_2}
$$

$$
+ \frac{e^{-ily_2} a_2 (-iy_2)^n \Gamma(\lambda_2)}{(-i\sigma_2 l)^{\lambda_2}} + O\left(\frac{e^{\bar{y}\Im l}}{|l|^{m'+3}} \right)
$$

$$
+ \frac{(-i)^n}{(il)^{m'+2}} \int_0^c e^{-il(-\sigma_2 y + y_2)} \frac{U_2^{(m'+2)}(y)}{(-\sigma_2)^{m'+2}}\, dy
$$

Since $y_1 = 0$ the expansion (d) in the case $h = 1$ becomes

(d') $(-\sigma_1 y)^n v(-\sigma_1 y) = a_1(-\sigma_1)^n y^{-1+\lambda_1+n} + (-\sigma_1 y)^n \phi_1(-\sigma_1 y).$

We will follow the same procedure as in part (1) except for a few minor differences. Since $0 \leq n' \leq n$ and $[y^n \phi_1(y)]^{(m'+1)} = [y^{n-n'} y^{n'} \phi_1(y)]^{(m'+1)}$ we have that $[y^n \phi_1(y)]^{(m'+1)}$ is absolutely continuous. This fact allows us to repeat the argument for part (1) with $h = 1$, m replaced by m', λ_h replaced by $\lambda_1 + n$, U_1 is defined to be the second term in the expansion (d'), and $a_h y_h^n$ replaced by $a_1(-\sigma_1)^n$. The outcome is

$$
(-i)^n e^{-ily_1} \int_0^c e^{i\sigma_1 ly}(-\sigma_1 y)^n v(-\sigma_1 y)\, dy
$$

$$
= \sigma_1 \sum_{j=0}^{m'+1} \frac{[(-iy)^n v(y)]^{(j)} e^{-ily}}{(il)^{j+1}} \Bigg|_{y=\bar{y}}
$$

$$
- \sigma_1 \sum_{j=0}^{m'+1} \frac{[(-iy)^n \phi_1(y)]^{(j)} e^{-ily}}{(il)^{j+1}} \Bigg|_{y=0}
$$

$$+ \frac{a_1 \Gamma(\lambda_1 + n)}{(-i\sigma_1 l)^{\lambda_1}(-l)^n} + O\left(\frac{e^{\bar{y}\Im l}}{|l|^{m'+3}}\right)$$

$$+ \frac{(-i)^n}{(il)^{m'+2}} \int_0^c e^{-il(-\sigma_1 y)} \frac{U_1^{(m'+2)}(y)}{(-\sigma_1)^{m'+2}} \, dy$$

The contributions from $y_1 = 0$ and y_2 are combined exactly as in part (1); the combined error terms are estimated in the exact same manner. $\square$

The Standard Assumption and Sharp Estimate. Now we will describe an assumption on the function v which we will term the *standard assumption*. This assumption will be sufficient, but by no means necessary, in order to assert that $V(k) = \int_{y_-}^{y_+} e^{-iky} v(y) \, dy$ satisfies sharp estimates of the form

$$|V^{(n)}(l)| \leq \frac{C_n e^{a\Im l}}{(1 + |l|)^{\min\{\lambda + n, \lambda'\}}},$$

where $n \geq 0, C_n > 0, a \in \mathbb{R}, \lambda > 0, \lambda \leq \lambda' \leq \infty$. These estimates will hold uniformly for l in the sectors $0 \leq L_0 \leq \pm \arg l \leq L_1 < \pi$. The constants $C_n, a, \lambda, \lambda'$ can depend on the sector (which includes a choice of the sign $\pm$). It will be convenient to define $\lambda_n = \min\{\lambda, \lambda' - n\}$ for each integer $n \geq 0$.

Suppose $y_- < y_0 < y_+$ and $m \geq -1$ is an integer. Suppose $m + 1 < \lambda_0 \leq m + 2$ is a real number. Suppose v is a function supported on $[y_-, y_+]$ such that $v^{(m)}$ is absolutely continuous on $[-1 + y_-, 1 + y_+]$, and $v^{(m+1)}$ is not absolutely continuous on the same interval. (If $m = -1$ this simply means that v is not absolutely continuous on $[-1 + y_-, 1 + y_+]$.) y_0 is the point within the support of v where the "worst" singularity occurs, and λ_0 is a measure of the severity of this singularity. There are two types of standard assumptions, one with $y_0 \neq 0$ and the other with $y_0 = 0$. First suppose $y_0 \neq 0$. Define $\bar{y}_\pm = \frac{1}{2}(y_0 + y_\pm)$. In this case *the standard assumption* is the following.

(1) $v^{(m+1)}$ is absolutely continuous on $[-1 + y_-, \bar{y}_-]$.
(2) $v(y) = (y - y_-)^{-1+\lambda_-}[a_- + O((y - y_-)^\epsilon)]$ for all $y \in (y_-, \bar{y}_-]$, where $a_- \neq 0, \lambda_- > m + 2$, and $\epsilon > 0$.
(3) $v^{(m+1)}$ is continuous on (y_-, y_0).
(4) $\{v(y) - [a_0 + a_1 \operatorname{sgn}(y - y_0)]|y - y_0|^{\lambda_0 - 1}\}^{(m+1)}$ is absolutely continuous on $[\bar{y}_-, \bar{y}_+]$, where $a_0, a_1 \in \mathbb{C}$ satisfy the non-degeneracy condition

$$a_0 \cos(\tfrac{\pi}{2}\lambda_0) - ia_1 \sin(\tfrac{\pi}{2}\lambda_0) \neq 0.$$

(5) $v^{(m+1)}$ is continuous on (y_0, y_+).
(6) $v(y) = (y_+ - y)^{-1+\lambda_+}[a_+ + O((y_+ - y)^\epsilon)]$ for all $y \in [\bar{y}_+, y_+)$, where $a_+ \neq 0, \lambda_+ > m + 2$, and $\epsilon > 0$.
(7) $v^{(m+1)}$ is absolutely continuous on $[\bar{y}_+, 1 + y_+]$.

When $y_0 = 0$ the standard assumption is more complicated. Assume there is a point $y_1 \in [y_-, y_+] \setminus \{y_0\}$ where (heuristically) the "second worst" singularity occurs. For definiteness, assume $y_0 < y_1 < y_+$. Define $\bar{y}_- = \frac{1}{2}(y_0 + y_-), \bar{y} =$

$\frac{1}{2}(y_0 + y_1), \bar{y}_+ = \frac{1}{2}(y_1 + y_+)$. Suppose $\lambda_1 > \lambda_0$ is a real number. Let $m' \geq m$ be the integer such that $m' + 1 < \lambda_1 \leq m' + 2$. In this case *the standard assumption* is the following.

(1') $v^{(m'+1)}$ is absolutely continuous on $[-1 + y_-, \bar{y}_-]$.

(2') $v(y) = (y - y_-)^{-1+\lambda_-}[a_- + O((y - y_-)^\epsilon)]$ for all $y \in (y_-, \bar{y}_-]$, where $a_- \neq 0$, $\lambda_- > m' + 2$, and $\epsilon > 0$.

(3') $v^{(m'+1)}$ is continuous on (y_-, y_0).

(4') $y^{m'-m}\{v(y) - [a_0 + a_1 \operatorname{sgn}(y)]|y|^{\lambda_0 - 1}\}^{(m'+1)}(y)$ is absolutely continuous for $y \in [\bar{y}_-, \bar{y}]$, where $a_0, a_1 \in \mathbb{C}$ satisfy the non-degeneracy condition stated in (4) above.

(5') $v^{(m'+1)}$ is continuous on (y_0, y_1).

(6') $\{v(y) - [b_0 + b_1 \operatorname{sgn}(y - y_1)]|y - y_1|^{\lambda_1 - 1}\}^{(m'+1)}$ is absolutely continuous on $[\bar{y}, \bar{y}_+]$, where $b_0, b_1 \in \mathbb{C}$ satisfy the non-degeneracy condition

$$b_0 \cos(\tfrac{\pi}{2}\lambda_1) - ib_1 \sin(\tfrac{\pi}{2}\lambda_1) \neq 0.$$

(7') $v^{(m'+1)}$ is continuous on (y_1, y_+).

(8') $v(y) = (y_+ - y)^{-1+\lambda_+}[a_+ + O((y_+ - y)^\epsilon)]$ for all $y \in [\bar{y}_+, y_+)$, where $a_+ \neq 0$, $\lambda_+ > m' + 2$, and $\epsilon > 0$.

(9') $v^{(m'+1)}$ is absolutely continuous on $[\bar{y}_+, 1 + y_+]$.

If the sector is $0 \leq L_0 \leq \pm \arg l \leq L_1 < \pi$, so that it lies in the upper half k plane (choose the upper sign) or the lower half k plane (choose the lower sign), then in our sharp estimate we always have $a = y_\pm$. If $L_0 > 0$ then $\lambda = \lambda_\pm$. If $L_0 > 0$ and $y_\pm \neq 0$ then by Lemma 1.5.1(1) we have $\lambda' = \lambda$, and hence $\min\{\lambda + n, \lambda'\} = \lambda_\pm$. If $L_0 > 0$ and $y_\pm = 0$ (i.e Lemma 1.5.1(2)) then we have $\lambda' = \infty$ so that $\min\{\lambda + n, \lambda'\} = \lambda_\pm + n$. If $L_0 = 0$ and $y_0 \neq 0$ then we have $\lambda = \lambda' = \lambda_0$. If $L_0 = 0$ and $y_0 = 0$ then $\lambda = \lambda_0$ and $\lambda' = \lambda_1$. In this last case the decay rate $\min\{\lambda + n, \lambda'\}$ can have a more complicated dependence on n. But in all these cases $\lambda_n = \min\{\lambda, \lambda' - n\}$ satisfies $0 \leq 1 + \lambda_{n+1} - \lambda_n \leq 1$ for all $n \geq 0$.

1.6 Terms of the expansion, for the contour C_j

In this section we will derive explicitly the terms of the expansion of

$$I(x, t; r, \rho, \gamma, V, C_j) \overset{\text{def}}{=} \int_{C_j} e^{-(-ikb + i\rho k^r)t} k^\gamma V(k) \, dk,$$

where $-J \leq j \leq K$ is fixed. This derivation is formal, so during it the reader may assume $b \neq 0$ is fixed. By comparing two successive terms we will be able to see how much this assumption can be relaxed. We will also derive explicit error terms for our expansion. The form of the terms (not the error terms) will show what sort of weighted estimates we must prove. The proof of the weighted error estimates can be found in the next section. Those estimates will be obtained without any assumption that b is constant, so the terms we derive here will be the correct ones in this more general situation as well.

As we saw in section 1.2, $C_j = C_j^+ \cup C_j^-$, and in the latter half of section 1.1 we derived some explicit formulae for the terms of the expansion of an integral over a contour *starting* at a saddle point. We will apply these formulae where the contour is either C_j^+ or $-C_j^-$. (Note that C_j^- *ends* at the saddle point.) The only difference in the expansions for these two cases will be a result of different choices of the phase of p_0.

Let $l = k_j(b)$ and $p(k;l) = i\rho(k^r - rl^{r-1}k)$. Define $\tilde{p}(l) \stackrel{\text{def}}{=} p(l;l) = -i\rho(r-1)l^r$ and let $p'(k;l)$ denote the derivative of $p(k;l)$ with respect to k. Then $p'(l;l) = 0$, and $p_0 = p''(l;l)/2 = i\rho r(r-1)l^{r-2}/2$. Thus $k = l$ is a simple saddle point, i.e. $\mu = 2$ in the notation of section 1.1, which we follow throughout. We have an expansion $p(k;l) - p(l;l) = (k-l)^2 \sum_{s=0}^\infty p_s(k-l)^s$. A short calculation shows that

$$\frac{p_j}{p_0} = \frac{1}{l^j} \prod_{h=1}^{j} \frac{r-h-1}{h+2}.$$

If we use this formula in Lemma 1.1.1 we obtain the following.

Lemma 1.6.1. *Suppose* $w = [p(k;l) - p(l;l)]^{1/2}$ *and* $k - l = \sum_{s=1}^\infty c_s w^s$. *Then for all* $s \geq 1$ *we have* $c_s = p_0^{-s/2} l^{1-s} \tilde{c}_s$ *and*

$$\tilde{c}_s = \frac{1}{s} \sum_{m=0}^{s-1} \left[\prod_{h=1}^{m} \left(-\frac{s}{2} - h + 1 \right) \right] \sum_{(\alpha_1,\dots,\alpha_s)} \prod_{j=1}^{s-1} \frac{1}{\alpha_j!} \left(\prod_{h=1}^{j} \frac{r-h-1}{h+2} \right)^{\alpha_j},$$

where the sum is over all s-tuples $(\alpha_1, \dots, \alpha_s)$ *of nonnegative integers satisfying the following restrictions*

$$\sum_{j=1}^{s} \alpha_j = m \qquad\qquad \sum_{j=1}^{s} j\alpha_j = s - 1.$$

For reference sake, the first few coefficients are:

$$\tilde{c}_1 = 1, \qquad \tilde{c}_2 = -\frac{r-2}{6}, \qquad \tilde{c}_3 = \frac{(r-2)(2r-1)}{72},$$

$$\tilde{c}_4 = -\frac{(r-2)(2r-1)(r+1)}{540}, \qquad \tilde{c}_5 = \frac{(r-2)(2r-1)(2r^2 + 19r + 2)}{17280},$$

$$\tilde{c}_6 = \frac{(r-2)(2r-1)(r+1)(2r^2 - 23r + 2)}{68040},$$

$$\tilde{c}_7 = -\frac{(r-2)(2r-1)(556r^4 - 1628r^3 - 9093r^2 - 1628r + 556)}{43545600}.$$

Proof. The only feature of the above result not immediately obvious is the dependence of c_s on l. When we substitute our expression for p_j/p_0 into the general formula in Lemma 1.1.1 we obtain $\sum_{j=1}^{s-1} j\alpha_j$ powers of $1/l$. But in all the s-tuples satisfying our conditions we have $\alpha_s = 0$. Thus there are $\sum_{j=1}^{s} j\alpha_j = s - 1$ powers of $1/l$ in each term; hence l^{1-s} can be factored out front. $\square$

Remarks. (1) Clearly, a tremendous simplification occurs when $r = 2$. In that case $p(k; l) = i\rho(k^2 - 2kl)$, $p(l; l) = -i\rho l^2$, so $p(k; l) - p(l; l) = i\rho(k - l)^2$; $p_0 = i\rho$, so $w = [p(k; l) - p(l; l)]^{1/2} = \sqrt{i\rho}(k - l)$. Thus no infinite series is needed to express w in terms of $k - l$.

(2) It also should be noted that the quantity $\tilde{c}_s$ which we introduced in the above lemma differs from the quantity given the same name in section 1.1. Henceforth we will consistently use this name to refer to what we have just defined. Maple procedures for computing $\tilde{c}_s$ explicitly can be obtained from the author upon request. Simple manipulations show that $\tilde{k} = \sum_{s=1}^{\infty} \tilde{c}_s \tilde{w}^s$, where $\tilde{k} = \frac{k-l}{l}$ and $\tilde{w} = \frac{w}{l\sqrt{p_0}}$ are the new variables introduced in section 1.1.

Next we must find the coefficients q_s in the Taylor series expansion $q(k) = k^\gamma V(k) = \sum_{s=0}^{\infty} q_s(k - l)^s$. Using Leibniz' formula for multiple derivatives of a product we get

$$q_s = l^{\gamma - s} \sum_{m=0}^{s} \binom{\gamma}{s - m} \frac{l^m V^{(m)}(l)}{m!},$$

where $\binom{\gamma}{s} = \frac{1}{s!} \prod_{h=1}^{s}(\gamma - h + 1)$. If we now expand $q(k)$ in powers of w, i.e. $q(k) = \sum_{s=0}^{\infty} b_s w^s$ (generically we have $\lambda = 1$), then we can apply Lemma 1.1.2 to find the coefficients b_s, $s \geq 0$.

Lemma 1.6.2. *For all $s \geq 0$ we have*

$$b_s = \frac{l^{\gamma - s}}{p_0^{s/2}} \sum_{m=0}^{s} m! \left[\sum_{n=0}^{m} \binom{\gamma}{m - n} \frac{l^n V^{(n)}(l)}{n!} \right] \sum_{(\alpha_1, \ldots, \alpha_{s+1})} \prod_{j=1}^{s} \frac{\tilde{c}_j^{\alpha_j}}{\alpha_j!}.$$

where the sum is over all $(s + 1)$-tuples $(\alpha_1, \ldots, \alpha_{s+1})$ of nonnegative integers such that

$$\sum_{j=1}^{s+1} \alpha_j = m \qquad \sum_{j=1}^{s+1} j\alpha_j = s.$$

Finally, if we define $v = w^2$ and $f(v) = \frac{1}{2} q(k) \frac{dk}{dw} v^{-1/2}$, then we need to compute the coefficients a_s in the expansion $f(v) = v^{-1/2} \sum_{s=0}^{\infty} a_s v^{s/2}$.

Lemma 1.6.3. *For all $s \geq 0$ we have*

$$a_s = \frac{l^{\gamma - s}}{p_0^{(s+1)/2}} \sum_{n=0}^{s} d_n^s(r, \gamma) l^n V^{(n)}(l)$$

where the coefficients $d_n^s(r, \gamma)$ depend only on r and γ, and are given in general by the formula

$$d_m^p(r, \gamma) = \frac{1}{2m!} \sum_{n=m}^{p} (p - n + 1) \tilde{c}_{p-n+1} \sum_{s=m}^{n} s! \binom{\gamma}{s - m} \sum_{(\alpha_1, \ldots, \alpha_{n+1})} \prod_{j=1}^{n} \frac{\tilde{c}_j^{\alpha_j}}{\alpha_j!},$$

where $0 \le m \le p$ are integers, and the third sum is over all $(n+1)$-tuples $(\alpha_1, \ldots, \alpha_{n+1})$ of nonnegative integers satisfying the conditions

$$\sum_{j=1}^{n+1} \alpha_j = s \qquad \sum_{j=1}^{n+1} j\alpha_j = n.$$

For reference sake the first several coefficients are explicitly given below:

$$d_0^0 = \frac{1}{2}.$$

$$d_1^1 = \frac{1}{2}, \qquad d_0^1 = \frac{\gamma}{2} - \frac{r-2}{6}.$$

$$d_2^2 = \frac{1}{4}, \qquad d_1^2 = \frac{\gamma}{2} - \frac{r-2}{4}, \qquad d_0^2 = \frac{\gamma^2}{2} - \frac{(r-1)\gamma}{4} + \frac{(r-2)(2r-1)}{48}.$$

$$d_3^3 = \frac{1}{12}, \qquad d_2^3 = \frac{\gamma}{4} - \frac{r-2}{6}, \qquad d_1^3 = \frac{\gamma^2}{4} - \frac{(4r-5)\gamma}{12} + \frac{(r-2)(r-1)}{12},$$

$$d_4^4 = \frac{1}{48}, \qquad d_3^4 = \frac{\gamma}{12} - \frac{5(r-2)}{72}, \qquad d_2^4 = \frac{\gamma^2}{8} - \frac{(5r-7)\gamma}{24} + \frac{5(r-2)(4r-5)}{288},$$

$$d_0^3 = \frac{\gamma^3}{12} - \frac{(2r-1)\gamma^2}{12} + \frac{r(r-1)\gamma}{12} - \frac{(r-2)(2r-1)(r+1)}{270}.$$

$$d_1^4 = \frac{\gamma^3}{12} - \frac{(5r-4)\gamma^2}{24} + \frac{(20r^2 - 35r + 14)\gamma}{144} - \frac{(r-2)(2r-1)(3r-2)}{288},$$

$$d_0^4 = \frac{\gamma^4}{48} - \frac{(5r-1)\gamma^3}{72} + \frac{(20r^2 - 5r - 4)\gamma^2}{288} - \frac{(r-1)(6r^2 + 7r - 2)\gamma}{288}$$
$$+ \frac{(r-2)(2r-1)(2r^2 + 19r + 2)}{6912}.$$

Proof. Use Lemma 1.1.3 and Lemma 1.6.2 and interchange the order of the summations. A Maple procedure which computes these coefficients for arbitrary values of m and p can be obtained from the author upon request. $\square$

Remarks. (1) This formula for $d_m^p(r, \gamma)$ simplifies dramatically when $r = 2$, yielding $d_m^p(2, \gamma) = \frac{1}{2m!}\binom{\gamma}{p-m}$.

(2) This same formula expressing $d_m^p(r, \gamma)$ in terms of $\tilde{c}_s$ can be derived independently from the relation $(1 + \tilde{k})^\gamma \tilde{k}^n \frac{dk}{d\tilde{w}} = \sum_{s=n}^{\infty} [2n! d_n^s(r, \gamma)] \tilde{w}^s$, where $\tilde{k}$ and $\tilde{w}$ are the variables we have discussed in earlier remarks. The verification involves Taylor's Theorem, Faà di Bruno's formula, and the Binomial expansion.

(3) We will abbreviate $d_m^p(r, \gamma)$ as d_m^p, or as $d_m^p(\gamma)$ when the dependence on γ is important.

Thus as in section 1.1 we have the expansion

$$\int_{\pm C_j^\pm} e^{-p(k;l)t} k^\gamma V(k)\, dk \sim e^{-\tilde{p}(l)t} \sum_{s=0}^{\infty} \Gamma\left(\frac{s+1}{2}\right) \frac{l^{\gamma-s}}{(p_0 t)^{(s+1)/2}} \sum_{n=0}^{s} d_n^s l^n V^{(n)}(l),$$

where the difference between the case $+C_j^+$ and $-C_j^-$ is in the choice of the phase of p_0. We will be more specific about such choices of phase later.

Asymptotic character of the expansion. Now we will investigate the degree to which b can be allowed to vary and still have the expansion we just wrote down retain its asymptotic character. We will index terms by pairs (s, n), $0 \leq n \leq s$; thus we think of the expansion as a double sum. For this purpose we consider the quotient of the (s_2, n_2) term by the (s_1, n_1) term (modulo nonzero constants which do not depend on x, t):

$$Q = \frac{d_{n_2}^{s_2}(r, \gamma)}{d_{n_1}^{s_1}(r, \gamma)} \frac{V^{(n_2)}(l)}{V^{(n_1)}(l)} \frac{|l|^{n_2 - \frac{r}{2} s_2 - (n_1 - \frac{r}{2} s_1)}}{t^{(s_2 - s_1)/2}}.$$

A glance at Lemma 1.6.3 makes clear that it is possible for $d_n^s(r, \gamma)$ to vanish, but for most values of r and γ this does not happen. The meaning of the rare event $d_n^s(r, \gamma) = 0$, which we will call an *anomaly*, is that the (s, n) term vanishes, and hence becomes an inappropriate basis of comparison when assessing the size of other terms. This phenomenon does not affect the asymptotic nature of the series as a whole. But it does make it hard to state and prove a single sharp asymptotic result. The anomaly forces upon us a different form of error estimate; these take on a variety of different forms, depending on where in the expansion the anomaly occurs. Generally, our method of dealing with anomalies is to ignore them; our error estimates will be sharp when there is no anomaly. Even when an anomaly is present, by including more terms than necessary one can arrive at the sharp error estimate. This estimate will follow from the estimates we will prove (sharp when there is no anomaly) as well as estimates of the extra terms which do not vanish.

If r and γ are fixed, it seems to be that a pair (s, n) such that $d_n^s(r, \gamma) = 0$ is rare. An exception to this observation is the case $r = 2$ and $\gamma \geq 0$ is an integer. In that case $d_n^s(2, \gamma) = \frac{1}{2n!}\binom{\gamma}{s-n} = 0$ for all $s - n > \gamma$. We could call such an occurrence a *super anomaly*, since a large number of anomalies occur for the same value of r and γ. It would not do to ignore the super anomalies $r = 2$, $\gamma \geq 0$ an integer, because they happen to coincide with cases of applied interest. Therefore we will discuss them separately (see below and section 4.1).

In the above quotient, we have $|l| \propto |b|^{1/(r-1)}$. We define $\zeta = xt^{-\beta}$, so that $b = \zeta t^{\beta - 1}$. In this discussion we will use β to describe a definite relationship between x and t; hence it can differ from the particular β involved in the definition of the outer region (which always satisfies $\frac{1}{r} < \beta < 1$). We distinguish three different limits of interest for this quotient as $t \to \infty$.

The case where $b \to 0^{\pm}$. Thus $|l| \to 0$ and $\beta < 1$. We may as well assume $V^{(n_1)}(0) \neq 0$, since the the contrary case is another sort of initial data dependent anomaly, which we will also ignore. Therefore we have that the quotient behaves like

$$Q \propto |\zeta|^{\frac{1}{r-1}[(n_2 - n_1) - \frac{r}{2}(s_2 - s_1)]} t^{\frac{-1}{2(r-1)}[2(1-\beta)(n_2 - n_1) + (r\beta - 1)(s_2 - s_1)]}.$$

Suppose $\zeta \neq 0$ is fixed. Then the terms, indexed by (s, n), are ordered asymptotically by the value of $2(1-\beta)n + (r\beta - 1)s$; a larger value of this quantity leads to a more rapidly decaying term as $t \to \infty$. When $\beta > \frac{1}{r}$ and for all $N > 0$ there are only finitely many terms (s,n) satisfying $2(1-\beta)n + (r\beta-1)s \leq N$. This means that in this case one can obtain asymptotic improvement by including more terms, and hence the asymptotic character of the expansion is confirmed. If $0 \leq \beta \leq \frac{1}{r}$ and there is no super-anomaly then there are infinitely many terms satisfying $2(1-\beta)n + (r\beta-1)s \leq N$. This signals a breakdown in the asymptotic character of the expansion. This is why we defined the outer regions the way we did. This is the phenomenon of non-uniformity of the outer expansion with respect to the parameter b.

If, however, we are in the super-anomalous case $r = 2$ and $\gamma \geq 0$ is an integer, then even when $0 \leq \beta \leq \frac{1}{r}$ and for all $N > 0$ there are only finitely many (non-vanishing) terms (s,n) satisfying $2(1-\beta)n + (r\beta-1)s \leq N$. Thus the effect of the super-anomaly is to allow the outer expansion to be uniformly valid. We will not address the appropriate error estimates for the super-anomalous case in this chapter; we save this for section 4.1.

The case where b is a nonzero constant independent of t. This means that $\beta = 1$ and $\zeta = b$. Here we must ignore a new type of anomaly by assuming $V^{(n_1)}(l) \neq 0$. Thus we have

$$Q \propto \frac{1}{t^{(s_2-s_1)/2}}$$

so that the terms are asymptotically well-ordered according to the value of s as $t \to \infty$.

The case where $b \to \pm\infty$. Thus $|l| \to \infty$ and $\beta > 1$. In order to analyze this case we assume $V(k) = \hat{v}(k)$, where $v(x)$ satisfies the standard assumptions stated near the end of section 1.5. Then we will be able to apply Lemma 1.5.1 and 1.5.2 to analyze the leading-order behavior of $V^{(n)}(l)$ as $l \to \infty$, $\arg l = S_j^\alpha$. These results imply that for each choice of j and α there are real constants a, λ_n and complex constants $c_n \neq 0$ such that for each integer $n \geq 0$ we have

$$V^{(n)}(l) = \frac{c_n e^{-ial}}{l^{n+\lambda_n}} + o\left(\frac{e^{a\Im l}}{|l|^{n+\lambda_n}}\right)$$

as $l \to \infty$, along the ray $\arg l = S_j^\alpha$. The numbers λ_n satisfy $0 \leq 1+\lambda_{n+1}-\lambda_n \leq 1$ for all $n \geq 0$ and $\lambda_0 > 0$. Using this result in the general expression for the quotient Q we obtain

$$Q \propto |\zeta|^{\frac{-1}{r-1}[(\lambda_{n_2}-\lambda_{n_1})+\frac{r}{2}(s_2-s_1)]} t^{\frac{-1}{2(r-1)}[2(\beta-1)(\lambda_{n_2}-\lambda_{n_1})+(r\beta-1)(s_2-s_1)]}.$$

Suppose first of all that $\zeta \neq 0$ is fixed and $\beta < \infty$. Then the terms are ordered asymptotically by the value of $2(\beta-1)\lambda_n + (r\beta-1)s$; larger values of this quantity lead to more rapidly decaying terms as $t \to \infty$. In general we have that $\lambda_n \geq \lambda_0 - n$ for all $n \geq 0$; also recall that $0 \leq n \leq s$. So if $r \geq 2$ and for all $N > 0$ there are at most finitely many terms (s,n) satisfying $2(\beta-1)\lambda_n+(r\beta-1)s \leq N$. Thus when $r \geq 2$ we can always gain an asymptotic improvement by adding more terms.

But when $1 < r < 2$ and $\beta \geq \frac{1}{2-r}$ it is possible that there are infinitely many terms such that $2(\beta - 1)\lambda_n + (r\beta - 1)s \leq N$. For example in the common situation where $\lambda_n = \lambda_0 - n$ for all $n \geq 0$ this is true. Thus $|x|$ cannot grow too rapidly in comparison to t without the expansion becoming asymptotically disordered. This signals the appearance of another pair of asymptotic regions which must be considered when $1 < r < 2$. There are special assumptions on $v(x)$ which imply that the sequence λ_n is bounded below (see section 1.5); for such $v(x)$ the expansion remains asymptotically well-ordered for all $\beta > 1$. We do not wish to discuss the complication of the appearance of these new regions further. We will avoid it by assuming $r \geq 2$ as we stated previously.

The value $r = 2$ is the borderline between the cases where the expansion is always asymptotically well-ordered as $t \to \infty$ and the cases where it can fail to be for sufficiently large β. We have seen that the outer expansion is asymptotically well-ordered as $t \to \infty$, ζ fixed, $\beta < \infty$, when $r = 2$. But now suppose $t \geq 1$ is fixed, and consider the asymptotic ordering of the terms as $|x| \to \infty$ (or $|\zeta| \to \infty$). From the above expression for Q we can see that in this limit the terms are ordered according to the value of $\lambda_n + \frac{r}{2}s$. When $r > 2$ and for all $N > 0$ there are at most finitely many terms (s, n) satisfying $\lambda_n + \frac{r}{2}s \leq N$. Thus when $r > 2$ the terms of the outer expansion are asymptotically well-ordered in this additional sense. But when $r = 2$ this might no longer be true; i.e. infinitely many terms of the outer expansion might decay at the same rate as $|x| \to \infty$, $t \geq 1$ fixed. We will return to this issue in sections 4.1 and 4.2.

To summarize, we have shown that for v satisfying the standard assumptions, the terms of the expansion we have computed form a doubly indexed asymptotic scale in the entire outer region as $t \to \infty$ provided $r \geq 2$. When $r > 2$ these terms also form an asymptotic scale as $|x| \to \infty$ for $t \geq 1$ fixed. This extra sense fails when $r = 2$.

Error terms. In the above we saw that in order to include all the terms decaying more slowly than a certain rate, we must specify the value of β, which in the outer region can take on any of the values $\beta > \frac{1}{r}$. In our treatment of the asymptotics over the contour $\tilde{C}$ we formulated our finite expansions and error terms in a β dependent manner. Here, however, we will depart from that practice and order our expansion as if $\beta = 1$. Hence we will consider our expansion to be a single sum over the index s. We lose no generality by doing this, and we gain a greater similarity in appearance between our expansion and the familiar form of Laplace expansions.

Consequently the sth term of the outer expansion involves the sum $\sum_{n=0}^{s} d_n^s l^n V^{(n)}(l)$. The most slowly decaying term as $l \to \infty$ in this sum is the one with $n = s$; since $d_s^s = 1/(2s!)$ this term does not vanish. On the other hand if we ignore the anomaly $d_0^s = 0$ we obtain a "generically" sharp bound of the following form

$$\left| \sum_{n=0}^{s} d_n^s l^n V^{(n)}(l) \right| \leq \frac{C_s e^{a\Re l}}{(1 + |l|)^{\lambda_s}},$$

where constants $C_s > 0, a \in \mathbb{R}, \lambda_s$ depend on the function v and on r and γ.

The optimal result we can prove is that the error committed by approximating the contour integral by the first N terms of the expansion is big O of the usual size of the $N + 1$st term. Suppose $N \geq 0$ is an integer and we define

$$E_N(x,t)$$

$$= \int_{\pm C_j^{\pm}} e^{-p(k;l)t} k^{\gamma} V(k)\, dk - e^{-\tilde{p}(l)t} \sum_{s=0}^{N-1} \Gamma\left(\frac{s+1}{2}\right) \frac{l^{\gamma-s}}{(p_0 t)^{\frac{s+1}{2}}} \sum_{n=0}^{s} d_n^s l^n V^{(n)}(l).$$

In preparation for the proof of estimates of this quantity, we now must derive expressions for it which are suitable for the estimation process. We want to express $E_N(x,t)$ in terms of Taylor remainders for analytic functions which do not depend on b (or l or p_0). This will not be possible entirely, since it will turn out that we must expand $V(k)$ about $k = l$; we will compensate for this l dependence with our detailed knowledge of the behavior of V. In Lemma 1.6.3 we computed the coefficients of the convergent expansion of $k^{\gamma} V(k)\frac{dk}{dw} = 2\sum_{s=0}^{\infty} a_s w^s$. Since this is an analytic function of w, this same power series can be found in a variety of different ways. First we will rearrange the series in a manner which suggests a more convenient means of obtaining the series for our present purposes.

$$k^{\gamma} V(k)\frac{dk}{dw} = 2\sum_{s=0}^{\infty} a_s w^s$$

$$= 2\sum_{s=0}^{\infty} \left(\frac{l^{\gamma-s}}{p_0^{(s+1)/2}} \sum_{n=0}^{s} d_n^s l^n V^{(n)}(l)\right) w^s$$

$$= \sum_{n=0}^{\infty} \frac{V^{(n)}(l)}{n!} \frac{l^{\gamma+n}}{\sqrt{p_0}} \sum_{s=n}^{\infty} (2n! d_n^s) \left(\frac{w}{l\sqrt{p_0}}\right)^s.$$

This suggests that we first expand $V(k)$ in its Taylor series at $k = l$:

$$V(k) = \sum_{n=0}^{N-1} \frac{V^{(n)}(l)}{n!}(k-l)^n + R_N(k;l).$$

Thus we have

$$k^{\gamma} V(k)\frac{dk}{dw} = \sum_{n=0}^{N-1} \frac{V^{(n)}(l)}{n!} k^{\gamma}(k-l)^n \frac{dk}{dw} + k^{\gamma} R_N(k;l)\frac{dk}{dw}.$$

Using the chain rule we have

$$\frac{dk}{dw} = \frac{dk}{d\tilde{k}}\frac{d\tilde{k}}{d\tilde{w}}\frac{d\tilde{w}}{dw} = l\frac{d\tilde{k}}{d\tilde{w}}\frac{1}{l\sqrt{p_0}} = \frac{1}{\sqrt{p_0}}\frac{d\tilde{k}}{d\tilde{w}}.$$

Here we are using the new variables $k = l(1 + \tilde{k})$ and $\tilde{w} = \frac{w}{l\sqrt{p_0}}$ which were introduced in section 1.1. These variables are preferred since $\tilde{k}$ is an analytic function of $\tilde{w}$ which is independent of l:

$$(1 + \tilde{k})^r - 1 - r\tilde{k} = \binom{r}{2}\tilde{w}^2.$$

Expressing everything in these variables as much as is possible we get

$$k^\gamma V(k)\frac{dk}{dw} = \sum_{n=0}^{N-1} \frac{V^{(n)}(l)}{n!}\frac{l^{\gamma+n}}{\sqrt{p_0}}(1 + \tilde{k})^\gamma \tilde{k}^n \frac{d\tilde{k}}{d\tilde{w}} + \frac{l^\gamma}{\sqrt{p_0}}(1 + \tilde{k})^\gamma \frac{d\tilde{k}}{d\tilde{w}} R_N(k; l).$$

If we compare this with our rearranged series above we obtain the following Taylor series expansion

$$(1 + \tilde{k})^\gamma \tilde{k}^n \frac{d\tilde{k}}{d\tilde{w}} = \sum_{s=n}^{N-1} (2n!d_n^s)\tilde{w}^s + R_N^n(\tilde{w}).$$

The Taylor remainder $R_N^n(\tilde{w})$ will be properly defined along the entire contour of integration when we show (see the next section) that the local analytic function $(1 + \tilde{k})^\gamma \tilde{k}^n \frac{d\tilde{k}}{d\tilde{w}}$ can be continued analytically along the ray $\tilde{w} = \frac{w}{l\sqrt{p_0}}$, $w \geq 0$. Substituting this into the previous equation we get

$$k^\gamma V(k)\frac{dk}{dw} = \sum_{n=0}^{N-1} \frac{V^{(n)}(l)}{n!}\frac{l^{\gamma+n}}{\sqrt{p_0}}\left[\sum_{s=n}^{N-1}(2n!d_n^s)\tilde{w}^s\right] + \frac{l^\gamma}{\sqrt{p_0}}(1 + \tilde{k})^\gamma \frac{d\tilde{k}}{d\tilde{w}} R_N(k; l)$$

$$+ \sum_{n=0}^{N-1} \frac{V^{(n)}(l)}{n!}\frac{l^{\gamma+n}}{\sqrt{p_0}}R_N^n(\tilde{w}).$$

Our expansion is based on integrating this expansion term-by-term. If we multiply both sides of this equation by $e^{-p(k;l)t} = e^{-\tilde{p}(l)t}e^{-vt}$ and by $\frac{dw}{dv} = \frac{1}{2\sqrt{v}}$ and integrate over $v \in [0, \infty)$ we obtain

$$E_N(x, t) = \frac{e^{-\tilde{p}(l)t}l^\gamma}{2\sqrt{p_0}} \int_0^\infty e^{-vt}(1 + \tilde{k})^\gamma \frac{d\tilde{k}}{d\tilde{w}} R_N(k; l)\frac{dv}{\sqrt{v}}$$

$$+ \frac{e^{-\tilde{p}(l)t}l^\gamma}{2\sqrt{p_0}} \sum_{n=0}^{N-1} \frac{V^{(n)}(l)}{n!}l^n \int_0^\infty e^{-vt}R_N^n(\tilde{w})\frac{dv}{\sqrt{v}}.$$

We would like to show that this quantity is big O of the $N + 1$st term of the expansion, i.e. the term with $s = N$

$$e^{-\tilde{p}(l)t}\Gamma\left(\frac{N+1}{2}\right)\frac{l^{\gamma-N}}{(p_0 t)^{(N+1)/2}}\sum_{n=0}^N d_n^N l^n V^{(n)}(l).$$

More precisely, since we wish to ignore anomalies, we want to prove that $E_N(x,t)$ is big O of

$$e^{-\tilde{p}(l)t} \frac{l^{\gamma-N}}{(p_0 t)^{(N+1)/2}} \frac{e^{a\Im l}}{(1+|l|)^{\lambda_N}},$$

where $\lambda_N = \min\{\lambda, \lambda' - N\}$. Thus our goal is to prove the following $N+1$ estimates:

(e_N)

$$\sup_{t \geq 1} \sup_{(-1)^\alpha \zeta \geq \epsilon} \frac{|l|^N (|p_0|t)^{\frac{N+1}{2}} (1+|l|)^{\lambda_N}}{e^{a\Im l} \sqrt{|p_0|}} \left| \int_0^\infty e^{-vt} (1+\tilde{k})^\gamma \frac{d\tilde{k}}{d\tilde{w}} R_N(k;l) \frac{dv}{\sqrt{v}} \right| < \infty,$$

(e_n)

$$\sup_{t \geq 1} \sup_{(-1)^\alpha \zeta \geq \epsilon} \frac{|l|^N (|p_0|t)^{\frac{N+1}{2}} (1+|l|)^{\lambda_N}}{e^{a\Im l} \sqrt{|p_0|}} \left| l^n V^{(n)}(l) \int_0^\infty e^{-vt} R_N^n(\tilde{w}) \frac{dv}{\sqrt{v}} \right| < \infty,$$

where $n = 0, \ldots, N-1$. Here $\zeta = xt^{-\beta}$ is the natural coordinate in the outer regions ($\frac{1}{r} < \beta < 1$), and $\alpha = 0, 1$ correspond to the regions R_+ and R_- respectively. The choice of the phase of p_0, which controls whether the contour of integration is $+C_j^+$ or $-C_j^-$, is implicit in these estimates.

1.7 Error Control, for the contour C_j

In this section we present the proofs of the estimates (e_n), $n = 0, 1, \ldots, N$, stated at the end of the previous section. The first step is to study the analytic mapping $\tilde{w} \mapsto \tilde{k}$. Define $g(\tilde{k}) = (1+\tilde{k})^r - 1 - r\tilde{k}$. We need to show that g is a one-to-one function on a certain domain, the image under g of which we need to compute as well. Then we will have $\tilde{k} = g^{-1}(r(r-1)\tilde{w}^2/2)$ for all $\tilde{w}$ such that $\tilde{w}^2$ lies in this image.

First we will define a curve in the $\tilde{k}$ plane which will form part of the boundary of the domain of g. For $b > 0$ consider what we have called C_0^-. This is a contour of steepest ascent starting at infinity in the valley $\theta = V_1 = \frac{3\pi}{2r}$ and ending at $l = k_0(b) = \left| \frac{b}{r\rho} \right|^{\frac{1}{r-1}}$. If $1 < r \leq 3$ this contour is contained in the sector $\arg(k) = \theta \in [0, V_1)$ (i.e. there is no overshoot, see Lemma 1.2.4B). If $r > 3$ there is an overshoot of angular width $\epsilon_0 > 0$, a number defined in Lemma 1.2.4B(1). In this case the contour is contained in the sector $\theta \in [0, V_1 + \epsilon_0]$. A detailed description of this contour can be found in Lemmas 1.2.4B and 1.2.5B. Every point $k = Re^{i\theta}$ on this contour satisfies the polar equation (**)

$$l^{r-1} R \cos(\theta) - \frac{1}{r} R^r \cos(r\theta) = \frac{r-1}{r} l^r.$$

Thus $1 + \tilde{k} = k/l = (R/l)e^{i\theta}$ satisfies the polar equation

$$\left(\frac{R}{l}\right) \cos(\theta) - \frac{1}{r} \left(\frac{R}{l}\right)^r \cos(r\theta) = \frac{r-1}{r}.$$

Thus by scaling C_0^-, which depends on $b > 0$, by l and then subtracting 1 we get a contour in the $\tilde{k}$ plane which is independent of $b > 0$. It is oriented, starting at infinity at $\arg(1 + \tilde{k}) = \theta = V_1$ and ending at $\tilde{k} = 0$. Call this contour C^-. C^- coincides with C_0^- shifted by 1 when $b = r\rho$ (i.e. $l = 1$).

Define C^+ to be the nonnegative real $\tilde{k}$ axis, oriented starting at $\tilde{k} = 0$ and ending at infinity. If we suppose that both C^+ and C^- contain the point at infinity then $C = C^- \cup C^+$ is a Jordan curve on the Riemann sphere S^2. Let D be the connected component of $S^2 \setminus C$ containing $\tilde{k} = 1 + i\epsilon$ for all sufficiently small $\epsilon > 0$.

Lemma 1.7.1. *Let $U \subset S^2$ be the open sector $0 < \arg z < 3\pi/2$. Then $g \colon \overline{D} \to \overline{U}$ is a homeomorphism, and $g^{-1} \colon U \to D$ is holomorphic. g^{-1} continues analytically across the ray $\arg z = 3\pi/2$.*

Proof. First we will show that g is injective when restricted to C. Clearly g maps C^+ injectively onto itself, taking 0 to 0 and ∞ to ∞. To understand how g maps C^- recall that the relation $g(\tilde{k}) = \frac{r(r-1)}{2} \tilde{w}^2$ was derived as being equivalent to $\varphi(k) - \varphi(l) = -w^2$ where $\varphi(k) = ikb - i\rho k^r$, $\tilde{k} = k/l - 1$, $\tilde{w} = \dfrac{w}{\sqrt{i\rho r(r-1)lr/2}}$ and $w \geq 0$. When $b = r\rho$, so that $l = 1$, we have $g(\tilde{k}) = \frac{r(r-1)}{2} \tilde{w}^2 = w^2/(i\rho)$, where $w^2 = \varphi(1) - \varphi(1 + \tilde{k})$. As $\tilde{k}$ traces through C^-, $k = 1 + \tilde{k}$ traces through C_0^-, which is a steepest ascent path for $\Re\varphi(k)$; there is a one-to-one correspondence between points $k \in C_0^- \setminus \{1\}$ and positive real numbers w. Since $\rho > 0$ we have that g maps C^- injectively onto the ray $\arg z = 3\pi/2$. Thus g maps C injectively onto the topological boundary of U in S^2. Clearly $g \colon \overline{D} \to S^2$ is continuous and g is holomorphic on D.

Let $h \colon S^2 \to S^2$ be the unique Möbius transformation such that $h(0) = 1$, $h(\infty) = -1$, and $h(e^{-i\pi/4}) = \infty$. Thus $h(\overline{U}) \subset \mathbb{C} = S^2 \setminus \{\infty\}$. Apply Proposition 2.8.14 on page 200 of [BG] to the composition $f = h \circ g$ to show that it determines a homeomorphism of $\overline{D}$ onto $h(\overline{U})$, and $f^{-1} \colon h(U) \to D$ is holomorphic. Thus $g = h^{-1} \circ f$ is a homeomorphism of $\overline{D}$ onto $\overline{U}$, and $g^{-1} = f^{-1} \circ h \colon U \to D$ is holomorphic.

To show that g^{-1} continues analytically across the ray $\arg z = 3\pi/2$ it suffices to show that $g'(\tilde{k}) \neq 0$ for $\tilde{k} \in C^- \setminus \{0, \infty\}$. This is obvious when $1 < r \leq 3$ because then C^- is contained in the sector $\theta = \arg(1 + \tilde{k}) \in [0, V_1)$ and $V_1 = \frac{3\pi}{2r} < \frac{2\pi}{r} < \frac{2\pi}{r-1}$. (If $g'(\tilde{k}) = 0$ then $\tilde{k} = e^{2\pi i n/(r-1)}$ for some integer n.) If $r > 3$ then C^- is contained in the sector $\theta = \arg(1 + \tilde{k}) \in [0, V_1 + \epsilon_0]$, and we will show that $V_1 + \epsilon_0 < \frac{2\pi}{r} < \frac{2\pi}{r-1}$. Lemma 1.2.4B implies $\epsilon_0 < \tilde{\epsilon}_0$, and the definition of $\tilde{\epsilon}_0$ implies that $\tilde{\epsilon}_0 \leq \frac{\pi}{2r}$. Therefore $V_1 + \epsilon_0 < \frac{3\pi}{2r} + \frac{\pi}{2r} = \frac{2\pi}{r}$. $\square$

Let $S\{\ln(1 + \tilde{k})\}$ denote the Riemann surface for the function $\tilde{k} \mapsto \ln(1 + \tilde{k})$. Topologically, it is a plane; it is an infinite-sheeted covering space of $\mathbb{C} \setminus \{-1\}$. Likewise let $S\{z^{1/2}\}$ denote the Riemann surface for the function $z \mapsto z^{1/2}$. Topologically, it is a sphere; it is a two-sheeted covering space of the Riemann sphere, with branch points $z = 0$ and $z = \infty$. Each of these Riemann surfaces can be equipped with polar coordinates centered at the branch point. Define the operation of complex conjugation on such surfaces by the rule $re^{i\theta} \mapsto re^{-i\theta}$. We associate with $D \subset \mathbb{C}$ the corresponding region (denoted also by D) in the principle branch of $S\{\ln(1 + \tilde{k})\}$; $\tilde{k}$ then denotes a point in D on the surface $S\{\ln(1+\tilde{k})\}$. We associate with $U \subset \mathbb{C}$ the corresponding region (denoted also by U) in the principle branch of $S\{z^{1/2}\}$; z then denotes a point in U on the surface $S\{z^{1/2}\}$. We denote by $\overline{D}$ the closure of D in the Riemann surface $S\{\ln(1+\tilde{k})\}$. We denote by $\overline{U}$ the closure of D in the Riemann surface $S\{z^{1/2}\} \setminus \{\infty\}$. We can consider $g \colon \overline{D} \to \overline{U}$ to be a homeomorphism, and $g^{-1} \colon U \to D$ to be a holomorphic mapping between Riemann surfaces.

Since $g^{-1}(z)$ is real when z is real and nonnegative, we can extend g^{-1} using Schwartz reflection for mappings between the Riemann surfaces $S\{z^{1/2}\} \setminus \{\infty\}$ and $S\{\ln(1 + \tilde{k})\}$. Now we must demonstrate that this extended mapping g^{-1} is sufficient to deal with all the steepest descent contours C_j, $-J \le j \le K$. That means we must show that $\arg(\tilde{w}^2) = \arg\left(\frac{w^2}{l^2 p_0}\right) = -2\arg(l) - \arg(p_0)$ is always in the interval $[-3\pi/2, 3\pi/2]$. $\arg(l) = S_j^\alpha$ and because of the relation $\arg(t) + \arg(p_0) + 2\omega^\pm = 0$, where $\omega^\pm$ are the angles along which $C_j^\pm$ depart from the saddle points $l = k_j(b)$, we can choose the argument of p_0. From Lemmas 1.2.5A(3) and 1.2.5B(3) we have that

$$\omega^\pm = \frac{1}{2} S_j^\alpha + (-1)^\alpha \left[\frac{\pi}{4} \mp \frac{\pi}{2}\right].$$

Since $t > 0$, $\arg t = 0$ and $\arg(p_0) = -2\omega^\pm$ and therefore

$$\arg(p_0) = -S_j^\alpha - (-1)^\alpha \left[\frac{\pi}{2} \mp \pi\right].$$

So $\arg(\tilde{w}^2) = -2S_j^\alpha + S_j^\alpha + (-1)^\alpha \left[\frac{\pi}{2} \mp \pi\right] = -S_j^\alpha + (-1)^\alpha \left[\frac{\pi}{2} \mp \pi\right]$. Thus we want to show

(1) $-S_j^1 + \frac{\pi}{2} \in [0, \frac{3\pi}{2}]$ for $-J \le j \le -1$.
(2) $-S_j^0 + \frac{3\pi}{2} \in [0, \frac{3\pi}{2}]$ for $0 \le j \le K$.
(3) $-S_j^1 - \frac{3\pi}{2} \in [-\frac{3\pi}{2}, 0]$ for $-J \le j \le -1$.
(4) $-S_j^0 - \frac{\pi}{2} \in [-\frac{3\pi}{2}, 0]$ for $0 \le j \le K$.

Since $S_j^1 \in [-\frac{\pi}{2}, 0)$ for $-J \le j \le -1$ and $S_j^0 \in [0, \frac{\pi}{2}]$ for $0 \le j \le K$ we have the desired result. A consequence of this fact is that estimates of the single function g^{-1} can be used to estimate $k(v)$ for all values of b.

Consider the sector $\arg \tilde{w} \in [-\frac{3\pi}{4}, \frac{3\pi}{4}]$, which we will denote by W. The mapping $\tilde{w} \mapsto r(r-1)\tilde{w}^2/2$ takes W into the domain of the extended mapping g^{-1}, which sits inside $S\{z^{1/2}\} \setminus \{\infty\}$. A consequence of the previous lemma is that the function $\tilde{k}(\tilde{w}) = g^{-1}(r(r-1)\tilde{w}^2/2)$ is well-defined and analytic for $\tilde{w}$ in this sector. Furthermore, this function can be analytically continued across the rays $\arg \tilde{w} = \pm\frac{3\pi}{4}$. In fact, since the derivative of the inverse map $\tilde{k} \mapsto \tilde{w}$ at $\tilde{k} = 0$ is 1, the mapping $\tilde{w} \mapsto \tilde{k}$ is also analytic in a sufficiently small neighborhood of $\tilde{w} = 0$. The image of W under the mapping $\tilde{w} \mapsto \tilde{k}$ is the domain of the Schwartz extension of the function g. Our next lemma gives some fundamental estimates of this mapping and its derivative.

Lemma 1.7.2. *Suppose $r > 1$.*

(1) *There is a positive number δ, depending only on r, such that*

$$|1 + \tilde{k}(\tilde{w})| \geq \delta$$

for all $\tilde{w} \in W$.

(2) *There is a positive number C, depending only on r, such that*

$$|\tilde{k}(\tilde{w})| \leq \frac{C|\tilde{w}|}{(1 + |\tilde{w}|)^{1-2/r}}.$$

for all $\tilde{w} \in W$.

(3) *There is a positive number C', depending only on r, such that*

$$\left| \frac{d\tilde{k}}{d\tilde{w}}(\tilde{w}) \right| \leq \frac{C'}{(1 + |\tilde{w}|)^{1-2/r}}$$

for all $\tilde{w} \in W$.

Proof. (1) This follows from the polar equation for the contour $\mathcal{C}^-$: $1 + \tilde{k} = (R/l)e^{i\theta}$, $\theta \in [0, V_1)$ when $1 < r \leq 3$ or $\theta \in [0, V_1 + \epsilon_0]$ when $r > 3$. $R(\theta)$ is a positive-valued continuous function.

(2) It is clear that $|g^{-1}(z)| \leq C_1 |z|^{1/2}$ for z sufficiently small and that $|g^{-1}(z)| \leq C_2 |z|^{1/r}$ for $|z|$ sufficiently large. Because of the compactness of the remaining domain and the continuity of $g^{-1}(z)|z|^{-1/2}(1 + |z|^{1/2})^{1-2/r}$ over that compact set we obtain the global estimate

$$|g^{-1}(z)| \leq \frac{C_3 |z|^{1/2}}{(1 + |z|^{1/2})^{1-2/r}},$$

valid for all z in the domain of g^{-1}. Applying this estimate with $z = r(r-1)\tilde{w}^2/2$ yields the result.

(3) Actually, it is easy to see that $1/2|z|^{1/r} \leq |g^{-1}(z)| \leq 2|z|^{1/r}$ for sufficiently large $|z|$. Computing we have

$$\frac{d\tilde{k}}{d\tilde{w}}(\tilde{w}) = \frac{r(r-1)\tilde{w}}{g'(g^{-1}(r(r-1)\tilde{w}^2/2))}.$$

Since $g'(\tilde{k}) = r[(1 + \tilde{k})^{r-1} - 1] \sim r\tilde{k}^{r-1}$ for $|\tilde{k}|$ large we have

$$\left| \frac{d\tilde{k}}{d\tilde{w}}(\tilde{w}) \right| \leq \frac{C_1 |\tilde{w}|}{|g^{-1}(r(r-1)\tilde{w}^2/2)|^{r-1}} \leq \frac{C_2 |\tilde{w}|}{|\tilde{w}^2|^{(r-1)/r}} = \frac{C_2}{|\tilde{w}|^{1-2/r}}$$

when $|\tilde{w}|$ is large. Since $\frac{d\tilde{k}}{d\tilde{w}}(0) = 1$ and $\frac{d\tilde{k}}{d\tilde{w}}(\tilde{w})$ is continuous in a neighborhood of $\tilde{w} = 0$, it is bounded there. Again by compactness and continuity we obtain the desired result. $\square$

Now we are ready to prove the error estimate (e_N).

Lemma 1.7.3. *Suppose $r > 3/2$, $\rho > 0$, and $N \geq 0$ is an integer. Suppose $-J \leq j \leq K$, $p_0 = i\rho r(r-1)l^{r-2}/2$, $l = k_j(b) = \left|\frac{b}{r\rho}\right|^{\frac{1}{r-1}} e^{iS_j^\alpha}$, $b = \zeta t^{\beta-1}$, $\frac{1}{r} < \beta < 1$, $\gamma > -1$, and $\epsilon > 0$. Suppose $k = l(1 + \tilde{k})$, $\tilde{k}$ is a function of $\tilde{w}$ as discussed in the previous two lemmata, $\tilde{w} = \frac{w}{l\sqrt{p_0}}$, and $v = w^2$. In the definition of $\tilde{w}$, let $\sqrt{p_0}$ be computed using the choice $\arg(p_0) = -S_j^\alpha - (-1)^\alpha \left[\frac{\pi}{2} \mp \pi\right]$, which depends on $\alpha \in \{0,1\}$, $(-1)^\alpha = \mathrm{sgn}(b)$, and whether we are integrating over C_j^+ (use upper sign) or over C_j^- (use lower sign). Suppose $V(k)$ is an entire function which satisfies*

$$|V^{(N)}(k)| \leq \frac{C_N e^{a\Im k}}{(1 + |k|)^{N + \lambda_N}},$$

uniformly for k in the smallest closed sector $L_0 \leq \arg k \leq L_1$ containing the contour $C_j^\pm$, where $C_N > 0, a \in \mathbb{R}, \lambda_N \in \mathbb{R}$ are constants. Define the Taylor remainder

$$R_N(k; l) = V(k) - \sum_{n=0}^{N-1} \frac{V^{(n)}(l)}{n!}(k - l)^n.$$

Then we have

$$\sup_{t \geq 1} \sup_{(-1)^\alpha \zeta \geq \epsilon} \frac{|l|^N (|p_0|t)^{\frac{N+1}{2}}(1 + |l|)^{\lambda_N}}{e^{a\Im l}\sqrt{|p_0|}} \left| \int_0^\infty e^{-vt}(1 + \tilde{k})^\gamma \frac{d\tilde{k}}{d\tilde{w}} R_N(k; l) \frac{dv}{\sqrt{v}} \right| < \infty.$$

This is asserted for both C_j^+ and C_j^- for all $-J \leq j \leq -1$ when $b < 0$ and $0 \leq j \leq K$ when $b > 0$.

Remarks. (1) If $\lambda_N = \min\{\lambda, \lambda' - N\}$, where $\lambda > 0, \lambda' > 0$ as was the case in section 1.5, then we have $\lambda_N > -N$. Our proof works in particular when $\lambda_0 = 0$, which allows us to cover the important special case where $V(k) = 1$ for all k.

(2) When $3/2 < r < 2$ the result we have stated involving the supremum over all x satisfying $(-1)^\alpha x t^{-\beta} \geq \epsilon$ is true, but unless we also have $|x|t^{-1/(2-r)} \to 0$ this estimate of the error term does not represent any improvement over the size of the Nth term.

Proof. If we introduce the new variable of integration $y = vt$ and take the supremum inside the integral we find that we must show that

$$\sup_{t \geq 1} \sup_{(-1)^\alpha \zeta \geq \epsilon} |l|^N (|p_0|t)^{N/2} e^{-a\Im l}(1 + |l|)^{\lambda_N} \left| (1 + \tilde{k})^\gamma \frac{d\tilde{k}}{d\tilde{w}} R_N(k; l) \right|$$

is an integrable function of $y \in (0, \infty)$ when multiplied by $e^{-y}y^{-1/2}$. Let us start by estimating the Taylor remainder ($N \geq 1$) using its integral representation

$$R_N(k; l) = \frac{(k - l)^N}{(N - 1)!} \int_0^1 (1 - z)^{N-1} V^{(N)}(l(1 - z) + kz)\, dz.$$

By assumption we have

$$|V^{(N)}(l(1-z)+kz)| \le \frac{C_N e^{a\Im[l+z(k-l)]}}{(1+|l+z(k-l)|)^{N+\lambda_N}} = \frac{C_N e^{a\Im(l)} e^{az\Im(k-l)}}{(1+|l|\cdot|1+z\tilde{k}|)^{N+\lambda_N}}.$$

This estimate also holds when $N = 0$. When $\lambda_N + N \ge 0$ we estimate as follows. An elementary calculus exercise shows that

$$\frac{1}{1+|l|\cdot|1+z\tilde{k}|} \le \max\left\{1, \frac{1}{|1+z\tilde{k}|}\right\} \frac{1}{1+|l|}.$$

When $3/2 < r \le 3$ we have that $V_1 < \pi$ and hence $1 + \tilde{k}$ is confined to a sector of angular width strictly less than π. This is also true when $r > 3$ since in that case $V_1 < \pi/2$ and by Lemma 1.2.4B(1) $V_1 + \epsilon_0 < \pi/2$. Thus by Lemma 1.7.2(1) we have that there is a positive constant δ' depending only on $r > 3/2$ such that $\inf_{z\in[0,1]} |1+z\tilde{k}| \ge \delta'$. (This fails when $r = 3/2$, as one can see from the figure for $r = 3/2$, $b > 0$, at the end of section 1.2.) Since $N + \lambda_N \ge 0$ we have

$$\frac{1}{(1+|l|\cdot|1+z\tilde{k}|)^{N+\lambda_N}} \le \max\left\{1, \frac{1}{\delta'}\right\}^{N+\lambda_N} \frac{1}{(1+|l|)^{N+\lambda_N}}.$$

When $\lambda_N + N < 0$, note that $1 + |l + z(k - l)| \le (1 + |l|)(1 + z|k - l|)$, so that

$$[1+|l+z(l-l)|]^{-\lambda_N - N} \le (1+|l|)^{-\lambda_N - N}(1+z|k-l|)^{-\lambda_N - N}.$$

Thus, regardless of the sign of $N + \lambda_N$, we have

$$[1+|l+z(l-l)|]^{-\lambda_N - N} \le (1+|l|)^{-\lambda_N - N}(1+z|k-l|)^{-\lambda'}.$$

where $\lambda' = \min\{\lambda_N + N, 0\}$. (This λ' is not related to the one in section 1.5.)
   On the other hand by Lemma 1.7.2(2)

$$|k-l| = |l\tilde{k}| \le C|l|\cdot|\tilde{w}|^{2/r} = C|l|\left|\frac{2w^2}{r(r-1)l^r}\right|^{1/r} = C'v^{1/r} = \frac{C'y^{1/r}}{t^{1/r}} \le C'y^{1/r},$$

since $t \ge 1$. Thus $e^{az\Im(k-l)} \le e^{|a|\cdot|k-l|} \le e^{Cy^{1/r}}$. Using the same estimate in Lemma 1.6.2(2) in a different way yields

$$|k-l| = |l|\cdot|\tilde{k}| \le C|l|\cdot|\tilde{w}| = C\left|\frac{lw}{l\sqrt{p_0}}\right| = C\frac{\sqrt{v}}{\sqrt{|p_0|}} = C\frac{\sqrt{y}}{\sqrt{|p_0|t}}.$$

Therefore $|k-l|^N \le Cy^{N/2}(|p_0|t)^{-N/2}$. Putting these estimates together we get

$$|l|^N(|p_0|t)^{N/2}e^{-a\Im l}(1+|l|)^{\lambda_N}|R_N(k;l)| \le C'y^{N/2}(1+y)^{-\frac{\lambda'}{r}}e^{Cy^{1/r}}.$$

In this estimate the constants C and C' are independent of $t \geq 1$ and $(-1)^\alpha \zeta \geq \epsilon$. To finish the proof notice that by Lemma 1.7.2(2)(3) we have

$$|1 + \tilde{k}|^\gamma \left| \frac{d\tilde{k}}{d\tilde{w}} \right| \leq \frac{C(1 + |\tilde{w}|)^{\frac{2}{r} \max\{0, \gamma\}}}{(1 + |\tilde{w}|)^{1 - 2/r}} = C(1 + |\tilde{w}|)^{\frac{2}{r}[1 + \max\{0, \gamma\}] - 1}.$$

When this exponent is positive we need to bound $|\tilde{w}|$ as it becomes large. But

$$|\tilde{w}| = \frac{\sqrt{2}w}{\sqrt{\rho r(r-1)}|l|^{r/2}} = \frac{c\sqrt{v}}{|b|^{\frac{r}{2(r-1)}}} = \frac{c\sqrt{y}}{|\zeta t^{\beta - 1} t^{\frac{r-1}{r}}|^{\frac{r}{2(r-1)}}} \leq \frac{c'\sqrt{y}}{t^{(\beta - \frac{1}{r})\frac{r}{2(r-1)}}} \leq c'\sqrt{y},$$

since $|\zeta| \geq \epsilon$, $\beta > 1/r$ and $t \geq 1$. Thus

$$|1 + \tilde{k}|^\gamma \left| \frac{d\tilde{k}}{d\tilde{w}} \right| \leq C(1 + c'\sqrt{y})^{\max\{0, \frac{2}{r}[1 + \max\{0, \gamma\}] - 1\}}.$$

Since $(1 + y)^\sigma e^{Cy^{1/r}}$ is integrable over $(0, \infty)$ when multiplied by $e^{-y}y^{-1/2}$ we are done. $\square$

The estimates (e_n), $n = 0, 1, \ldots, N - 1$ can also now be proved.

Lemma 1.7.4. *Adopt the notation and assumptions of the previous lemma with the following modifications. Suppose $0 \leq n \leq N - 1$ is an integer and the entire function $V(k)$ satisfies*

$$|V^{(n)}(l)| \leq \frac{C_n e^{a\Im l}}{(1 + |l|)^{n + \lambda_n}}$$

for all l on the ray $\arg l = S_j^\alpha$ containing the saddle point k_j, where $C_n > 0$, a, and $\lambda_n \geq \lambda_N$ are real numbers. Define the Taylor remainder

$$R_N^n(\tilde{w}) = (1 + \tilde{k})^\gamma \tilde{k}^n \frac{d\tilde{k}}{d\tilde{w}} - \sum_{s=n}^{N-1} (2n! d_n^s)\tilde{w}^s$$

for all $\tilde{w}$ in the domain of the mapping $\tilde{w} \mapsto \tilde{k}$ described earlier. Then

$$\sup_{t \geq 1} \sup_{(-1)^\alpha \zeta \geq \epsilon} \frac{|l|^N (|p_0|t)^{\frac{N+1}{2}} (1 + |l|)^{\lambda_N}}{e^{a\Im l}\sqrt{|p_0|}} \left| l^n V^{(n)}(l) \int_0^\infty e^{-vt} R_N^n(\tilde{w}) \frac{dv}{\sqrt{v}} \right| < \infty.$$

Remark. If $\lambda_n = \min\{\lambda, \lambda' - n\}$, where $\lambda > 0, \lambda' > 0$ as was the case in section 1.5, then we have that $\lambda_n \geq \lambda_N$ when $n \leq N$. This relation also holds in the important special case $V(k) = 1$ for all k, where $\lambda_0 = 0$.

Proof. By our assumptions on the function $V^{(n)}(l)$ we have

$$e^{-a\Im l}(1 + |l|)^{\lambda_N}|l|^n|V^{(n)}(l)| \leq \frac{C_n|l|^n}{(1 + |l|)^{n + \lambda_n - \lambda_N}} \leq C_n,$$

since $\lambda_n - \lambda_N \geq 0$. Thus it suffices to prove

$$\sup_{t \geq 1} \sup_{(-1)^\alpha \zeta \geq \epsilon} \frac{|l|^N (|p_0|t)^{(N+1)/2}}{\sqrt{|p_0|}} \left| \int_0^\infty e^{-vt} R_N^n(\tilde{w}) \frac{dv}{\sqrt{v}} \right| < \infty.$$

If we introduce the new variable of integration $y = vt$ and take the supremum inside the integral we find that it suffices to show that

$$\sup_{t \geq 1} \sup_{(-1)^\alpha \zeta \geq \epsilon} |l|^N (|p_0|t)^{N/2} |R_N^n(\tilde{w})|$$

is an integrable function of $y \in (0, \infty)$ when multiplied by $e^{-y} y^{-1/2}$. Since

$$\tilde{w} = \frac{w}{l\sqrt{p_0}} = \frac{\sqrt{v}}{l\sqrt{p_0}} = \frac{\sqrt{y}}{l\sqrt{p_0 t}}$$

we see that it suffices to show that

$$\sup_{t \geq 1} \sup_{(-1)^\alpha \zeta \geq \epsilon} \left| \frac{R_N^n(\tilde{w})}{\tilde{w}^N} \right|$$

is an integrable function of $y \in (0, \infty)$ when multiplied by $e^{-y} y^{(N-1)/2}$. Since $R_N^n(\tilde{w})$ is the Taylor remainder of a fixed (independent of t and ζ) analytic function defined on the domain of the mapping $\tilde{w} \mapsto \tilde{k}$, the quotient $|R_N^n(\tilde{w}) \tilde{w}^{-N}|$ is bounded by an absolute constant on the part of the domain where $|\tilde{w}| \leq 1$. When $|\tilde{w}| \geq 1$ we estimate as follows:

$$\left| \frac{R_N^n(\tilde{w})}{\tilde{w}^N} \right| \leq \left| \frac{(1+\tilde{k})^\gamma \tilde{k}^n}{\tilde{w}^N} \frac{d\tilde{k}}{d\tilde{w}} \right| + \sum_{s=n}^{N-1} |2n! d_n^s| \cdot |\tilde{w}|^{s-N}$$

$$\leq C(1 + |\tilde{w}|)^{\frac{2}{r}(\max\{0,\gamma\}+n+1)-N-1} + \sum_{s=n}^{N-1} |2n! d_n^s|$$

Here we used the estimates from Lemma 1.7.2(2) and (3). When $\frac{2}{r}(\max\{0, \gamma\} + n + 1) - N - 1 > 0$ we must estimate $|\tilde{w}|$ as was done near the end of the proof of Lemma 1.7.3, i.e.

$$\sup_{t \geq 1} \sup_{(-1)^\alpha \zeta \geq \epsilon} |\tilde{w}| \leq c' \sqrt{y},$$

where the constant $c' > 0$ depends on ρ, r, ϵ. Since $e^{-y}(1 + y)^\sigma y^{(N-1)/2}$ is integrable for $y \in (0, \infty)$, we are done. $\quad\square$

Now we have completed the proof of the following.

Theorem 1.7.5. *Suppose* $r \geq 2$, $\rho > 0$, $\gamma > -1$, $\frac{1}{r} < \beta < 1$, $\epsilon > 0$, *and* $N \geq 0$ *and* $-J \leq j \leq K$ *are integers, where* $J = \left[\frac{r+1}{4}\right]$ *and* $K = \left[\frac{r-1}{4}\right]$. *Define the parameter* $b = x/t$, $x \in \mathbb{R}$, $t \geq 1$, *and* $(-1)^\alpha = \mathrm{sgn}(b)$, $\alpha \in \{0,1\}$. *Assume* $\alpha = 1$ *when* $j < 0$ *and* $\alpha = 0$ *when* $j \geq 0$. *Suppose* $C_j^\pm$ *is the contour described in section 1.2 depending on the parameter* b *and passing through the saddle point* $l = k_j(b) = \left|\frac{b}{r\rho}\right|^{\frac{1}{r-1}} e^{iS_j^\alpha}$, $S_j^\alpha = \frac{\pi}{r-1}(2j+\alpha)$. *Suppose* $L_0 \leq \arg k \leq L_1$ *is the smallest closed sector containing the contour* $C_j^\pm$ *in its entirety (see section 1.2). Suppose* $V(k)$ *is an entire function satisfying the bounds*

$$|V^{(n)}(k)| \leq \frac{C_n e^{a\Im k}}{(1+|k|)^{n+\lambda_n}},$$

uniformly for k *in the sector* $L_0 \leq \arg k \leq L_1$, *where* $n = 0, 1, \ldots, N$, $C_n > 0$, $a \in \mathbb{R}$, $\lambda_n \geq \lambda_N$ *are constants (see section 1.5). Define the* N-*term asymptotic approximation*

$$A_N^\pm(x,t) = e^{-p(l;l)t} \sum_{s=0}^{N-1} \Gamma\left(\frac{s+1}{2}\right) \frac{l^{\gamma-s}}{(p_0 t)^{(s+1)/2}} \sum_{n=0}^{s} d_n^s l^n V^{(n)}(l),$$

where $p(k; l) = i\rho(k^r - rl^{r-1}k)$, $p_0 = i\rho r(r-1)l^{r-2}/2$, $\arg(p_0) = -S_j^\alpha - (-1)^\alpha \left[\frac{\pi}{2} \mp \pi\right]$, *and* d_n^s *are certain explicit functions of* r *and* γ *described in Lemma 1.6.1 and Lemma 1.6.3. Define the* Nth *weight function*

$$W_N(x,t) = \exp\left[-(r-1)\rho\left|\frac{x}{r\rho t}\right|^{\frac{r}{r-1}} t\sin(|S_j^\alpha|) + a\left|\frac{x}{r\rho t}\right|^{\frac{1}{r-1}} \sin(S_j^\alpha)\right] \cdot$$

$$\cdot \frac{(1+|x/t|)^{-\frac{\lambda_N}{r-1}}}{t^{\frac{N+1}{2}} |x/t|^{\frac{1}{r-1}[\frac{r}{2}(N+1)-\gamma-1]}}.$$

Then we have the asymptotic result

$$\sup_{t \geq 1} \sup_{(-1)^\alpha xt^{-\beta} \geq \epsilon} \frac{1}{W_N(x,t)} \left| \pm \int_{C_j^\pm} e^{-p(k;l)t} k^\gamma V(k)\, dk - A_N^\pm(x,t) \right| < \infty.$$

Remarks. (1) The theorem should be read consistently throughout with either the upper sign or the lower sign.

(2) An equivalent weight function to the one defined above is the one given in section 1.6 in terms of l and p_0, namely

$$\exp\{-\Re[\bar{p}(l)]t + a\Im l\} \frac{|l|^{\gamma-N}(1+|l|)^{-\lambda_N}}{(|p_0|t)^{(N+1)/2}}.$$

If the assumed estimate on $V^{(n)}(k)$ is sharp on the ray containing the saddle point, then this weight function has the same (ignoring anomalies) asymptotic decay rate as the $N+1$st term of the asymptotic expansion. We have already seen (see section 1.6) that under reasonable hypotheses the terms of our asymptotic expansion form an asymptotic sequence. Hence we have proved the optimal result.

If we combine the contributions of C_j^+ and C_j^- there is some cancellation. Thus we get the following.

Corollary 1.7.6. *Adopt the assumptions and notation of the previous theorem, with the following modifications. Suppose* $\arg p_0 = -S_j^\alpha + (-1)^\alpha \frac{\pi}{2}$, *and* N *is even, i.e.* $N = 2M$ *where* $M \geq 0$ *is an integer. If* $j \neq 0$ *then we assume the same estimate holds for* $V^{(n)}(k)$ *as in the previous theorem except assume it holds uniformly on the smallest closed sector containing the contour* C_j *(note that* $\Im k$ *has a definite sign in this sector). If* $j = 0$ *we assume the entire function* $V(k)$ *satisfies the estimate*

$$|V^{(n)}(k)| \leq \frac{C_n e^{\max\{a_- \Im k, a_+ \Im k\}}}{(1+|k|)^{n+\lambda_n}},$$

uniformly for k *in the smallest closed sector* $L_0 \leq \arg k \leq L_1$ *containing the contour* C_0, *where* $n = 0, 1, \ldots, N$, $C_n > 0, a_- \leq a_+ \in \mathbb{R}$, *and* $\lambda_n \geq \lambda_N$ *are constants. Define*

$$A_M(x,t) = 2e^{-p(l;l)t} \sum_{s=0}^{M-1} \Gamma\left(s + \frac{1}{2}\right) \frac{l^{\gamma-2s}}{(p_0 t)^{s+1/2}} \sum_{n=0}^{2s} d_n^{2s} l^n V^{(n)}(l),$$

and

$$W_{2M}(x,t) = \exp\left[-(r-1)\rho \left|\frac{x}{r\rho t}\right|^{\frac{r}{r-1}} t \sin(|S_j^\alpha|) + a\left|\frac{x}{r\rho t}\right|^{\frac{1}{r-1}} \sin(S_j^\alpha)\right] \cdot$$
$$\cdot \frac{(1+|x/t|)^{-\frac{\lambda_{2M}}{r-1}}}{t^{M+\frac{1}{2}}|x/t|^{\frac{1}{r-1}[r(M+\frac{1}{2})-\gamma-1]}},$$

where $a \sin(S_j^\alpha) = \max\{a_- \sin(S_j^\alpha), a_+ \sin(S_j^\alpha)\}$. *Then we have the asymptotic result*

$$\sup_{t \geq 1} \sup_{(-1)^\alpha xt^{-\beta} \geq \epsilon} \frac{1}{W_{2M}(x,t)} \left| \int_{C_j} e^{-p(k;l)t} k^\gamma V(k)\, dk - A_M(x,t) \right| < \infty.$$

CHAPTER 2

EXPANSION IN THE INNER REGION, MATCHING

As we saw in sections 1.3 and 1.6, it was necessary to restrict attention to $(x,t) \in R_{(-1)^a} = \{(x,t) \in \mathbb{R} \times [1,\infty) \mid (-1)^a xt^{-\beta} > \epsilon\}$ with $\beta > 1/r$ and $0 < \epsilon < 1$. If we allowed $\beta = 1/r$ then we found that infinitely many of the terms of the asymptotic expansions we derived decayed at the same rate. We had no way of telling if those infinitely many terms formed a convergent series or not, and if convergent we had no way of writing down the sum of that series. Thus $\beta = 1/r$ is in some sense a wall over which one cannot go (using the same ideas); we are at a loss over how to obtain successively improving asymptotic approximations to $I(x,t)$.

The problem is not the coalescence of the saddle points *per se*. The saddle points coalesce when $x = \zeta t^\beta$ and $t \to \infty$ provided $\beta < 1$, and yet when $\beta > 1/r$ the expansion remains ordered and valid as an asymptotic approximation. The problem is that when $\beta = 1/r$ the contribution of the integral over each contour of steepest ascent/descent no longer is dominated by the immediate neighborhood of the saddle point; the integral over each segment of the contour makes an equally important asymptotic contribution. Define the variable $\xi = xt^{-1/r}$. This variable will play a similar role in the inner region (to be defined below) as $b = x/t$ did in the outer regions.

2.1 Expansion and the error term

If $\frac{1}{r} < \beta < 1$ and $0 < \epsilon < 1$ we can define the *inner region* R_0 as follows:

$$R_0 = \{(x,t) \in \mathbb{R} \times [1,\infty) \mid |x|t^{-\beta} < \epsilon^{-1}\}.$$

We still employ the notation $\zeta = xt^{-\beta}$ for the natural variable in the inner region. Notice that the *overlap region* $R_0 \cap R_{(-1)^a}$ can be described by the inequalities $\epsilon < (-1)^a \zeta < \epsilon^{-1}$. We seek another method of generating an expansion of $I(x,t)$ which will work throughout the inner region, and which will yield an expansion which can be matched to the expansion we have already obtained in the outer regions, provided $\epsilon < (-1)^a \zeta < \epsilon^{-1}$. Our approach in the outer regions was to expand $V(k)$ at $k = 0$ (for the contour $\tilde{C}$) or at $k = k_j(b)$ (for the contour C_j). Since in the inner region the saddle points always coalesce as $t \to \infty$ at $k = 0$, one strategy (which turns out to be the correct one) is to forget about expanding $V(k)$ at the saddle points, and simply expand it at the fixed point $k = 0$ for all the contours:

$$V(k) = \sum_{m=0}^{N-1} \frac{V^{(m)}(0)}{m!} k^m + R_N(k;0).$$

Letting $C(b)$ denote any of the contours $\tilde{C}, C_j, C^+_{-J}, C^+_K$ (the last two only being relevant when $r = 4J - 1$ or $r = 4K + 1$ respectively) we have

$$\int_{C(b)} e^{ikx - i\rho k^r t} k^\gamma V(k)\, dk = \sum_{m=0}^{N-1} \frac{V^{(m)}(0)}{m!} \int_{C(b)} e^{ikx - i\rho k^r t} k^{\gamma + m}\, dk$$

$$+ \int_{C(b)} e^{ikx - i\rho k^r t} k^\gamma R_N(k; 0)\, dk.$$

In the outer regions, and more particularly when $b = x/t$ was a nonzero constant, each of the integrals $\int_{\tilde{C}(b)} e^{ikx - i\rho k^r t} k^{\gamma + m}\, dk$ and $\int_{\pm C^\pm_j(b)} e^{ikx - i\rho k^r t} k^\gamma (k - l)^m\, dk$ gave rise to an asymptotic expansion obtained by re-expressing the integrand in the new variable $v \geq 0$, which measured the difference in the real part of the phase function $\varphi(k) = ikb - i\rho k^r$ along the contour. The integrand was expanded in powers of v, an acknowledgment of the fact that the neighborhood of the contour where $v = 0$ was of prime asymptotic importance, and then the resulting expansion was integrated term-by-term, yielding the desired asymptotic expansion of the integral. But as we have noted above, when we move to the wall delimiting the outer region by supposing ξ is a nonzero constant ($\zeta = \xi$ when $\beta = 1/r$), we find each of these term-by-term integrals (indexed by the integer $s \geq 0$) decay in t at the same rate, namely $t^{-(1+\gamma+m)/r}$:

$$\frac{1}{|b|^s} \frac{1}{(|b|t)^{1+\gamma+m+s(r-1)}} = \frac{1}{|\frac{x}{t}|^s} \frac{1}{(|x|)^{s(r-1)}} \frac{1}{(|\xi|t^{1/r})^{1+\gamma+m}} = \frac{1}{|\xi|^{1+\gamma+m+sr}} \frac{1}{t^{\frac{1+\gamma+m}{r}}},$$

$$e^{-p(l)t} \frac{|l|^{\gamma+m-s}}{(|p_0|t)^{\frac{s+1}{2}}} \propto e^{-c|b|^{\frac{r}{r-1}}t} \frac{|l|^{1+\gamma+m}}{(|l|^r t)^{\frac{s+1}{2}}} \propto e^{-c|\xi|^{\frac{r}{r-1}}} \frac{|b|^{\frac{1+\gamma+m}{r-1}}}{(|b|^{\frac{r}{r-1}}t)^{\frac{s+1}{2}}} = \frac{f_s(\xi)}{t^{\frac{1+\gamma+m}{r}}}.$$

To go beyond this wall into the inner region, it makes sense to re-express the integral being expanded in terms of the variable ξ:

$$\int_{C(b)} e^{ikx - i\rho k^r t} k^{\gamma + m}\, dk = \frac{1}{t^{(1+\gamma+m)/r}} \int_{C(\xi)} e^{i\omega\xi - i\rho\omega^r} \omega^{\gamma + m}\, d\omega$$

where we have changed variables $\omega = kt^{1/r}$. $C(\xi)$ is defined to be the contour in the ω-plane such that $\omega \in C(\xi)$ if and only if $k \in C(b)$. In order to see that this definition makes sense independent of the value of t we examine the polar equations for the contours $\tilde{C}(b)$ and $C_j(b)$:

$$\tilde{C}(b); \qquad bR\cos(\theta) - \rho R^r \cos(r\theta) = 0,$$

$$C_j(b): \qquad bR\cos(\theta) - \rho R^r \cos(r\theta) = (-1)^\alpha (r - 1)\rho \left|\frac{b}{r\rho}\right|^{\frac{r}{r-1}} \cos(S^\alpha_j),$$

where $k = Re^{i\theta}$. If $\omega = Qe^{i\theta}$ then $Q = Rt^{1/r}$. If we multiply both sides of these polar equations by t and re-express in terms of ξ and Q we get

$$\tilde{C}(\xi); \qquad \xi Q\cos(\theta) - \rho Q^r \cos(r\theta) = 0,$$

$$C_j(\xi): \qquad \xi Q\cos(\theta) - \rho Q^r \cos(r\theta) = (-1)^\alpha (r - 1)\rho \left|\frac{\xi}{r\rho}\right|^{\frac{r}{r-1}} \cos(S^\alpha_j).$$

Thus the polar expressions of $C(\xi)$ are the same as described in section 1.2 except k is replaced by ω and b is replaced by ξ.

We take the view that $A(\xi; r, \rho, \gamma, C) \stackrel{\text{def}}{=} \int_C e^{i\omega\xi - i\rho\omega^r} \omega^\gamma \, d\omega$ is some sort of special function, well-known and studied in certain cases, and as such is an acceptable component in the terms of an asymptotic expansion. Thus we have the N-term expansion (with error term)

$$\int_{C(b)} e^{ikx - i\rho k^r t} k^\gamma V(k) \, dk = \sum_{m=0}^{N-1} \frac{V^{(m)}(0)}{m!} \frac{A(\xi; r, \rho, \gamma + m, C(\xi))}{t^{(1+\gamma+m)/r}}$$

$$+ \frac{1}{t^{(1+\gamma+N)/r}} \int_{C(\xi)} e^{i\omega\xi - i\rho\omega^r} \omega^{\gamma+N} \frac{R_N(k; 0)}{k^N} \bigg|_{k = \omega t^{-1/r}} \, d\omega.$$

In the inner region both of the limits $\xi \to 0$ and $(-1)^\alpha \xi \to \infty$ are possible. Thus although it is clear for ξ fixed that the terms of this expansion form an asymptotic scale, we need that to be also the case throughout the inner region. Thus we need to understand the asymptotic behavior of $A(\xi; r, \rho, \gamma + m, C(\xi))$ as ξ tends to infinity. Since the contour $C(\xi)$ depends on ξ, we need to assume $(-1)^\alpha \xi \geq 0$ so that the contour is well-defined. We will find out what happens when $(-1)^\alpha \xi \to 0^+$ later.

To find the asymptotics of $A(\xi; r, \rho, \gamma + m, C(\xi))$ as $(-1)^\alpha \xi \to \infty$ it will be useful to first re-express this quantity as an integral over a fixed contour. If $\omega \in C(\xi)$ then define $\omega = \eta |\xi|^{1/(r-1)}$. It turns out that η lies on a single contour denoted by $C((-1)^\alpha)$. To see this refer back to the polar equations of $\check{C}(\xi)$ and $C_j(\xi)$. If $\eta = S e^{i\theta}$ then $Q = S|\xi|^{1/(r-1)}$. Plugging this equation into these polar equations and dividing both sides by $|\xi|^{r/(r-1)}$ yields

$$\check{C}((-1)^\alpha): \qquad (-1)^\alpha S \cos(\theta) - \rho S^r \cos(r\theta) = 0,$$

$$C_j((-1)^\alpha): \qquad (-1)^\alpha S \cos(\theta) - \rho S^r \cos(r\theta) = (-1)^\alpha \frac{(r-1)}{(r^r \rho)^{\frac{1}{r-1}}} \cos(S_j^\alpha).$$

Thus $C((-1)^\alpha)$ is described in polar coordinates in section 1.2 if k is replaced by η and b is replaced by $(-1)^\alpha$. If we perform this change of variables in the integral defining $A(\xi; r, \rho, \gamma + m, C(\xi))$ we get

$$A(\xi; r, \rho, \gamma + m, C(\xi)) = |\xi|^{\frac{1+\gamma+m}{r-1}} \int_{C((-1)^\alpha)} e^{i[\eta(-1)^\alpha - \rho\eta^r]|\xi|^{\frac{r}{r-1}}} \eta^{\gamma+m} \, d\eta.$$

This last integral is the same as the one we studied in section 1.1-1.7 where $|\xi|^{\frac{r}{r-1}}$ plays the role of t, $(-1)^\alpha$ the role of b, and $V(k)$ is replaced by the constant 1. The following lemma gives the full asymptotic expansion of this quantity, c.f. [SSS], [F].

Lemma 2.1.1. Define $\tilde{p}(l) = -i\rho(r-1)l^r$, $p_0(l) = i\rho\frac{r(r-1)}{2}l^{r-2}$ and $l_0 = (r\rho)^{-\frac{1}{r-1}} e^{iS_j^\alpha}$. The following are asymptotic expansions of $A(\xi; r, \rho, \gamma + m, C(\xi))$

as $(-1)^\alpha \xi \to \infty$ for the indicated choices of the contour $C(\xi)$.

(1) $\tilde{C}(\xi)$:

$$\sum_{s=0}^{\infty} \frac{\Gamma(1+\gamma+m+sr)}{s!} \left(\frac{\rho}{i}\right)^s \left(\frac{i}{\xi}\right)^{1+\gamma+m+sr}$$

In the special case where $\tilde{C}(\xi)$ is a finite line segment (because it hits a saddle point) it might be advantageous to replace the coefficient $\frac{1}{s!}\Gamma(1+\gamma+m+sr)$ by

$$\frac{\Gamma(1+\gamma+m+sr)}{s!} - \binom{\gamma+m+sr}{s}\Gamma\left(1+\gamma+m+s(r-1),(r-1)\rho\left|\frac{\xi}{r\rho}\right|^{\frac{r}{r-1}}\right).$$

(2) $C_j(\xi)$, $-J \le j < 0$ $(\xi < 0, \alpha = 1)$ or $0 \le j \le K$ $(\xi > 0, \alpha = 0)$:

$$e^{-\tilde{p}(l_0)|\xi|^{r/(r-1)}} \sum_{s=0}^{\infty} \frac{\Gamma(s+\frac{1}{2})2d_0^{2s}(\gamma+m)l_0^{1+\gamma+m}}{[l_0^2 p_0(l_0)]^{s+\frac{1}{2}}} |\xi|^{\frac{1}{r-1}[1+\gamma+m-r(s+\frac{1}{2})]}$$

where $\arg[p_0(l_0)] = -S_j^\alpha + (-1)^\alpha \frac{\pi}{2}$.

(3) C_{-J}^+ $(r = 4J - 1, \xi < 0, \alpha = 1)$ or C_K^+ $(r = 4K + 1, \xi > 0, \alpha = 0)$:

$$e^{-\tilde{p}(l_0)|\xi|^{r/(r-1)}} \sum_{s=0}^{\infty} \Gamma\left(\frac{s+1}{2}\right) \frac{l_0^{1+\gamma+m} d_0^s(\gamma+m)}{[l_0^2 p_0(l_0)]^{(s+1)/2}} |\xi|^{\frac{1}{r-1}[1+\gamma+m-\frac{r}{2}(s+1)]}$$

where $\arg[p_0(l_0)] = 0$.

Proof. (1) Theorem 1.4.9. (2) Corollary 1.7.6. (3) Theorem 1.7.5. $\quad\square$

Expressing A in terms of special functions. First we introduce standard contours which do not depend on ξ. We will then employ Cauchy's theorem to change the contour of integration to these standard ones when $(-1)^\alpha \xi \le 1$. Recall we defined Γ_θ to be the ray $\arg k = \theta$ oriented from 0 to infinity. Thus we could replace $C_j(\xi)$ by $\Lambda_j \overset{\text{def}}{=} (-1)^\alpha(\Gamma_{V_j} - \Gamma_{V_{j+1}})$. By Cauchy's Theorem we have

$$A(\xi; r, \rho, \gamma, C_j(\xi)) = A(\xi; r, \rho, \gamma, \Lambda_j).$$

Also when $\xi < 0$ and $r \ne 4J - 1$ or $\xi > 0$ and $r \ne 4K + 1$ then $\tilde{C}(\xi)$ could be replaced by a ray $\tilde{\Gamma} \overset{\text{def}}{=} \Gamma_{\tilde{V}}$, where $\tilde{V}$ was defined after Lemma 1.2.2B. In those cases we have

$$A(\xi; r, \rho, \gamma, \tilde{C}(\xi)) = A(\xi; r, \rho, \gamma, \tilde{\Gamma}).$$

In the case $\xi < 0, r = 4J - 1$ (resp. $\xi > 0, r = 4K + 1$), the contour $\tilde{C}(\xi)$ ends at a saddle point whose position depends on ξ. In these cases we could replace the contour $\tilde{C}(\xi) \cup C_{-J}^+(\xi)$ (resp. $\tilde{C}(\xi) \cup C_K^+(\xi)$) with $\tilde{\Gamma} = \Gamma_{\tilde{V}}$. Then we would have

$$A(\xi; r, \rho, \gamma, \tilde{C}(\xi) \cup C_{-J}^+(\xi)) = A(\xi; r, \rho, \gamma, \tilde{\Gamma}),$$
$$A(\xi; r, \rho, \gamma, \tilde{C}(\xi) \cup C_K^+(\xi)) = A(\xi; r, \rho, \gamma, \tilde{\Gamma}).$$

This is a legitimate thing to do in the inner region, but it is the wrong thing to do if one wants the expansion to be uniformly valid. See section 3.1.

We will express $A(\xi; r, \rho, \gamma, C)$ in terms of a special function defined by

$$F_0(r, \delta; y) = \int_0^\infty e^{-\frac{1}{r}\sigma^r + y\sigma} \sigma^{\delta-1} \, d\sigma, \qquad r > 1, \delta > 0, y \in \mathbb{C}.$$

$F_0(r, \delta; y)$ is an entire function of y, which can also be expressed in terms of Faxén's integral $\mathrm{Fi}(\alpha, \delta; y)$, see [O1], page 332:

$$F_0(r, \delta; y) = r^{\delta/r-1} \mathrm{Fi}(\tfrac{1}{r}, \tfrac{\delta}{r}; yr^{1/r}).$$

When $r \geq 3$ is an odd integer we will also need the incomplete form of this function:

$$\mathbb{F}_0(r, \delta; y, z) = \int_0^z e^{-\frac{1}{r}\sigma^r + y\sigma} \sigma^{\delta-1} \, d\sigma, \qquad r > 1, \delta > 0, y \in \mathbb{C}, z \in S\{\sigma^{\delta-1}\}.$$

The contour of integration can be taken to be the straight line segment connecting 0 and z lying in the Riemann surface $S\{\sigma^{\delta-1}\}$. Later we will define $F_n(r, \delta; y)$, $\mathbb{F}_n(r, \delta; y, z)$, and $\mathbb{F}_n^c(r, \delta; y, z)$ for $n \geq 1$ an integer; see section 3.1.

In the following lemma we express $A(\xi; r, \rho, \gamma, C)$ for $C = \tilde{C}(\xi), \tilde{\Gamma}$ and Γ_{V_0} using the above special functions. For similar expressions when $C = C_j(\xi)$ or $C_{-J}^+(\xi), C_K^+(\xi)$, see Lemma 3.1.3 (where $n = 0$).

Lemma 2.1.2. *Suppose* $(-1)^\alpha \xi \geq 0$. *Then*

$$A(\xi; r, \rho, \gamma, \Gamma_{V_0}) = \left(\frac{e^{iV_0}}{(r\rho)^{1/r}}\right)^{1+\gamma} F_0\left(r, 1+\gamma; \frac{i\xi e^{iV_0}}{(r\rho)^{1/r}}\right)$$

$$A(\xi; r, \rho, \gamma, \tilde{\Gamma}) = \left(\frac{e^{i\tilde{V}}}{(r\rho)^{1/r}}\right)^{1+\gamma} F_0\left(r, 1+\gamma; \frac{i\xi e^{i\tilde{V}}}{(r\rho)^{1/r}}\right)$$

If $r = 4J - 1$ *and* $\alpha = 1$ *or if* $r = 4K + 1$ *and* $\alpha = 0$ *then*

$$A(\xi; r, \rho, \gamma, \tilde{C}(\xi)) = \left(\frac{e^{-(-1)^\alpha i \frac{\pi}{2}}}{(r\rho)^{1/r}}\right)^{1+\gamma} \mathbb{F}_0\left(r, 1+\gamma; y, e^{(-1)^\alpha i\pi} y^{\frac{1}{r-1}}\right),$$

where $y = \frac{|\xi|}{(r\rho)^{1/r}}$.

Proof. Parameterize the contour Γ_{V_j} as Re^{iV_j}, $R \geq 0$, and then change variables $\sigma = (r\rho)^{1/r}R$. □

Later we will see that when $r = 2$ or 3 this function is related to other well-known special functions.

Asymptotic Character of the Expansion. In order to check that each term of our expansion represents an asymptotic improvement over the previous terms we must consider the quotient of the $(N+1)$st term by the Nth term (ignoring multiplicative factors which do not depend on ξ or t)

$$Q_N = \frac{A(\xi; r, \rho, \gamma + N, C(\xi))}{t^{(1+\gamma+N)/r}} \cdot \frac{t^{(1+\gamma+N-1)/r}}{A(\xi; r, \rho, \gamma + N - 1, C(\xi))}$$

The factor $V^{(N)}(0)/V^{(N-1)}(0)$ has been omitted, even though we obviously need $V^{(N-1)}(0) \neq 0$ in order for the comparison of the sizes of the two terms to make sense. We can distinguish three different limits of interest in the inner region.

The case where $(-1)^\alpha \xi \to 0^+$. Let Γ^α denote any of these standard contours $\tilde{\Gamma}$ or Λ_j, where $(-1)^\alpha \xi > 0$. $A(\xi; r, \rho, \gamma + N, \Gamma^\alpha)$ is an entire function of ξ, so $A(0; r, \rho, \gamma + N, \Gamma^\alpha)$ is well-defined, and is nonzero (the integrand does not oscillate). As $(-1)^\alpha \xi \to 0^+$ we have that

$$Q_N \sim \frac{A(0; r, \rho, \gamma + N, \Gamma^\alpha)}{t^{1/r} A(0; r, \rho, \gamma + N - 1, \Gamma^\alpha)}.$$

Clearly this tends to zero as $t \to \infty$.

In the case $\xi < 0, r = 4J - 1$ (resp. $\xi > 0, r = 4K + 1$) we would also like to know how our expansions over $\tilde{C}(\xi)$, $C^+_{-J}(\xi)$, and $C^+_K(\xi)$ behave by themselves (not combined with each other). First, in this case $\tilde{C}(\xi)$ is a path of length proportional to $|\xi|^{1/(r-1)}$, so when we change variables in the integral $\omega = \eta |\xi|^{1/(r-1)}$ as we did before Lemma 2.1.1 we find that

$$A(\xi; r, \rho, \gamma + m, \tilde{C}(\xi)) = |\xi|^{\frac{1+\gamma+m}{r-1}} [c + o(1)], \qquad (-1)^\alpha \xi \to 0^+,$$

where $c \neq 0$ is a constant depending only on r and ρ. Thus for $\tilde{C}(\xi)$ we have

$$Q_N = \frac{A(\xi; r, \rho, \gamma + N, \tilde{C}(\xi))}{t^{1/r} A(\xi; r, \rho, \gamma + N - 1, \tilde{C}(\xi))} \sim \frac{c|\xi|^{\frac{1+\gamma+N}{r-1}}}{t^{1/r} |\xi|^{\frac{1+\gamma+N-1}{r-1}}} \propto \frac{|\xi|^{\frac{1}{r-1}}}{t^{1/r}},$$

which tends to zero as $|\xi| \to 0$, $t \to \infty$. Since $1 + \gamma + m \geq 1 + \gamma > 0$ we also see that $A(\xi; r, \rho, \gamma + m, \tilde{C}(\xi)) \to 0$ as $(-1)^\alpha \xi \to 0^+$. Therefore if $C^\alpha(\xi)$ represents either $C^+_{-J}(\xi)$ or $C^+_K(\xi)$, we have

$$Q_N = \frac{A(\xi; r, \rho, \gamma + N, C^\alpha(\xi))}{t^{1/r} A(\xi; r, \rho, \gamma + N - 1, C^\alpha(\xi))}$$

$$= \frac{A(\xi; r, \rho, \gamma + N, \tilde{\Gamma}) - A(\xi; r, \rho, \gamma + N, \tilde{C}(\xi))}{t^{1/r} [A(\xi; r, \rho, \gamma + N - 1, \tilde{\Gamma}) - A(\xi; r, \rho, \gamma + N - 1, \tilde{C}(\xi))]}$$

$$\sim \frac{A(0; r, \rho, \gamma + N, \tilde{\Gamma})}{t^{1/r} A(0; r, \rho, \gamma + N - 1, \tilde{\Gamma})},$$

which also tends to zero as $t \to \infty$, $(-1)^\alpha \xi \to 0^+$.

The case where $(-1)^\alpha \xi > 0$ is fixed. Here we have

$$Q_N = \frac{A(\xi; r, \rho, \gamma + N, C(\xi))}{t^{1/r} A(\xi; r, \rho, \gamma + N - 1, C(\xi))}.$$

There is an anomaly if $A(\xi; r, \rho, \gamma + N - 1, C(\xi)) = 0$, i.e the Nth term is abnormally small (it vanishes!). Lemma 2.1.1 shows that it cannot happen if $(-1)^\alpha \xi$ is sufficiently large; it cannot happen when ξ is sufficiently near (but not equal to) zero either. It may never happen but it is certainly very rare. We will ignore any such anomalies by only proving estimates which will be sharp when the anomaly does not occur. Clearly then the quotient vanishes in the limit as $t \to \infty$.

The case where $(-1)^\alpha \xi \to \infty$. Here Lemma 2.1.1 gives us exactly the information we need. We begin with $\tilde{C}(\xi)$:

$$Q_N = \frac{A(\xi; r, \rho, \gamma + N, \tilde{C}(\xi))}{t^{1/r} A(\xi; r, \rho, \gamma + N - 1, \tilde{C}(\xi))} \sim \frac{c|\xi|^{1+\gamma+N-1}}{t^{1/r}|\xi|^{1+\gamma+N}} \propto \frac{1}{t^{1/r}|\xi|},$$

which tends to zero as $t \to \infty$, $|\xi| \to \infty$. On the other hand, if $C^\alpha(\xi)$ denotes $C_j(\xi)$, or $C_{-J}^+(\xi)$ (when $\xi < 0, r = 4J - 1$), or $C_K^+(\xi)$ (when $\xi > 0, r = 4K + 1$), then

$$Q_N = \frac{A(\xi; r, \rho, \gamma + N, C^\alpha(\xi))}{t^{1/r} A(\xi; r, \rho, \gamma + N - 1, C^\alpha(\xi))} \sim \frac{c|\xi|^{\frac{1}{r-1}[1+\gamma+N-\frac{r}{2}]}}{t^{1/r}|\xi|^{\frac{1}{r-1}[1+\gamma+N-1-\frac{r}{2}]}} \propto \frac{|\xi|^{\frac{1}{r-1}}}{t^{1/r}}.$$

But $|\xi|^{1/(r-1)} t^{-1/r} = |xt^{-1/r} t^{-(r-1)/r}|^{1/(r-1)} = |xt^{-1}|^{1/(r-1)} = |\zeta t^{\beta-1}|^{1/(r-1)}$, which tends to zero as $t \to \infty$ since $|\zeta| \le \epsilon^{-1}$ and $\beta < 1$. So there is a definite outer limit to the region of validity of the inner expansion for the contour $C^\alpha(\xi)$. But we have shown that throughout the inner region, the terms of the inner expansion form an asymptotic scale.

Weighted Error Estimates. Because the terms of our expansion form an asymptotic scale, the appropriate weight function for the error of an N-term expansion will faithfully reflect the behavior of the $(N+1)$st term of the expansion. This weight function will depend on which contour is being investigated. Consider first the contour $\tilde{C}(\xi)$. Define the weight function

$$W_N(x, t) = \frac{1}{[t^{1/r}(1 + |\xi|)]^{1+\gamma+N}}.$$

When $\xi < 0, r = 4J - 1$, or $\xi > 0, r = 4K + 1$, this weight function does not faithfully reflect the behavior of the $(N+1)$st term as $(-1)^\alpha \xi \to 0^+$. In these cases we should replace the weight function by the following:

$$W_N(x, t) = \left[\frac{|\xi|^{1/(r-1)}}{t^{1/r}(1 + |\xi|)^{r/(r-1)}}\right]^{1+\gamma+N}.$$

We can use the same weight function for any of the contours $C_j(\xi), C_j^+(\xi), C_j^-(\xi)$, $-J \leq j \leq K$. It is the following:

$$W_N(x,t) = \exp\left\{-(r-1)\rho\left|\frac{\xi}{r\rho}\right|^{\frac{r}{r-1}}\sin(|S_j^\alpha|)\right\}\frac{(1+|\xi|)^{\frac{1}{r-1}[1+\gamma+N-\frac{r}{2}]}}{t^{\frac{1+\gamma+N}{r}}}.$$

For any of these choices of $C(b)$ and with the corresponding weight function $W_N(x,t)$, the weighted error estimate we must prove is:

$$\sup_{t\geq 1}\ \sup_{0\leq(-1)^\alpha xt^{-\beta}\leq\epsilon-1}\frac{t^{-\frac{1+\gamma+N}{r}}}{W_N(x,t)}\left|\int_{C(\xi)}e^{i\omega\xi-i\rho\omega^r}\omega^{\gamma+N}\frac{R_N(k;0)}{k^N}\right|_{k=\omega t^{-\frac{1}{r}}}d\omega\right| < \infty.$$

2.2 Error Control for the contour $\tilde{C}$

We will consider first the case where the contour $\tilde{C}(\xi)$ does not hit a saddle point. So if $\xi < 0$ we assume $r \neq 4J - 1$ and if $\xi > 0$ we assume $r \neq 4K + 1$. We will come back to the case where $\tilde{C}(\xi)$ does hit a saddle point later in this section.

Theorem 2.2.1. *Suppose $r > 1$, $\rho > 0$, $\gamma > -1$, and $\frac{1}{r} < \beta < 1$. Suppose $N \geq 0$ is an integer, $b = x/t$, and $\xi = xt^{-1/r}$; let $(-1)^\alpha = \operatorname{sgn}(x)$, $\alpha \in \{0,1\}$. If $\xi < 0$ we assume $r \neq 4J - 1$ and if $\xi > 0$ we assume $r \neq 4K + 1$. Let $\tilde{C}(b), \tilde{C}((-1)^\alpha), \tilde{\Gamma}$ be the contours described in the previous section. Suppose $V(k)$ is an entire function satisfying the estimate*

$$|V^{(N)}(k)| \leq \frac{C_N e^{a\Im k}}{(1+|k|)^{\lambda_N+N}}$$

uniformly for all k in the smallest closed sector containing $\tilde{C}((-1)^\alpha)$, where $C_N > 0, a \in \mathbb{R}, \lambda_N \in \mathbb{R}$ are constants. Define

$$W_N(x,t) = \frac{1}{[t^{1/r}(1+|\xi|)]^{1+\gamma+N}}.$$

and

$$A_N(x,t) = \sum_{m=0}^{N-1}\frac{V^{(m)}(0)}{m!}\frac{A(\xi;r,\rho,\gamma+m,\tilde{\Gamma})}{t^{(1+\gamma+m)/r}},$$

where $A(\xi;r,\rho,\gamma+m,\tilde{\Gamma})$ is the special function defined by

$$A(\xi;r,\rho,\gamma+m,\tilde{\Gamma}) \stackrel{\text{def}}{=} \int_{\tilde{\Gamma}}e^{i\omega\xi-i\rho\omega^r}\omega^{\gamma+m}\,d\omega.$$

Then we have

$$\sup_{t\geq 1}\ \sup_{(-1)^\alpha\xi\geq 0}\frac{1}{W_N(x,t)}\left|\int_{\tilde{C}(b)}e^{(ikb-i\rho k^r)t}k^\gamma V(k)\,dk - A_N(x,t)\right| < \infty.$$

Proof. As we saw in the previous section we must show that

$$\sup_{t\geq 1}\ \sup_{(-1)^\alpha\xi\geq 0}(1+|\xi|)^{1+\gamma+N}\left|\int_{\tilde{C}(\xi)}e^{i\omega\xi-i\rho\omega^r}\omega^{\gamma+N}\left(\frac{R_N(k;0)}{k^N}\right)d\omega\right|<\infty.$$

We will bound this quantity in two different ways, depending on whether $|\xi|\geq 1$ or $|\xi|<1$. First consider the case where $|\xi|\geq 1$. We basically follow the proof of Lemma 1.4.3. If $v=-[i\omega\xi-i\rho\omega^r]$, where $\omega\in\tilde{C}(\xi)$, then $v\geq 0$ is a real number. As in section 1.1 we define $\tilde{\omega}=(\frac{\xi}{iv})\omega-1$ and $\tilde{v}=\frac{\rho}{\xi}(\frac{iv}{\xi})^{r-1}$, and consequently $\tilde{\omega}(1+\tilde{\omega})^{-r}=\tilde{v}$. Hence we can apply all the results of section 1.4 where ω plays the role of k, $\tilde{\omega}$ plays the role of $\tilde{k}$, and ξ plays the role of b. In particular, we can consider ω to be a function of v (and ξ) and by Lemma 1.4.2 we have the estimate $|\omega(v)|\leq Cv^{1/r}$, where the positive constant C is independent of ξ. Also we have the formula (see section 1.3)

$$\omega^{\gamma+N}\frac{d\omega}{dv}=\frac{i}{\xi}\left(\frac{iv}{\xi}\right)^{\gamma+N}\frac{(1+\tilde{\omega})^{1+\gamma+N}}{1-(r-1)\tilde{\omega}}.$$

By Lemmas 1.4.1 and 1.4.2 the last fraction on the right is bounded independently of ξ and $v>0$. By the integral form of Taylor remainders (for $N\geq 1$) we have

$$\frac{R_N(k;0)}{k^N}=\frac{1}{(N-1)!}\int_0^1(1-z)^{N-1}V^{(N)}\left(\frac{\omega z}{t^{1/r}}\right)dz.$$

By our assumptions on V we have

$$\left|\frac{R_N(k;0)}{k^N}\right|\leq\frac{1}{(N-1)!}\sup_{0\leq z\leq 1}\left|V^{(N)}\left(\frac{\omega z}{t^{1/r}}\right)\right|\leq C'_N\sup_{0\leq z\leq 1}\frac{e^{azt^{-1/r}\Im\omega}}{(1+|\omega|zt^{-1/r})^{\lambda_N+N}}.$$

When $\lambda_N+N\geq 0$ this is bounded by $Ce^{C'v^{1/r}}$, since $t\geq 1$, where the positive constants C,C' are independent of t and ξ. If $\lambda_N+N<0$ then

$$(1+|\omega|zt^{-1/r})^{-\lambda_N-N}\leq C(1+v)^{-(\lambda_N+N)/r},$$

and hence $|R_N(k;0)/k^N|$ is bounded by $C(1+v)^{-(\lambda_N+N)/r}e^{C'v^{1/r}}$. In either case this quantity is bounded by $C(1+v)^{-\lambda'/r}e^{C'v^{1/r}}$, where $\lambda'=\min\{\lambda_N+N,0\}$. The same estimate holds when $N=0$. Thus in the integral we must estimate, after we change variables to v we obtain

$$(1+|\xi|)^{1+\gamma+N}\left|\int_{\tilde{C}(\xi)}e^{i\omega\xi-i\rho\omega^r}\omega^{\gamma+N}\left(\frac{R_N(k;0)}{k^N}\right)d\omega\right|$$

$$\leq C\int_0^\infty e^{-v}v^{\gamma+N}(1+v)^{-\lambda'/r}e^{C'v^{1/r}}dv$$

where the constants C,C' are independent of $(-1)^\alpha\xi\geq 1$ and $t\geq 1$. This finishes the argument when $|\xi|\geq 1$.

The change of contour from $\tilde{C}(\xi)$ to $\tilde{\Gamma}$ is allowed for any $(-1)^\alpha \xi > 0$, but it becomes most useful when $1 \geq (-1)^\alpha \xi > 0$. The justification of the change of contour in the integral comprising the error term is similar to the argument given near the end of section 1.2. Let V_f be the argument of center of the valley into which $\tilde{C}(\xi)$ tends as $\omega \to \infty$. Then $\tilde{\Gamma}$ is parameterized by $\omega = Q e^{i V_f}$, $Q \geq 0$. For such ω we have $e^{-i\rho\omega^r} = e^{-\rho Q^r}$ and $|e^{i\omega\xi}| = e^{-Q|\xi|\sin|V_f|} \leq 1$. We can estimate the Taylor remainder in the same way as above except that $\Im\omega = Q\sin(V_f)$, so that for all $N \geq 0$

$$\left| \frac{R_N(k;0)}{k^N} \right| \leq C(1+Q)^{-\lambda'} e^{|a|Q\sin|V_f|}.$$

Thus we have that

$$(1+|\xi|)^{1+\gamma+N} \left| \int_{\tilde{C}(\xi)} e^{i\omega\xi - i\rho\omega^r} \omega^{\gamma+N} \left(\frac{R_N(k;0)}{k^N} \right) d\omega \right|$$

$$\leq 2^{1+\gamma+N} \left| \int_{\tilde{\Gamma}} e^{i\omega\xi - i\rho\omega^r} \omega^{\gamma+N} \left(\frac{R_N(k;0)}{k^N} \right) d\omega \right|$$

$$\leq C \int_0^\infty e^{-\rho Q^r} Q^{\gamma+N} (1+Q)^{-\lambda'} e^{|a|Q\sin|V_f|} \, dQ$$

as long as $0 \leq (-1)^\alpha \xi \leq 1$, where $C > 0$ does not depend on $t \geq 1$ or on ξ. This completes the proof for the case $0 \leq (-1)^\alpha \xi \leq 1$. $\square$

Now we turn our attention to the case where $\tilde{C}(\xi)$ hits a saddle point.

Theorem 2.2.2. *Suppose $r > 1$, $\rho > 0$, $\gamma > -1$, and $\frac{1}{r} < \beta < 1$. Suppose $N \geq 0$ is an integer, $b = x/t$, and $\xi = xt^{-1/r}$; let $(-1)^\alpha = \operatorname{sgn}(x)$, $\alpha \in \{0,1\}$. If $\xi < 0$ we assume $r = 4J - 1$ and if $\xi > 0$ we assume $r = 4K + 1$. Let $\tilde{C}(b), \tilde{C}(\xi)$ be the contours described in the previous section. Suppose $V(k)$ is an entire function satisfying the estimate*

$$|V^{(N)}(k)| \leq \frac{C_N e^{a\Im k}}{(1+|k|)^{\lambda_N + N}}$$

uniformly for all k in the ray $k = (-1)^\alpha iR, R \geq 0$, where $C_N > 0, a \in \mathbb{R}, \lambda_N \in \mathbb{R}$ are constants. Define

$$W_N(x,t) = \left[\frac{|\xi|^{1/(r-1)}}{t^{1/r}(1+|\xi|)^{r/(r-1)}} \right]^{1+\gamma+N}.$$

and

$$A_N(x,t) = \sum_{m=0}^{N-1} \frac{V^{(m)}(0)}{m!} \frac{A(\xi; r, \rho, \gamma+m, \tilde{C}(\xi))}{t^{(1+\gamma+m)/r}},$$

where $A(\xi; r, \rho, \gamma + m, \tilde{C}(\xi))$ is defined by

$$A(\xi; r, \rho, \gamma + m, \tilde{C}(\xi)) \overset{\text{def}}{=} \int_{\tilde{C}(\xi)} e^{i\omega\xi - i\rho\omega^r} \omega^{\gamma + m} \, d\omega.$$

Then we have

$$\sup_{t \geq 1} \sup_{(-1)^\alpha \xi \geq 0} \frac{1}{W_N(x, t)} \left| \int_{\tilde{C}(b)} e^{(ikb - i\rho k^r)t} k^\gamma V(k) \, dk - A_N(x, t) \right| < \infty.$$

Proof. Again the proof will be split up according as $|\xi| \geq 1$ or $|\xi| \leq 1$. First, let us consider the case $|\xi| \geq 1$. We must show that

$$\sup_{t \geq 1} \sup_{(-1)^\alpha \xi \geq 1} |\xi|^{1 + \gamma + N} \left| \int_{\tilde{C}(\xi)} e^{i\omega\xi - i\rho\omega^r} \omega^{\gamma + N} \left(\frac{R_N(k; 0)}{k^N} \right) d\omega \right| < \infty.$$

This is proved as in the proof of the previous theorem, with the necessary modifications for the fact that $v \in [0, v_0]$, where $v_0 = (r - 1)\rho \left| \frac{\xi}{r\rho} \right|^{\frac{r}{r-1}}$, as in the proof of Lemma 1.4.6.

Now suppose $|\xi| \leq 1$. We must show that

$$\sup_{t \geq 1} \sup_{0 \leq (-1)^\alpha \xi \leq 1} \frac{1}{|\xi|^{\frac{1 + \gamma + N}{r - 1}}} \left| \int_{\tilde{C}(\xi)} e^{i\omega\xi - i\rho\omega^r} \omega^{\gamma + N} \left(\frac{R_N(k; 0)}{k^N} \right) d\omega \right| < \infty.$$

We change variables $\omega = \eta |\xi|^{\frac{1}{(r-1)}}$, where $\eta \in \tilde{C}((-1)^\alpha)$. The contour $\tilde{C}((-1)^\alpha)$ is an oriented segment of the imaginary axis starting at $\eta = 0$ and ending at $\eta = (-1)^\alpha i(r\rho)^{-1/(r-1)}$. Thus we must bound

$$\left| \int_0^{(-1)^\alpha i(r\rho)^{-1/(r-1)}} e^{i[\eta(-1)^\alpha - \rho\eta^r]|\xi|^{\frac{r}{r-1}}} \eta^{\gamma + m} \left(\frac{R_N(k; 0)}{k^N} \right) d\eta \right|.$$

We parameterize the contour of integration as $\eta = (-1)^\alpha iS$, $S \in [0, (r\rho)^{-1/(r-1)}]$. We have $i[\eta(-1)^\alpha - \rho\eta^r] = -S + \rho S^r$. Thus $e^{i[\eta(-1)^\alpha - \rho\eta^r]|\xi|^{\frac{r}{r-1}}} \leq 1$ for $S \in [0, (r\rho)^{-1/(r-1)}]$. Also we have the bound

$$\left| \frac{R_N(k; 0)}{k^N} \right| \leq C_N' \sup_{0 \leq z \leq 1} \frac{e^{azt^{-1/r}\Im\omega}}{(1 + |\omega|zt^{-1/r})^{\lambda_N + N}} \leq C_N' \frac{e^{|a|t^{-1/r}|\xi|^{1/(r-1)}S}}{(1 + S|\xi|^{1/(r-1)}t^{-1/r})^{\lambda'}}$$

$$\leq C_N' \frac{e^{|a|S}}{(1 + S)^{\lambda'}},$$

where $\lambda' = \min\{\lambda_N + N, 0\}$, since $|\xi| \leq 1$ and $t \geq 1$. Thus the above integral is bounded by

$$C \int_0^{(r\rho)^{-1/(r-1)}} S^{\gamma + N}(1 + S)^{-\lambda'} e^{|a|S} \, dS < \infty,$$

where $C > 0$ is independent of $t \geq 1$ and ξ. $\quad\square$

The reader will notice that we never mentioned the assumption $(-1)^\alpha \zeta \leq \epsilon^{-1}$ in the statements or proofs of the last two theorems. Thus these two theorems hold without this assumption.

2.3 Error Control for the contour C_j

First we will prove a result for the contour $C_j^\pm$, which will apply directly to C_{-J}^+ and C_K^+, and can be used in a simple way to obtain the desired result for the contours C_j and $\tilde{C} \cup C_{-J}^+$ and $\tilde{C} \cup C_K^+$.

Theorem 2.3.1. *Suppose* $r \geq 2$, $\rho > 0$, $\gamma > -1$, $\frac{1}{r} < \beta < 1$, $0 < \epsilon < 1$, *and* $N \geq 0$ *and* $-J \leq j \leq K$ *are integers, where* $J = [\frac{r+1}{4}]$ *and* $K = [\frac{r-1}{4}]$. *Define the parameters* $b = x/t$, $\xi = xt^{-1/r}$, $\zeta = xt^{-\beta}$, $x \in \mathbb{R}$, $t \geq 1$, *and* $(-1)^\alpha = \text{sgn}(x)$, $\alpha \in \{0,1\}$. *Assume* $j < 0$ *when* $\alpha = 1$ *and* $j \geq 0$ *when* $\alpha = 0$. *Suppose* $C_j^\pm(b)$ *is the contour described in section 1.2 depending on the parameter* b *and passing through the saddle point* $k_j(b) = \left|\frac{b}{r\rho}\right|^{\frac{1}{r-1}} e^{iS_j^\alpha}$, $S_j^\alpha = \frac{\pi}{r-1}(2j + \alpha)$. *Suppose* $L_0 \leq \arg k \leq L_1$ *is the smallest closed sector containing the contour* $C_j^\pm(b)$ *in its entirety (see section 1.2). Suppose* $V(k)$ *is an entire function satisfying the bounds*

$$|V^{(N)}(k)| \leq \frac{C_N e^{a\Im k}}{(1 + |k|)^{N + \lambda_N}},$$

uniformly for k *in the sector* $L_0 \leq \arg k \leq L_1$, *where* $C_N > 0, a \in \mathbb{R}, \lambda_N \in \mathbb{R}$ *are constants. Define the* N-*term asymptotic approximation*

$$A_N^\pm(x,t) = \sum_{m=0}^{N-1} \frac{V^{(m)}(0)}{m!} \frac{A(\xi; r, \rho, \gamma + m, C_j^\pm(\xi))}{t^{(1+\gamma+m)/r}},$$

where the contour $C_j^\pm(\xi)$ *is described in section 2.1 and*

$$A(\xi; r, \rho, \gamma + m, C_j^\pm(\xi)) \stackrel{\text{def}}{=} \int_{C_j^\pm(\xi)} e^{i\omega\xi - i\rho\omega^r} \omega^{\gamma+m} \, d\omega.$$

Define the Nth *weight function*

$$W_N(x,t) = \exp\left\{-(r-1)\rho\left|\frac{\xi}{r\rho}\right|^{\frac{r}{r-1}} \sin(|S_j^\alpha|)\right\} \frac{(1 + |\xi|)^{\frac{1}{r-1}[1+\gamma+N-\frac{r}{2}]}}{t^{\frac{1+\gamma+N}{r}}}.$$

Then we have the asymptotic result

$$\sup_{t \geq 1} \sup_{0 \leq (-1)^\alpha\zeta \leq \epsilon-1} \frac{1}{W_N(x,t)} \left|\left|\int_{C_j^\pm(b)} e^{(ikb-i\rho k^r)t} k^\gamma V(k) \, dk - A_N^\pm(x,t)\right|\right| < \infty.$$

Proof. Let $X_N(\xi) = W_N(x,t)t^{\frac{1+\gamma+N}{r}}$. Using our expressions for the error term we find that we must show that

$$\sup_{t \geq 1} \sup_{0 \leq (-1)^\alpha\zeta \leq \epsilon-1} \frac{1}{X_N(\xi)} \left|\int_{C_j^\pm(\xi)} e^{i\omega\xi - i\rho\omega^r} \omega^{\gamma+N} \left(\frac{R_N(k;0)}{k^N}\right) d\omega\right| < \infty.$$

The proof of this divides into the two cases $(-1)^\alpha \xi \geq 1$ and $0 \leq (-1)^\alpha \xi \leq 1$. Suppose to begin with that $(-1)^\alpha \xi \geq 1$. Define $p(\omega; \xi) = -[i\omega\xi - i\rho\omega^r]$ and $l(\xi) = \left|\frac{\xi}{r\rho}\right|^{\frac{1}{r-1}} e^{iS_j^\alpha}$. Then

$$e^{i\omega\xi - i\rho\omega^r} = e^{-p(l(\xi);\xi)} e^{-[p(\omega;\xi) - p(l(\xi);\xi)]} = e^{-p(l(\xi);\xi)} e^{-v},$$

where $v = p(\omega; \xi) - p(l(\xi); \xi)$. Notice that (see section 1.1)

$$-p(l(\xi); \xi) = (r-1)\rho \left|\frac{\xi}{r\rho}\right|^{\frac{r}{r-1}} (-1)^\alpha i e^{iS_j^\alpha}$$

$$= (r-1)\rho \left|\frac{\xi}{r\rho}\right|^{\frac{r}{r-1}} (-1)^\alpha [-\sin(S_j^\alpha) + i\cos(S_j^\alpha)].$$

Because of the assumed relation between j and α we have $(-1)^\alpha S_j^\alpha = |S_j^\alpha|$. Hence

$$\left|e^{-p(l(\xi);\xi)}\right| = \exp\left\{-(r-1)\rho \left|\frac{\xi}{r\rho}\right|^{\frac{r}{r-1}} \sin(|S_j^\alpha|)\right\}.$$

As we have seen, if $\omega \in C_j^\pm(\xi)$ then v is real and nonnegative. We can apply all the relationships and estimates of section 1.7 where k is replaced by ω and b is replaced by ξ. Thus $\omega = l(\xi)(1 + \tilde\omega)$, $\tilde\omega = \frac{w}{l(\xi)\sqrt{p_0(l(\xi))}}$, where as usual $p_0(l) = i\rho r(r-1)l^{r-2}/2$, $v = w^2$, and $(1 + \tilde\omega)^r - 1 - r\tilde\omega = r(r-1)\tilde\omega^2/2$. Therefore we have

$$\omega^{\gamma+N} = l(\xi)^{\gamma+N}(1 + \tilde\omega)^{\gamma+N},$$

and hence the bound

$$|\omega|^{\gamma+N} \leq \left|\frac{\xi}{r\rho}\right|^{\frac{\gamma+N}{r-1}} (1 + |\tilde\omega|)^{\gamma+N}.$$

As in our discussion of the error term in section 1.6 we have

$$\frac{d\omega}{dv} = \frac{1}{2\sqrt{v}\sqrt{p_0(l(\xi))}} \frac{d\tilde\omega}{d\tilde{w}}.$$

By Lemma 1.7.2 we have

$$\left|\frac{d\tilde\omega}{d\tilde{w}}\right| \leq \frac{C}{(1 + |\tilde{w}|)^{1-2/r}} \leq C,$$

where C depends only on r. As in Theorem 2.2.1 we have the bound

$$\left|\frac{R_N(k;0)}{k^N}\right| \leq C'_N \sup_{0 \leq z \leq 1} \frac{e^{azt^{-1/r}\Im\omega}}{(1 + |\omega|zt^{-1/r})^{\lambda_N+N}} \leq C'_N \frac{e^{|a|\cdot|l(\xi)|t^{-1/r} + |a|\cdot|l(\xi)\tilde\omega|}}{(1 + |l(\xi)|t^{-1/r} + |l(\xi)\tilde\omega|)^{\lambda'}},$$

where $\lambda' = \min\{\lambda_N + N, 0\}$. Putting these together we have the estimate

$$
\frac{1}{X_N(\xi)} \left| \int_{C_j^\pm(\xi)} e^{i\omega\xi - i\rho\omega^r} \omega^{\gamma+N} \left(\frac{R_N(k;0)}{k^N} \right) d\omega \right|
$$

$$
\leq \frac{C|\xi|^{\frac{\gamma+N}{r-1}}}{(1+|\xi|)^{\frac{1}{r-1}[1+\gamma+N-\frac{r}{2}]}} \int_0^\infty e^{-v} \frac{(1+|\tilde\omega|)^{\gamma+N}}{\sqrt{v}|\xi|^{\frac{r-2}{2(r-1)}}} \frac{e^{|a| \cdot |l(\xi)| t^{-1/r} + |a| \cdot |l(\xi)\tilde\omega|}}{(1+|l(\xi)|t^{-1/r} + |l(\xi)\tilde\omega|)^{\lambda'}} dv
$$

Since $|\xi| \geq 1$ the factor involving ξ is bounded independently of ξ. Also

$$
\frac{|l(\xi)|}{t^{1/r}} = c\frac{|\xi|^{1/(r-1)}}{t^{1/r}} = c|xt^{-\frac{1}{r}}t^{-\frac{r-1}{r}}|^{1/(r-1)} = c|\zeta t^{\beta-1}|^{1/(r-1)} \leq ce^{-1/(r-1)},
$$

since $\beta < 1$, $t \geq 1$ and $|\zeta| \leq \epsilon^{-1}$. By Lemma 1.7.2, $|\tilde\omega| \leq C|\tilde w|^{2/r}$. Recall $\tilde w = \frac{w}{l(\xi)\sqrt{p_0(l(\xi))}}$. But $l(\xi)^2 p_0(l(\xi)) = i\rho r(r-1)l(\xi)^r/2$. So

$$
|\tilde\omega| \leq C' \frac{|w|^{2/r}}{(|\xi|^{\frac{r}{2(r-1)}})^{2/r}} = C'\frac{v^{1/r}}{|\xi|^{1/(r-1)}} \leq C'v^{1/r},
$$

since $|\xi| \geq 1$. Using the same estimate we also have

$$
|l(\xi)\tilde\omega| \leq C''|\xi|^{1/(r-1)} \frac{v^{1/r}}{|\xi|^{1/(r-1)}} \leq C''v^{1/r}.
$$

If $\gamma + N > 0$ we have $(1+|\tilde\omega|)^{\gamma+N} \leq (1+C'v^{1/r})^{\gamma+N}$. Otherwise we have $(1+|\tilde\omega|)^{\gamma+N} \leq 1$. Set $\gamma' = \max\{0, \gamma+N\}$. Thus the quantity we are estimating is bounded by

$$
Ce^{|a|ce^{-1/(r-1)}} \int_0^\infty \frac{e^{-v}}{\sqrt{v}} (1+C'v^{1/r})^{\gamma'} (1+ce^{-1/(r-1)} + C''v^{1/r})^{-\lambda'} e^{|a|C''v^{1/r}} dv
$$

where the constants $C, C', C'' > 0$ are independent of ξ and t. This finishes the proof when $|\xi| \geq 1$.

Now suppose $0 \leq (-1)^\alpha \xi \leq 1$. We must show that

$$
\sup_{t \geq 1} \sup_{0 \leq (-1)^\alpha \xi \leq 1} \left| \int_{C_j^\pm(\xi)} e^{i\omega\xi - i\rho\omega^r} \omega^{\gamma+N} \left(\frac{R_N(k;0)}{k^N} \right) d\omega \right| < \infty.
$$

Let V_f be the argument of the asymptotic center of the valley containing the contour $C_j^\pm(\xi)$ at infinity. We can use Cauchy's Theorem to change the contour of integration from $C_j^\pm(\xi)$ to $\Lambda_j(\xi)$, where $\Lambda_j(\xi)$ consists of the union of two pieces. One piece is a circular arc $\omega = |l(\xi)|e^{i\theta}$ where θ ranges between S_j^α and V_f. The second piece is $\omega = Qe^{iV_f}$, $Q \geq |l(\xi)|$. Since $k = \omega/t^{1/r}$, the quotient $|R_N(k;0)/k^N|$ is bounded independently of t and ξ and ω in the first piece. Hence the integral over the first piece is bounded independently of t and

ξ. On the second piece we have $e^{-i\rho\omega^r} = e^{-\rho Q^r}$ and $|e^{i\omega\xi}| = e^{-(-1)^\alpha|\xi|Q\sin(V_f)}$. This last exponential is bounded by 1 except for the contour $C_0^+(\xi)$, since there $\alpha = 0$, and $V_f = -\frac{\pi}{2r} < 0$. In all cases we have the exponentially growing bound $|e^{i\omega\xi}| \leq e^{Q\sin|V_f|}$. We also have the estimate (as usual, $\lambda' = \min\{\lambda_N + N, 0\}$)

$$\left|\frac{R_N(k;0)}{k^N}\right| \leq C_N' \sup_{0 \leq z \leq 1} \frac{e^{azt^{-1/r}\Im\omega}}{(1+|\omega|zt^{-1/r})^{\lambda_N+N}} \leq C_N' \frac{e^{|a|Q\sin|V_f|}}{(1+Q)^{\lambda'}}.$$

Thus the integral over the second piece is bounded by

$$C \int_{|l(\xi)|}^\infty e^{-\rho Q^r} Q^{\gamma+N} \frac{e^{(1+|a|)Q\sin|V_f|}}{(1+Q)^{\lambda'}} dQ \leq C \int_0^\infty e^{-\rho Q^r} Q^{\gamma+N} \frac{e^{(1+|a|)Q\sin|V_f|}}{(1+Q)^{\lambda'}} dQ$$

where $C > 0$ is independent of t and ξ. $\square$

Now we can combine this result with itself to obtain our result for the contour C_j.

Corollary 2.3.2. *Suppose* $r \geq 2$, $\rho > 0$, $\gamma > -1$, $\frac{1}{r} < \beta < 1$, $0 < \epsilon < 1$, *and* $N \geq 0$ *and* $-J \leq j \leq K$ *are integers, where* $J = \left[\frac{r+1}{4}\right]$ *and* $K = \left[\frac{r-1}{4}\right]$. *Define the parameters* $b = x/t$, $\xi = xt^{-1/r}$, $\zeta = xt^{-\beta}$, $x \in \mathbb{R}$, $t \geq 1$, *and* $(-1)^\alpha = \mathrm{sgn}(x)$, $\alpha \in \{0,1\}$. *Assume* $j < 0$ *when* $\alpha = 1$ *and* $j \geq 0$ *when* $\alpha = 0$. *Suppose* $C_j(b)$ *is the contour described in section 1.2 depending on the parameter* b *and passing through the saddle point* $k_j(b) = \left|\frac{b}{r\rho}\right|^{\frac{1}{r-1}} e^{iS_j^\alpha}$, $S_j^\alpha = \frac{\pi}{r-1}(2j+\alpha)$. *Suppose* $L_0 \leq \arg k \leq L_1$ *is the smallest closed sector containing the contour* $C_j(b)$ *in its entirety (see section 1.2). Suppose* $V(k)$ *is an entire function satisfying the bounds*

$$|V^{(N)}(k)| \leq \frac{C_N e^{\max\{a_-\Im k, a_+\Im k\}}}{(1+|k|)^{N+\lambda_N}},$$

uniformly for k *in the sector* $L_0 \leq \arg k \leq L_1$, *where* $C_N > 0, a_- \leq a_+ \in \mathbb{R}, \lambda_N \in \mathbb{R}$ *are constants. Define the N-term asymptotic approximation*

$$A_N(x,t) = \sum_{m=0}^{N-1} \frac{V^{(m)}(0)}{m!} \frac{A(\xi;r,\rho,\gamma+m,\Lambda_j)}{t^{(1+\gamma+m)/r}},$$

where the contour Λ_j *is described in section 2.1 and*

$$A(\xi;r,\rho,\gamma+m,\Lambda_j) \stackrel{\text{def}}{=} \int_{\Lambda_j} e^{i\omega\xi-i\rho\omega^r} \omega^{\gamma+m} d\omega.$$

Define the Nth weight function

$$W_N(x,t) = \exp\left\{-(r-1)\rho \left|\frac{\xi}{r\rho}\right|^{\frac{r}{r-1}} \sin(|S_j^\alpha|)\right\} \frac{(1+|\xi|)^{\frac{1}{r-1}[1+\gamma+N-\frac{r}{2}]}}{t^{\frac{1+\gamma+N}{r}}}.$$

Then we have the asymptotic result

$$\sup_{t \geq 1} \sup_{0 \leq (-1)^\alpha \zeta \leq \epsilon - 1} \frac{1}{W_N(x,t)} \left| \int_{C_j(b)} e^{(ikb - i\rho k^r)t} k^\gamma V(k)\, dk - A_N(x,t) \right| < \infty.$$

Proof. Combine the results of the previous theorem for C_j^+ and C_j^-, and then change contours from $C_j = C_j^- \cup C_j^+$ to Λ_j as discussed in section 2.1.   □

Let C^α denote either C_{-J}^+ when $\alpha = 1$ and $r = 4J - 1$ or C_K^+ when $\alpha = 0$ and $r = 4K + 1$. Theorem 2.3.1 says that

$$\int_{C^\alpha(b)} e^{(ikb - i\rho k^r)t} k^\gamma V(k)\, dk = \sum_{m=0}^{N-1} \frac{V^{(m)}(0)}{m!} \frac{A(\xi; r, \rho, \gamma + m, C^\alpha(\xi))}{t^{(1+\gamma+m)/r}}$$

$$+ O\left(\exp\left\{ -(r-1)\rho \left| \frac{\xi}{r\rho} \right|^{\frac{r}{r-1}} \right\} \frac{(1+|\xi|)^{\frac{1}{r-1}[1+\gamma+N-\frac{r}{2}]}}{t^{\frac{1+\gamma+N}{r}}} \right)$$

in the inner region. On the other hand, Theorem 2.2.2 says that

$$\int_{\tilde{C}(b)} e^{(ikb - i\rho k^r)t} k^\gamma V(k)\, dk = \sum_{m=0}^{N-1} \frac{V^{(m)}(0)}{m!} \frac{A(\xi; r, \rho, \gamma + m, \tilde{C}(\xi))}{t^{(1+\gamma+m)/r}}$$

$$+ O\left(\left[\frac{|\xi|^{1/(r-1)}}{t^{1/r}(1+|\xi|)^{r/(r-1)}} \right]^{1+\gamma+N} \right)$$

also in the inner region. We saw in section 2.1 that $A(\xi; r, \rho, \gamma + m, \tilde{C}(\xi) \cup C^\alpha(\xi)) = A(\xi; r, \rho, \gamma + m, \tilde{\Gamma})$. So adding these two results and changing contours as indicated we obtain the result

$$\int_{\tilde{C}(b) \cup C^\alpha(b)} e^{(ikb - i\rho k^r)t} k^\gamma V(k)\, dk = \sum_{m=0}^{N-1} \frac{V^{(m)}(0)}{m!} \frac{A(\xi; r, \rho, \gamma + m, \tilde{\Gamma})}{t^{(1+\gamma+m)/r}}$$

$$+ O\left(\frac{1}{[t^{1/r}(1+|\xi|)]^{1+\gamma+N}} \right)$$

in the inner region. Notice that the weight function occurring inside the big O in this result is the equivalent to the maximum of the weight functions in the two component results, and is the same weight function which appears in Theorem 2.2.1. Thus we have the following.

Corollary 2.3.3. *Theorem 2.2.1 actually holds with $\tilde{C}(b)$ replaced by $\tilde{\Gamma}$ without the assumption $r \neq 4J - 1$ when $x < 0$ and $r \neq 4K + 1$ when $x > 0$. However, when this assumption is not true, we must assume that the estimate on $V(k)$ holds uniformly on the smallest closed sector containing the contour $\tilde{C}((-1)^\alpha) \cup C^\alpha((-1)^\alpha)$, where C^α denotes either C_{-J}^+ when $\alpha = 1$ and $r = 4J - 1$ or C_K^+ when $\alpha = 0$ and $r = 4K + 1$.*

2.4 Matching

Since the outer regions overlap with the inner region, and we have obtained sharp asymptotic results holding throughout each region, it simply must be the case that when we compare the inner expansion and the outer expansions in the overlap regions we will find that they are in fact effectively the same expansion. But in order to provide a check on our expansions, as well as to provide insight into methods for generating uniformly valid expansions, we will perform this matching between the inner and outer expansions. We will match expansions corresponding to the same contour of integration.

Inner and Outer Expansions for the Contour $\tilde{C}$. Suppose as usual $\frac{1}{r} < \beta < 1$, $0 < \epsilon < 1$, $\zeta = xt^{-\beta}$, and $\epsilon \le (-1)^\alpha \zeta \le \epsilon^{-1}$. First we will re-express the outer and inner expansions in the variables ζ and t. The outer expansion is (Theorem 1.4.9)

$$
\int_{\tilde{C}(b)} e^{(ikb - i\rho k^r)t} k^\gamma V(k)\, dk
$$

$$
= \sum_{m=0}^{N-1} \sum_{s=0}^{s_m - 1} \frac{V^{(m)}(0)}{m!s!} \Gamma(1 + \gamma + m + sr) \left(\frac{\rho t}{i}\right)^s \left(\frac{i}{\zeta t^\beta}\right)^{1 + \gamma + m + sr}
$$

$$
+ O\left(|\zeta t^\beta|^{-(1 + \gamma + N)}\right),
$$

where for each $m = 0, \ldots, N - 1$, $s_m \ge 0$ is the smallest integer such that $m + s_m(r - 1/\beta) \ge N$. The inner expansion is (Theorems 2.2.1 and 2.2.2)

$$
\int_{\tilde{C}(b)} e^{(ikb - i\rho k^r)t} k^\gamma V(k)\, dk = \sum_{m=0}^{N-1} \frac{V^{(m)}(0)}{m!} \frac{A(\zeta t^{\beta - 1/r}; r, \rho, \gamma + m, \tilde{C}(\zeta t^{\beta - 1/r}))}{t^{(1 + \gamma + m)/r}}
$$

$$
+ O\left(|\zeta t^\beta|^{-(1 + \gamma + N)}\right),
$$

where we have adjusted the form of the error term to fit the overlap region $\epsilon \le (-1)^\alpha \zeta \le \epsilon^{-1}$ (the case where $\tilde{C}(b)$ hits a saddle point has the same weight function in this overlap region as does the case where it does not). In order for these two to match we must expand the special function $A(\zeta t^{\beta - 1/r}; r, \rho, \gamma + m, \tilde{C}(\zeta t^{\beta - 1/r}))$ as $t \to \infty$. By Lemma 2.1.1(1) we have

$$
A(\zeta t^{\beta - 1/r}; r, \rho, \gamma + m, \tilde{C}(\zeta t^{\beta - 1/r}))
$$

$$
= \sum_{s=0}^{s_m - 1} \frac{\Gamma(1 + \gamma + m + sr)}{s!} \left(\frac{\rho}{i}\right)^s \left(\frac{i}{\zeta t^{\beta - 1/r}}\right)^{1 + \gamma + m + sr}
$$

$$
+ O\left(|\zeta t^{\beta - 1/r}|^{-(1 + \gamma + m + s_m r)}\right).
$$

Inserting this into the inner expansion, we find agreement with the outer expansion; and because of the way we chose s_m we see that all the error terms can be absorbed into a single big O term, the one present in the outer expansion. Since in Theorems 2.2.1 and 2.2.2 the restriction to the inner region was not necessary, we see that the inner expansion contains the outer expansion as well, and hence is an uniformly valid expansion, i.e. for all $(-1)^\alpha x \ge 0$, $t \ge 1$.

Inner and Outer Expansions for the Contour C_j. As usual, we assume $-J \le j < 0$ when $x < 0$ and $0 \le j \le K$ when $x > 0$. The outer expansion (Corollary 1.7.6) in terms of the variables $b = \frac{x}{t}$ and t is

$$
\int_{C_j} e^{-p(k)t} k^\gamma V(k)\, dk
$$

$$
= 2e^{-p(l)t} \sum_{s=0}^{M-1} \Gamma\left(s+\frac{1}{2}\right) \frac{l^{1+\gamma}}{[l^2 p_0(l)t]^{s+1/2}} \sum_{n=0}^{2s} d_n^{2s}(\gamma) l^n V^{(n)}(l)
$$

$$
+ O\left(\exp\left[-(r-1)\rho \left| \frac{b}{r\rho} \right|^{\frac{r}{r-1}} t \sin(|S_j^\alpha|) + a \left| \frac{b}{r\rho} \right|^{\frac{1}{r-1}} \sin(S_j^\alpha) \right] \cdot \right.
$$

$$
\left. \cdot \frac{|b|^{\frac{1+\gamma}{r-1}}(1+|b|)^{-\frac{\lambda_2 M}{r-1}}}{t^{M+\frac{1}{2}} |b|^{\frac{r}{r-1}(M+\frac{1}{2})}} \right),
$$

where as usual $l = \left| \frac{b}{r\rho} \right|^{\frac{1}{r-1}} e^{iS_j^\alpha}$, $p(k) = -(ikb - i\rho k^r) = i\rho(k^r - rl^{r-1}k)$, and $p_0(l) = i\rho r(r-1)l^{r-2}/2$. In order to facilitate the comparison with the inner expansion, let us define $l_0 = (r\rho)^{-\frac{1}{r-1}} e^{iS_j^\alpha}$, so that $l = |b|^{\frac{1}{r-1}} l_0$. For precision, define $\tilde{p}(l) = -i\rho(r-1)l^r$; note that $\tilde{p}(l) = p(l)$, but $\tilde{p}(l) = \tilde{p}(l_0)|b|^{\frac{r}{r-1}}$ and $p(l_0)$ is ambiguous. Thus $p(l)t = \tilde{p}(l_0)|b|^{\frac{r}{r-1}}t = \tilde{p}(l_0)|\xi|^{\frac{r}{r-1}}$, where as usual $\xi = xt^{-1/r}$. Also $l^2 p_0(l)t = l_0^2 p_0(l_0)|b|^{\frac{r}{r-1}}t = l_0^2 p_0(l_0)|\xi|^{\frac{r}{r-1}}$. Using this notation we can rewrite the above in terms of the variables b and ξ:

$$
\int_{C_j} e^{-p(k)t} k^\gamma V(k)\, dk
$$

$$
= e^{-\tilde{p}(l_0)|\xi|^{\frac{r}{r-1}}} \sum_{s=0}^{M-1} \Gamma\left(s+\frac{1}{2}\right) \frac{l^{1+\gamma}}{[l_0^2 p_0(l_0)|\xi|^{\frac{r}{r-1}}]^{s+1/2}} \sum_{n=0}^{2s} 2n! d_n^{2s}(\gamma) l^n \frac{V^{(n)}(l)}{n!}
$$

$$
+ O\left(\exp\left[-(r-1)\rho \left| \frac{\xi}{r\rho} \right|^{\frac{r}{r-1}} \sin(|S_j^\alpha|) \right] \frac{|b|^{\frac{1+\gamma}{r-1}}}{|\xi|^{\frac{r}{r-1}(M+\frac{1}{2})}} \right).
$$

Since $b = \zeta t^{\beta-1}$, we have $b \to 0$ in the overlap region, so we have adjusted the weight function to take that into account.

On the other hand, in the overlap region the inner expansion is (Corollary 2.3.2)

$$
\int_{C_j(b)} e^{(ikb - i\rho k^r)t} k^\gamma V(k)\, dk = \sum_{m=0}^{N-1} \frac{V^{(m)}(0)}{m!} \frac{A(\xi; r, \rho, \gamma+m, \Gamma_j)}{t^{(1+\gamma+m)/r}}
$$

$$
+ O\left(\exp\left\{ -(r-1)\rho \left| \frac{\xi}{r\rho} \right|^{\frac{r}{r-1}} \sin(|S_j^\alpha|) \right\} \frac{|b|^{\frac{1}{r-1}(1+\gamma+N)}}{|\xi|^{\frac{r}{r-1}\frac{1}{2}}} \right).
$$

Since $\xi = \zeta t^{\beta-1/r}$ we have $|\xi| \to \infty$ in the overlap region. We have adjusted the weight function accordingly, and expressed it in terms of ξ and b.

In order to match these two expansions to each other, we need to expand $V^{(n)}(l)$ in powers of l (note that $l \to 0$ in the overlap region) in the outer expansion, and we need to expand $A(\xi; r, \rho, \gamma + m, \Gamma_j)$ as $(-1)^\alpha \xi \to \infty$ in the inner expansion. For the first, we have

$$l^n \frac{V^{(n)}(l)}{n!} = \sum_{m=n}^{N-1} \frac{V^{(m)}(0)}{m!} \binom{m}{n} l^m + O(|l|^N).$$

We assume $N \geq 2M - 1$ so that this expansion will apply for $n = 0, \ldots, 2s$, $s = 0, \ldots, M - 1$. Therefore

$$\sum_{n=0}^{2s} 2n! d_n^{2s}(\gamma) l^n \frac{V^{(n)}(l)}{n!} = \sum_{n=0}^{2s} 2n! d_n^{2s}(\gamma) \left[\sum_{m=n}^{N-1} \frac{V^{(m)}(0)}{m!} \binom{m}{n} l^m + O(|l|^N) \right]$$

$$= \sum_{m=0}^{N-1} \frac{V^{(m)}(0)}{m!} l^m \sum_{n=0}^{\min\{m, 2s\}} \binom{m}{n} 2n! d_n^{2s}(\gamma) + O(|l|^N).$$

Let $D(m, s, \gamma) = \sum_{n=0}^{\min\{m, 2s\}} \binom{m}{n} 2n! d_n^{2s}(\gamma)$. If we insert this into the outer expansion we get

(o)

$$\int_{C_j} e^{-p(k)t} k^\gamma V(k) \, dk$$

$$= e^{-\tilde{p}(l_0)|\xi|^{\frac{r}{r-1}}} \sum_{m=0}^{N-1} \frac{V^{(m)}(0)}{m!} \sum_{s=0}^{M-1} \Gamma\left(s + \frac{1}{2}\right) \frac{l_0^{1+\gamma+m} |b|^{\frac{1}{r-1}(1+\gamma+m)}}{[l_0^2 p_0(l_0) |\xi|^{\frac{r}{r-1}}]^{s+1/2}} D(m, s, \gamma)$$

$$+ e^{-\tilde{p}(l_0)|\xi|^{\frac{r}{r-1}}} \sum_{s=0}^{M-1} O\left(\frac{|b|^{\frac{1+\gamma+N}{r-1}}}{|\xi|^{\frac{r}{r-1}(s+\frac{1}{2})}} \right)$$

$$+ O\left(\exp\left[-(r-1)\rho \left| \frac{\xi}{r\rho} \right|^{\frac{r}{r-1}} \sin(|S_j^\alpha|) \right] \frac{|b|^{\frac{1+\gamma}{r-1}}}{|\xi|^{\frac{r}{r-1}(M+\frac{1}{2})}} \right).$$

Since $|b| \to 0$ and $|\xi| \to \infty$ we can replace all these error terms by

(e) $\quad \exp\left[-(r-1)\rho \left| \frac{\xi}{r\rho} \right|^{\frac{r}{r-1}} \sin(|S_j^\alpha|) \right] \frac{|b|^{\frac{1+\gamma}{r-1}}}{|\xi|^{\frac{r}{r-1}\frac{1}{2}}} \left[O\left(|b|^{\frac{N}{r-1}}\right) + O\left(|\xi|^{-\frac{rM}{r-1}}\right) \right].$

Applying Lemma 2.1.1(2) we have as $(-1)^\alpha \xi \to \infty$ that

$$A(\xi; r, \rho, \gamma + m, \Gamma_j)$$

$$= e^{-\tilde{p}(l_0)|\xi|^{r/(r-1)}} \sum_{s=0}^{M-1} \Gamma\left(s + \frac{1}{2}\right) \frac{l_0^{1+\gamma+m} 2d_0^{2s}(\gamma + m)}{[l_0^2 p_0(l_0)]^{s+\frac{1}{2}}} |\xi|^{\frac{1}{r-1}[1+\gamma+m-r(s+\frac{1}{2})]}$$

$$+ e^{-\tilde{p}(l_0)|\xi|^{r/(r-1)}} O\left(|\xi|^{\frac{1}{r-1}[1+\gamma+m-r(M+\frac{1}{2})]} \right).$$

Inserting this into the inner expansion we have

(i)

$$
\int_{C_j(b)} e^{(ikb - i\rho k^r)t} k^\gamma V(k)\, dk
$$

$$
= e^{-\tilde{p}(l_0)|\xi|^{\frac{r}{r-1}}} \sum_{m=0}^{N-1} \frac{V^{(m)}(0)}{m!} \sum_{s=0}^{M-1} \Gamma\left(s + \frac{1}{2}\right) \frac{l_0^{1+\gamma+m} 2 d_0^{2s}(\gamma+m)}{[l_0^2 p_0(l_0)]^{s+\frac{1}{2}}} \frac{|b|^{\frac{1}{r-1}(1+\gamma+m)}}{|\xi|^{\frac{r}{r-1}(s+\frac{1}{2})}}
$$

$$
+ e^{-\tilde{p}(l_0)|\xi|^{r/(r-1)}} \sum_{m=0}^{N-1} O\left(\frac{|b|^{\frac{1}{r-1}(1+\gamma+m)}}{|\xi|^{\frac{r}{r-1}(M+\frac{1}{2})}}\right)
$$

$$
+ O\left(\exp\left\{-(r-1)\rho \left|\frac{\xi}{r\rho}\right|^{\frac{r}{r-1}} \sin(|S_j^\alpha|)\right\} \frac{|b|^{\frac{1}{r-1}(1+\gamma+N)}}{|\xi|^{\frac{r}{r-1}\frac{1}{2}}}\right).
$$

We can replace all these error terms by the same error term we ended up with for the outer expansion, namely (e).

Comparing the expansions in (o) and (i) we find that they match provided $D(m, s, \gamma) = 2 d_0^{2s}(\gamma + m)$. This is a consequence of the following.

Lemma 2.4.1. *Suppose $m \geq 0, s \geq 0$ are integers and γ is a real number. Let $d_n^s(\gamma)$ be as defined in Lemma 1.6.3. Then*

$$
2 d_0^s(\gamma + m) = \sum_{n=0}^{\min\{m,s\}} \binom{m}{n} 2 n! \, d_n^s(\gamma).
$$

Proof. We could prove this directly from the formulae in Lemma 1.6.3, but it is easier to use the fact that

$$
(1 + \tilde{k})^\gamma \tilde{k}^n \frac{d\tilde{k}}{d\tilde{w}} = \sum_{s=n}^{\infty} 2 n! \, d_n^s(\gamma) \tilde{w}^s
$$

(see Remark (2) after Lemma 1.6.3). Multiply both sides by $\binom{m}{n}$ and sum from $n = 0, \ldots, m$. By the Binomial expansion, the left side becomes $(1 + \tilde{k})^{\gamma + m} \frac{d\tilde{k}}{d\tilde{w}} = \sum_{s=0}^{\infty} 2 d_0^s(\gamma + m) \tilde{w}^s$. If we interchange the order of summation on the right-hand-side and compare coefficients of $\tilde{w}^s$ we obtain the desired result. $\square$

In summary, in the overlap region both the outer and the inner expansions can be written as

$$
\int_{C_j(b)} e^{(ikb - i\rho k^r)t} k^\gamma V(k)\, dk
$$

$$
= e^{-\tilde{p}(l_0)|\xi|^{\frac{r}{r-1}}} \sum_{m=0}^{N-1} \frac{V^{(m)}(0)}{m!} \sum_{s=0}^{M-1} \Gamma\left(s + \frac{1}{2}\right) \frac{l_0^{1+\gamma+m} 2 d_0^{2s}(\gamma+m)}{[l_0^2 p_0(l_0)]^{s+\frac{1}{2}}} \frac{|b|^{\frac{1}{r-1}(1+\gamma+m)}}{|\xi|^{\frac{r}{r-1}(s+\frac{1}{2})}}
$$

$$
+ \exp\left[-(r-1)\rho \left|\frac{\xi}{r\rho}\right|^{\frac{r}{r-1}} \sin(|S_j^\alpha|)\right] \frac{|b|^{\frac{1+\gamma}{r-1}}}{|\xi|^{\frac{r}{r-1}\frac{1}{2}}} \left[O\left(|b|^{\frac{N}{r-1}}\right) + O\left(|\xi|^{-\frac{rM}{r-1}}\right)\right].
$$

Depending on the value of β, some of the terms in the double sum may be absorbed into the error term, once one expresses everything in the variables ζ and t. But since N and M can be chosen as large as necessary, we do not need to do this in order to see that matching has occurred.

Matching of the outer and inner expansions for the contours C_{-J}^+ and C_K^+ can be done in a very similar manner.

Matching between $x \leq 0$ and $x \geq 0$ in the Inner Region. Recall that the contour Γ_{V_0} is the ray $\arg k = V_0$ starting at $k = 0$ and ending at infinity. In section 1.2 we considered Γ_{V_0} to be the "original" contour of integration. But, depending on the sign of x, it was deformed to a new contour of integration Γ, which consisted of several components. We have now found the large t asymptotics of the integral over each of these components in the inner region. Suppose that r is not an odd integer. We do this only to reduce writing to a minimum. Every conclusion we draw will also be true when r is an odd integer. If we add the contributions of each of the component contours we find for $x \leq 0$ that

$$I(x,t)$$
$$\sim \sum_{n=0}^{\infty} \frac{V^{(n)}(0)}{n! t^{(1+\gamma+n)/r}} \left[A(\xi; r, \rho, \gamma + n, \tilde{C}(\xi)) + \sum_{j=1}^{J} A(\xi; r, \rho, \gamma + n, C_{-j}(\xi)) \right]$$

in the inner region as $t \to \infty$. Similarly, for $x \geq 0$ we have

$$I(x,t)$$
$$\sim \sum_{n=0}^{\infty} \frac{V^{(n)}(0)}{n! t^{(1+\gamma+n)/r}} \left[A(\xi; r, \rho, \gamma + n, \tilde{C}(\xi)) + \sum_{j=0}^{K} A(\xi; r, \rho, \gamma + n, C_j(\xi)) \right]$$

in the inner region as $t \to \infty$. But when $x \leq 0$ the contour $\tilde{C}(\xi) \cup \bigcup_{j=1}^{J} C_{-j}(\xi)$ can be changed to Γ_{V_0}. And when $x \geq 0$ the contour $\tilde{C}(\xi) \cup \bigcup_{j=0}^{K} C_j(\xi)$ can also be changed to Γ_{V_0}. Thus regardless of whether $x \leq 0$ or $x \geq 0$ the inner expansion for the "original" contour Γ_{V_0} is

$$I(x,t) \sim \sum_{n=0}^{\infty} \frac{V^{(n)}(0)}{n!} \frac{A(\xi; r, \rho, \gamma + n, \Gamma_{V_0})}{t^{(1+\gamma+n)/r}}.$$

The functions $A(\xi; r, \rho, \gamma + m, \Gamma_{V_0})$ are entire in ξ, and thus allow us to smoothly connect our expansions for $x \leq 0$ and $x \geq 0$. Asymptotics for $A(\xi; r, \rho, \gamma + m, \Gamma_{V_0})$ as $\xi \to -\infty$ or as $\xi \to \infty$ can be extracted from Lemma 2.1.1. Typically, the asymptotic behavior as $\xi \to \infty$ is dominated by the contribution of the contour C_0, whereas the asymptotic behavior as $\xi \to -\infty$ is dominated by the contribution of the contour $\tilde{C}$. Thus the function $A(\xi; r, \rho, \gamma + m, \Gamma_{V_0})$ does not decay rapidly in either direction.

CHAPTER 3

UNIFORMLY VALID EXPANSIONS AS $t \to \infty$

Since we have given a complete matched asymptotic description of each of the integrals comprising $I(x,t)$ it could be argued that a single uniformly valid expansion is unnecessary. However, we found that the expansions we derived for the contour $\tilde{C}$ in the inner region are already uniformly valid. Certainly possession of an uniform expansion for each of the other contours greatly simplifies the issue of which terms should be included in a finite expansion, and the asymptotic size of the error term corresponding to that finite expansion. Also, as we will see, an uniformly valid leading-order term is in some sense more accurate than a non-uniformly-valid term. After exploring these uniform expansions in section 3.1, we will give a complete uniformly valid expansion for $I(x,t)$ (including all its component contours) in section 3.2.

3.1 The contours C_j, C^+_{-J} and C^+_K

We have given expansions of

$$I(x,t;C_j) = \int_{C_j(b)} e^{ikx - i\rho k^r t} k^\gamma V(k) \, dk$$

in both the outer and inner regions, and have shown that these two expansions agree on the overlap of the outer and inner regions. These expansions have quite different appearances, and in the outer region the expansion takes the form of a double sum. The asymptotic ordering of these terms changes as one moves from the inner region, through the overlap region, and into the outer region. For example, consider "the" leading-order term with its error estimate. (See section 2.4 for an explanation of the notation.)

(1) In the inner region $0 \le (-1)^\alpha \zeta \le \epsilon^{-1}$ we have

$$I(x,t;C_j) = V(0)A(\xi;r,\rho,\gamma,\Lambda_j)t^{-(1+\gamma)/r}$$

$$+ O\left(\exp\left\{ -(r-1)\rho \left| \frac{\xi}{r\rho} \right|^{\frac{r}{r-1}} \sin(|S_j^\alpha|) \right\} \frac{(1+|\xi|)^{\frac{1}{r-1}[2+\gamma-\frac{r}{2}]}}{t^{\frac{2+\gamma}{r}}} \right).$$

(2) In the overlap region $\epsilon \le (-1)^\alpha \zeta \le \epsilon^{-1}$ we have

$$I(x,t;C_j) = V(0)e^{-\tilde{p}(l_0)|\xi|^{r/(r-1)}}\Gamma(\tfrac{1}{2})\frac{l_0^{1+\gamma}}{[l_0^2 p_0(l_0)]^{\frac{1}{2}}}\frac{|b|^{\frac{1}{r-1}(1+\gamma)}}{|\xi|^{\frac{r}{2(r-1)}}}$$

$$+ \exp\left\{ -(r-1)\rho \left| \frac{\xi}{r\rho} \right|^{\frac{r}{r-1}} \sin(|S_j^\alpha|) \right\} \frac{|b|^{\frac{1+\gamma}{r-1}}}{|\xi|^{\frac{r}{r-1}\frac{1}{2}}} \left[O\left(|b|^{\frac{1}{r-1}} \right) + O\left(|\xi|^{-\frac{r}{r-1}} \right) \right].$$

(3) In the outer region $\epsilon \leq (-1)^\alpha \zeta$ we have

$$I(x,t;C_j) = V(l_0|b|^{\frac{1}{r-1}})e^{-\tilde{p}(l_0)t|b|^{r/(r-1)}}\Gamma(\tfrac{1}{2})\frac{l_0^{1+\gamma}}{[l_0^2 p_0(l_0)t]^{\frac{1}{2}}}|b|^{\frac{1}{r-1}(1+\gamma-\frac{r}{2})}$$

$$+ \exp\left[-(r-1)\rho\left|\frac{b}{r\rho}\right|^{\frac{r}{r-1}}t\sin(|S_j^\alpha|) + a\left|\frac{b}{r\rho}\right|^{\frac{1}{r-1}}\sin(S_j^\alpha)\right] \cdot$$

$$\cdot O\left(\frac{|b|^{\frac{1}{r-1}(1+\gamma-\frac{3r}{2})}}{t^{3/2}(1+|b|)^{\frac{\Lambda_2}{r-1}}}\right).$$

In all three cases the error becomes comparable to the leading-order term as $\beta \to \frac{1}{r}^+$ and/or $\beta \to 1^-$. This is because what we have given as the leading-order term is actually only an approximation to "the real" leading-order term, an approximation whose accuracy deteriorates near the boundary of the region. In an attempt to avoid this deterioration of accuracy as much as possible within the above framework, it is natural to choose $\beta = 2/(r+1)$ so that $|b|^{\frac{1}{r-1}} = |\zeta|^{\frac{1}{r-1}}t^{-1/(r+1)}$ and $|\xi|^{-\frac{r}{r-1}} = |\zeta|^{-\frac{r}{r-1}}t^{-1/(r+1)}$. This is the optimal cross-over point between the inner and outer regions in the sense that the asymptotic "gap" between the leading-order term and the error is maximized if this cross-over point is chosen. (The "asymptotic gap" $g(\beta)$, depends on the choice of β, and is defined so that the error is $t^{-g(\beta)}O$(leading-order term). A larger gap means a smaller asymptotic error.) With this choice of β in the overlap region we may use the leading-order terms of (1), (2), or (3) above interchangeably, with an error of the same asymptotic size. But this does not mean that there is no better choice for the leading-order term in the overlap region—better in the sense that the associated error is asymptotically smaller. The problem with the leading-order terms of (1), (2), or (3) above is not that they decay at the wrong rate, or with the wrong coefficient, but that they are too simple in their dependence on the relevant variables to allow the asymptotic gap in the overlap region to be larger than $g(2/(r+1)) = 1/(r+1)$. This simplicity in functional dependence is a virtue in many contexts, so many that we have chosen in favor of this simplicity in our mode of presentation and proof (thus far). But already in the inner region we cannot avoid the need to evaluate the special function $A(\xi; r, \rho, \gamma, \Lambda_j)$ (most especially when ξ is a nonzero constant; when $(-1)^\alpha \xi \to 0^+$ the most relevant information is the Taylor series of $A(\xi; r, \rho, \gamma, \Lambda_j)$ at $\xi = 0$). It turns out that a better leading-order term can be chosen which is no more complicated in its functional dependence than $V(l_0|b|^{\frac{1}{r-1}})A(\xi; r, \rho, \gamma, \Lambda_j)t^{-(1+\gamma)/r}$.

An uniformly valid expansion of $I(x,t;C_j)$ as $t \to \infty$ is an expansion which works equally well in both the inner and outer regions. Naturally it must reduce to (1) in the inner region and to (3) in the outer region, but the "asymptotic gap" between the leading-order term and the associated error will be larger with the uniformly valid expansion. It turns out that when $\beta = 2/(r+1)$, if we use the uniformly valid leading-order term the gap is $g(\beta) = 2/(r+1)$, i.e. twice as large as the gap we found for (1), (2), or (3).

In section 2.1, when deriving the form of the expansion in the inner region, we noticed that the outer expansion could be derived by first expanding

$$\int_{C_j(b)} e^{ikx - i\rho k^r t} k^\gamma V(k)\, dk = \sum_{n=0}^{N-1} \frac{V^{(n)}(l)}{n!} \int_{C_j(b)} e^{ikx - i\rho k^r t} k^\gamma (k - l)^n\, dk$$
$$+ \int_{C_j(b)} e^{ikx - i\rho k^r t} k^\gamma R_N(k; l)\, dk$$

and then expanding further each of the integrals $\int_{C_j(b)} e^{ikx - i\rho k^r t} k^\gamma (k - l)^n\, dk$, since in the outer region the immediate neighborhood of the saddle point $k = l$ is of principal importance. But we also saw there that this further expansion produces no benefit when $\beta = 1/r$, since each of the resulting terms decays at the same rate. The key to the expansion in the inner region was to take the original expansion about $k = 0$ and to re-express the integral in terms of the variable ξ. This suggests that in order to get an uniformly valid expansion, one should keep the original expansion about $k = l$ and to re-express each of the integrals in the variable ξ:

$$\int_{C_j(b)} e^{ikx - i\rho k^r t} k^\gamma V(k)\, dk$$
$$= \sum_{n=0}^{N-1} \frac{V^{(n)}(l)}{n!} t^{-(1+\gamma+n)/r} \int_{C_j(\xi)} e^{i\omega\xi - i\rho\omega^r} \omega^\gamma (\omega - l_0 |\xi|^{\frac{1}{r-1}})^n\, d\omega$$
$$+ \int_{C_j(b)} e^{ikx - i\rho k^r t} k^\gamma R_N(k; l)\, dk.$$

This leads us to define the special function

$$A_n(\xi; r, \rho, \gamma, \Lambda_j) = \int_{\Lambda_j} e^{i\omega\xi - i\rho\omega^r} \omega^\gamma (\omega - l_0 |\xi|^{\frac{1}{r-1}})^n\, d\omega,$$

where $l_0 = (r\rho)^{\frac{-1}{r-1}} e^{iS_j^\alpha}$ and $0 \le (-1)^\alpha \xi$. Suppose $C^\alpha(\xi)$ denotes $C_{-J}^+(\xi)$ when $\alpha = 1$ and $r = 4J - 1$, or $C_K^+(\xi)$ when $\alpha = 0$ and $r = 4K + 1$. Let $A_n(\xi; r, \rho, \gamma, C^\alpha(\xi))$ be defined as above with Λ_j replaced by $C^\alpha(\xi)$. In the below what we say about $A_n(\xi; r, \rho, \gamma, \Lambda_j)$ applies equally to $A_n(\xi; r, \rho, \gamma, C^\alpha(\xi))$ with minor obvious modifications, which we only point out when they are substantial.

When $n = 0$ we have $A_0(\xi; r, \rho, \gamma, \Lambda_j) = A(\xi; r, \rho, \gamma, \Lambda_j)$. In fact, using the Binomial expansion, we can express $A_n(\xi; r, \rho, \gamma, \Lambda_j)$ in terms of $A(\xi; r, \rho, \gamma + q, \Lambda_j)$, $q = 0, 1, \dots, n$:

$$A_n(\xi; r, \rho, \gamma, \Lambda_j) = \sum_{q=0}^{n} \binom{n}{q} (-l_0 |\xi|^{\frac{1}{r-1}})^{n-q} A(\xi; r, \rho, \gamma + q, \Lambda_j).$$

Our next lemma records the asymptotic behavior of this special function.

Lemma 3.1.1. *If x is a real number let $((x))$ denote the smallest integer greater than or equal to x. Then for all $n \geq 0$ we have*

$$A_n(\xi;r,\rho,\gamma,\Lambda_j)$$

$$\sim e^{-\tilde{p}(l_0)|\xi|^{\frac{r}{r-1}}} (l_0|\xi|^{\frac{1}{r-1}})^{1+\gamma+n} \sum_{s=((\frac{n}{2}))}^{\infty} \Gamma(s+\tfrac{1}{2}) \frac{2n!d_n^{2s}(\gamma)}{[l_0^2 p_0(l_0)|\xi|^{\frac{r}{r-1}}]^{s+\frac{1}{2}}},$$

$$A_n(\xi;r,\rho,\gamma,C^{\alpha}(\xi))$$

$$\sim e^{-\tilde{p}(l_0)|\xi|^{\frac{r}{r-1}}} (l_0|\xi|^{\frac{1}{r-1}})^{1+\gamma+n} \sum_{s=n}^{\infty} \Gamma(\tfrac{s+1}{2}) \frac{n!d_n^s(\gamma)}{[l_0^2 p_0(l_0)|\xi|^{\frac{r}{r-1}}]^{\frac{s+1}{2}}},$$

as $(-1)^{\alpha}\xi \to \infty$.

Proof. Using Lemma 2.1.1(2) and the above formula we get

$$A_n(\xi;r,\rho,\gamma,\Lambda_j) = \sum_{q=0}^{n} \binom{n}{q}(-l_0|\xi|^{\frac{1}{r-1}})^{n-q} A(\xi;r,\rho,\gamma+q,\Lambda_j)$$

$$\sim \sum_{q=0}^{n} \binom{n}{q}(-l_0|\xi|^{\frac{1}{r-1}})^{n-q} e^{-\tilde{p}(l_0)|\xi|^{\frac{r}{r-1}}}.$$

$$\cdot (l_0|\xi|^{\frac{1}{r-1}})^{1+\gamma+q} \sum_{s=0}^{\infty} \Gamma(s+\tfrac{1}{2}) \frac{2d_0^{2s}(\gamma+q)}{[l_0^2 p_0(l_0)|\xi|^{\frac{r}{r-1}}]^{s+\frac{1}{2}}}$$

$$= e^{-\tilde{p}(l_0)|\xi|^{\frac{r}{r-1}}} (l_0|\xi|^{\frac{1}{r-1}})^{1+\gamma+n} \sum_{s=0}^{\infty} \Gamma(s+\tfrac{1}{2}) \frac{1}{[l_0^2 p_0(l_0)|\xi|^{\frac{r}{r-1}}]^{s+\frac{1}{2}}}.$$

$$\cdot \sum_{q=0}^{n}(-1)^{n-q} \binom{n}{q} 2d_0^{2s}(\gamma+q).$$

Thus the desired result is true provided

$$\sum_{q=0}^{n}(-1)^{n-q}\binom{n}{q} 2d_0^s(\gamma+q) = \begin{cases} 0 & s < n, \\ 2n!d_n^s(\gamma) & s \geq n. \end{cases}$$

To prove this identity we use Remark (2) after Lemma 1.6.3 in the following form

$$(1+\tilde{k})^{\gamma+q}\frac{d\tilde{k}}{d\tilde{w}} = \sum_{s=0}^{\infty} 2d_0^s(\gamma+q)\tilde{w}^s.$$

Multiply both sides by $(-1)^{n-q}\binom{n}{q}$ and sum in $q = 0,1,\ldots,n$. Using the Binomial expansion and interchanging the order of summation on the right-hand-side we obtain

$$(1+\tilde{k})^{\gamma}\tilde{k}^n \frac{d\tilde{k}}{d\tilde{w}} = \sum_{s=0}^{\infty}\left[\sum_{q=0}^{n}(-1)^{n-q}\binom{n}{q} 2d_0^s(\gamma+q)\right]\tilde{w}^s.$$

By Remark (2) after Lemma 1.6.3 the left-hand-side of the above equation is equal to $\sum_{s=n}^{\infty} [2n! d_n^s(\gamma)] \tilde{w}^s$. Comparing the coefficients of like powers of $\tilde{w}$ yields the identity.

The proof for the contour $C^\alpha(\xi)$ is very similar.   □

From the leading-term of this expansion we obtain the following sharp estimate.

Lemma 3.1.2.

(1) Suppose $d_n^{2S}(\gamma) \neq 0$, where $S = ((\frac{n}{2}))$. Then there is a positive constant C such that for all $(-1)^\alpha \xi \geq 0$ we have

$$|A_n(\xi; r, \rho, \gamma, \Lambda_j)|$$
$$\leq C \exp\left[-(r-1)\rho \left|\frac{\xi}{r\rho}\right|^{\frac{r}{r-1}} \sin(|S_j^\alpha|)\right] (1+|\xi|)^{\frac{1}{r-1}\{1+\gamma+n-r[((\frac{n}{2}))+\frac{1}{2}]\}}.$$

(2) Suppose $(-1)^\alpha \xi \geq 0$ and $r = 4J - 1$ when $\alpha = 1$ and $r = 4K + 1$ when $\alpha = 0$. Then there is a positive constant C such that

$$|A_n(\xi; r, \rho, \gamma, C^\alpha(\xi))| \leq C \exp\left[-(r-1)\rho \left|\frac{\xi}{r\rho}\right|^{\frac{r}{r-1}}\right] (1+|\xi|)^{\frac{1}{r-1}\{1+\gamma+n-r[\frac{n+1}{2}]\}}.$$

Remark. If n is even, then the restriction $0 \neq d_n^{2S}(\gamma) = d_n^n(\gamma)$ in (1) always holds. An example of a situation where it fails is $r = 2$, $\gamma = 0$, and n odd (a super-anomalous case).

Expression of A_n in terms of special functions. When $n \geq 0$ is an integer, define the special function

$$F_n(r, \delta; y) = \int_0^\infty e^{-\frac{1}{r}\sigma^r + y\sigma} \sigma^{\delta-1} (\sigma - y^{\frac{1}{r-1}})^n \, d\sigma, \qquad r > 1, \delta > 0, y \in S\{y^{\frac{1}{r-1}}\}.$$

In the special case $r = 2$, $F_n(2, \delta; y)$ is entire in y, and is mentioned on page 346 of Olver [O1]. When $r = 3$ and $\delta = 1$, $F_n(3, 1; y) = \pi Q i_n(y)$, a function which is discussed on page 354 of Olver [O1]. Using this special function we can express A_n. The benefit is that we can make contact with well-studied special functions in special cases, and in general the dependence on ρ and the contour will be simplified.

When $r \geq 3$ is an odd integer we will also need the incomplete forms of this function:

$$\mathbb{F}_n(r, \delta; y, z) = \int_0^z e^{-\frac{1}{r}\sigma^r + y\sigma} \sigma^{\delta-1} (\sigma - z)^n \, d\sigma, \qquad \delta > 0, y \in \mathbb{C}, z \in S\{\sigma^{\delta-1}\},$$

$$\mathbb{F}_n^c(r, \delta; y, z) = \int_z^\infty e^{-\frac{1}{r}\sigma^r + y\sigma} \sigma^{\delta-1} (\sigma - z)^n \, d\sigma, \qquad \delta > 0, y \in \mathbb{C}, z \in S\{\ln\sigma\}.$$

The contour of integration for the first integral can be taken as the line segment joining 0 and z on the Riemann surface $S\{\sigma^{\delta-1}\}$. The contour of integration in the second integral lies on the Riemann surface $S\{\ln \sigma\}$, avoids $\sigma = 0$, and tends to infinity with $|\arg \sigma| \leq \frac{\pi}{2r} - \epsilon$. Notice that when $|\arg y| < (r-1)\pi$ we have

$$F_n(r, \delta; y) = \mathbb{F}_n\left(r, \delta; y, y^{\frac{1}{r-1}}\right) + \mathbb{F}^c_n\left(r, \delta; y, y^{\frac{1}{r-1}}\right).$$

Unfortunately, these incomplete forms have generally received much less study, although when $n = 0$ and for r an integer they have been mentioned by Bleistein [B].

Lemma 3.1.3. *Suppose that $\alpha = 1$ when $-J \leq j \leq -1$ and $\alpha = 0$ when $0 \leq j \leq K$, and $(-1)^\alpha \xi \geq 0$. Then for all integers $n \geq 0$ we have*

$$A_n(\xi; r, \rho, \gamma, \Lambda_j) = (-1)^\alpha \left(\frac{e^{iV_j}}{(r\rho)^{1/r}}\right)^{1+\gamma+n} F_n\left(r, 1+\gamma; \frac{|\xi|e^{i(\frac{\pi}{2}+\alpha\pi+V_j)}}{(r\rho)^{1/r}}\right)$$

$$- (-1)^\alpha \left(\frac{e^{iV_{j+1}}}{(r\rho)^{1/r}}\right)^{1+\gamma+n} F_n\left(r, 1+\gamma; \frac{|\xi|e^{i(-\frac{3\pi}{2}+\alpha\pi+V_{j+1})}}{(r\rho)^{1/r}}\right),$$

$$A_n(\xi; r, \rho, \gamma, C^\alpha(\xi)) = \left(\frac{e^{i\tilde{V}}}{(r\rho)^{1/r}}\right)^{1+\gamma+n} \mathbb{F}^c_n\left(r, 1+\gamma; y, y^{\frac{1}{r-1}}\right),$$

where $y = \frac{i\xi e^{i\tilde{V}}}{(r\rho)^{1/r}}$, $\arg(i\xi) = (-1)^\alpha \pi/2$.

Proof. Similar to the proof of Lemma 2.1.2; some algebra shows that $S^\alpha_j - V_j = \frac{1}{r-1}[\frac{\pi}{2} + \alpha\pi + V_j]$ and $S^\alpha_j - V_{j+1} = \frac{1}{r-1}[-\frac{3\pi}{2} + \alpha\pi + V_{j+1}]$. $\square$

When $r \geq 2$ is an integer, Bleistein [B] has given an integration-by-parts procedure for deriving uniformly valid expansions which applies to the integral over the contour C_j we have expanded in this section. This procedure is as follows. First, one defines $V_0(k; b) = V(k)$. Then one proceeds inductively: suppose $V_m(k; b)$ is known for some integer $m \geq 0$. One finds complex numbers $a^m_0(b), a^m_1(b), \ldots, a^m_{r-1}(b)$ (depending on $b = x/t$) such that

$$\frac{V_m(k; b) - \sum_{h=0}^{r-1} a^m_h(b)k^h}{ik(r\rho k^{r-1} - b)} = W_m(k; b)$$

is analytic at $k = 0$ and at all of the $r-1$ saddle points $k = k_j(b)$, $j = 0, \ldots, r-2$. Thus we have

$$\int_{\Lambda_j} e^{ikx - i\rho k^r t} k^\gamma V_m(k; b)\, dk = \sum_{h=0}^{r-1} a^m_h(b) t^{-(1+\gamma+h)/r} \int_{\Lambda_j} e^{i\omega\xi - i\rho\omega^r} \omega^{\gamma+h}\, d\omega$$

$$+ \int_{\Lambda_j} e^{ikx - i\rho k^r t} k^{1+\gamma} i(r\rho k^{r-1} - b) W_m(k; b)\, dk.$$

The cleverness of this approach is evident when one integrates by parts in the error term to get

$$\frac{1}{t} \int_{\Lambda_j} e^{ikx - i\rho k^r t} k^\gamma V_{m+1}(k;b)\, dk$$

where $V_{m+1}(k;b) = (1+\gamma)W_m(k;b) + k\frac{\partial}{\partial k}W_m(k;b)$. Note that this integral is similar to the one we started with. We can combine these results to obtain an expansion of the form

$$\int_{\Lambda_j} e^{ikx - i\rho k^r t} k^\gamma V(k)\, dk \sim \sum_{h=0}^{r-1} \sum_{m=0}^{\infty} a_h^m(b) t^{-(1+\gamma+h)/r - m} A(\xi; r, \rho, \gamma + h, \Lambda_j).$$

By contrast, we have expanded the same integral in terms of $A_n(\xi; r, \rho, \gamma, \Lambda_j)$, which would seem (upon reflection upon the equation prior to Lemma 3.1.1) to involve an infinite sum $0 \le h < \infty$ instead of a finite sum $0 \le h < r$. However, this is not the case. This is because A_n (and F_n) satisfy recursion relations in the variable n which allow us to convert the infinite sum to a finite sum. For example, when $r = 2$ the recursion relations for F_n are given in Exercise 9.1, page 346 of Olver [O1]. These recursion relations are difficult to describe for general (integral) $r \ge 2$.

Our expansion technique (not entirely original to the author) yields an expansion which can be related to Bleistein's when r is an integer—but the two resulting expansions are not finite rearrangements of one another. In our expansion the functions of b are explicitly related to $V(k)$ and involve only one saddle point, whereas in Bleistein's method they are more complicated and involve $r-1$ saddle points. See sections 4.1 and 4.5 for comparisons of the two expansions when $r = 2$ and $r = 3$. When r is not an integer it is not clear how an integration by parts procedure is to be carried out. We have followed an approach which Olver tends to favor, and it appears to be more generally applicable.

The expansion we obtain for the integral over the contour $C^\alpha(b)$ also differs from the type of expansion obtained from the integration by parts approach in that ours has no terms arising from the endpoints of integration. However, like in [B], we do need to consider incomplete special functions. See section 4.5 for more details of this comparison in the case $r = 3$.

Asymptotic character of the expansion. Now we must look at the relative sizes of the terms of our proposed uniform expansion

$$\sum_{n=0}^{\infty} \frac{V^{(n)}(l_0|b|^{\frac{1}{r-1}})}{n!} t^{-(1+\gamma+n)/r} A_n(\xi; r, \rho, \gamma, \Lambda_j)$$

to see if it does what we desire, namely provide a succession of improving asymptotic approximations to the integral $I(x, t; C_j)$. Suppose we compare the sizes of the $n = N + 1$ term and the $n = N$ term. As usual, we are actually comparing size estimates for these terms, and not the terms themselves, so that we are ignoring *anomalies*. Then the quotient of the $n = N + 1$ term by the $n = N$ term is proportional to

$$\frac{t^{(1+\gamma+N)/r}}{t^{(2+\gamma+N)/r}} \frac{(1+|b|^{\frac{1}{r-1}})^{N+\lambda_N}}{(1+|b|^{\frac{1}{r-1}})^{N+1+\lambda_{N+1}}} \frac{(1+|\xi|)^{\frac{1}{r-1}\{2+\gamma+N-r[((\frac{N+1}{2}))+\frac{1}{2}]\}}}{(1+|\xi|)^{\frac{1}{r-1}\{1+\gamma+N-r[((\frac{N}{2}))+\frac{1}{2}]\}}}.$$

Suppose $N \geq 0$ is even. In the notation of sections 1.5 and 1.6 we have $\lambda_N = \min\{\lambda, \lambda' - N\}$, where $0 < \lambda \leq \lambda' \leq \infty$. Thus $\lambda_N \geq \lambda_{N+1} \geq \lambda_N - 1$, so that $1 + \lambda_{N+1} - \lambda_N \geq 0$. Also, since N is even, $((\frac{N}{2})) = \frac{N}{2}$ and $((\frac{N+1}{2})) = \frac{N}{2} + 1$. Thus this quotient is

$$\frac{1}{t^{1/r}} \frac{1}{(1+|b|^{\frac{1}{r-1}})^{1+\lambda_{N+1}-\lambda_N}} \frac{1}{(1+|\xi|)}.$$

Thus an even term and its successor are well-ordered asymptotically, both as $t \to \infty$ and as $|x| \to \infty$ for $t \geq 1$.

But if N is odd, $((\frac{N+1}{2})) = \frac{N+1}{2} = ((\frac{N}{2}))$, so that the quotient is

$$\frac{1}{t^{1/r}} \frac{1}{(1+|b|^{\frac{1}{r-1}})^{1+\lambda_{N+1}-\lambda_N}} (1+|\xi|)^{\frac{1}{r-1}}.$$

Thus only in the exceptional circumstance $\lambda_{N+1} = \lambda_N$ (see section 1.5) do we avoid asymptotic disordering of this pair of terms as $|b| \to \infty$. When $|b|$ is bounded and $t \to \infty$ the pair of terms is asymptotically well-ordered.

Fortunately, this misbehavior does not hurt the overall utility of the expansion, because when we compute the quotient of the $n = N+2$ term by the $n = N$ term we get (for both even and odd N)

$$\frac{t^{(1+\gamma+N)/r}}{t^{(3+\gamma+N)/r}} \frac{(1+|b|^{\frac{1}{r-1}})^{N+\lambda_N}}{(1+|b|^{\frac{1}{r-1}})^{N+2+\lambda_{N+2}}} \frac{(1+|\xi|)^{\frac{1}{r-1}\{3+\gamma+N-r[((\frac{N+2}{2}))+\frac{1}{2}]\}}}{(1+|\xi|)^{\frac{1}{r-1}\{1+\gamma+N-r[((\frac{N}{2}))+\frac{1}{2}]\}}}$$

$$= \frac{1}{t^{2/r}} \frac{1}{(1+|b|^{\frac{1}{r-1}})^{2+\lambda_{N+2}-\lambda_N}} \frac{1}{(1+|\xi|)^{\frac{r-2}{r-1}}}.$$

Since $2 + \lambda_{N+2} - \lambda_N \geq 0$ and $r \geq 2$, this pair of terms is always asymptotically well-ordered as $t \to \infty$. When $r > 2$ it is also asymptotically well-ordered as $(-1)^\alpha x \to \infty$, $t \geq 1$. This also holds when $r = 2$ and $2 + \lambda_{N+2} - \lambda_N > 0$. But in the typical case, $2 + \lambda_{N+2} - \lambda_N = 0$, $r = 2$, this additional sense of asymptotic well-ordering fails.

Thus if we include the terms $n = 0, 1, \ldots, N - 1$ in our expansion, and N is odd, we can be assured (formally) that the error is smaller than the last term included ($n = N - 1$). But it is not bounded by the first term omitted ($n = N$), but by the maximum of the $n = N$ and $n = N + 1$ terms. However, if N is even, then the error is bounded (formally) by the size of the first term omitted. This is the type of error estimate which we will soon prove. The other type (where N is odd) follows from it.

When we look at the asymptotic character of our expansion

$$\sum_{n=0}^{\infty} \frac{V^{(n)}(l_0|b|^{\frac{1}{r-1}})}{n!} t^{-(1+\gamma+n)/r} A_n(\xi; r, \rho, \gamma, C^{\alpha}(\xi))$$

for the integral over the contour $C^{\alpha}(b)$ we find that the story is much simpler. Because of the simpler dependence of the asymptotic behavior of $A_n(\xi; r, \rho, \gamma, C^{\alpha})$ on n (see Lemma 3.1.2) we find that the difference between the cases where N is even or odd disappears. The quotient of the $n = N + 1$ term by the $n = N$ term is

$$\frac{1}{t^{1/r}} \frac{1}{(1 + |b|^{\frac{1}{r-1}})^{1+\lambda_{N+1}-\lambda_N}} \frac{1}{(1 + |\xi|)^{\frac{r-2}{2(r-1)}}}.$$

Thus the terms are asymptotically well-ordered as $t \to \infty$. The asymptotic ordering in the sense $(-1)^{\alpha} x \to \infty$, $t \geq 1$ is the same as that present in the even terms for the contour C_j.

Error Control. The following result could be considered as a consequence of Corollary 1.7.6 and Corollary 2.3.2. However, we will give an independent proof to show how the ideas we have introduced earlier suffice for the case of an uniformly valid expansion.

Theorem 3.1.4. *Suppose $r \geq 2$, $\rho > 0$, $\gamma > -1$, and $N \geq 0$ and $-J \leq j \leq K$ are integers, where $J = \left[\frac{r+1}{4}\right]$ and $K = \left[\frac{r-1}{4}\right]$. Define the parameters $b = x/t$, $\xi = xt^{-1/r}$, $x \in \mathbb{R}$, $t \geq 1$, and $(-1)^{\alpha} = \text{sgn}(x)$, $\alpha \in \{0, 1\}$. Assume $j < 0$ when $\alpha = 1$ and $j \geq 0$ when $\alpha = 0$. Define $l_0 = (r\rho)^{\frac{-1}{r-1}} e^{iS_j^{\alpha}}$, where $S_j^{\alpha} = \frac{\pi}{r-1}(2j + \alpha)$. Suppose $C_j(b)$ is the contour described in section 1.2 depending on the parameter b and passing through the saddle point $l = k_j(b) = l_0 |b|^{\frac{1}{r-1}}$. Suppose $L_0 \leq \arg k \leq L_1$ is the smallest closed sector containing the contour $C_j(b)$ in its entirety (see section 1.2). Suppose, when N is even, that $V(k)$ is an entire function satisfying the bounds*

$$|V^{(N)}(k)| \leq \frac{C_N e^{\max\{a_- \Im k, a_+ \Im k\}}}{(1 + |k|)^{N+\lambda_N}},$$

uniformly for k in the sector $L_0 \leq \arg k \leq L_1$, where $C_N > 0, a_- \leq a_+ \in \mathbb{R}, \lambda_N \in \mathbb{R}$ are constants. If N is odd make a similar assumption on $V^{(N+1)}(k)$, where $0 \leq 1 + \lambda_{N+1} - \lambda_N \leq 1$. Define the N-term asymptotic approximation

$$A_N(x, t) = \sum_{n=0}^{N-1} \frac{V^{(n)}(l)}{n!} \frac{A_n(\xi; r, \rho, \gamma, \Lambda_j)}{t^{(1+\gamma+n)/r}},$$

where the contour Λ_j is described in section 2.1 and

$$A_n(\xi; r, \rho, \gamma, \Lambda_j) \overset{\text{def}}{=} \int_{\Lambda_j} e^{i\omega\xi - i\rho\omega^r} \omega^{\gamma} (\omega - l_0 |\xi|^{\frac{1}{r-1}})^n \, d\omega.$$

Suppose $a\sin(S_j^\alpha) = \max\{a_-\sin(S_j^\alpha), a_+\sin(S_j^\alpha)\}$. *If N is even, define the weight function*

$$W_N(x,t) = \exp\left\{-(r-1)\rho\left|\frac{\xi}{r\rho}\right|^{\frac{r}{r-1}}\sin(|S_j^\alpha|) + a\left|\frac{b}{r\rho}\right|^{\frac{1}{r-1}}\sin(S_j^\alpha)\right\} \cdot$$

$$\cdot\,\frac{(1+|\xi|)^{\frac{1}{r-1}\{1+\gamma+N-\frac{r}{2}(N+1)\}}}{t^{\frac{1+\gamma+N}{r}}(1+|b|)^{\frac{N+\lambda_N}{r-1}}}.$$

When N is odd, define the weight function as follows:

$$W_N(x,t) = \exp\left\{-(r-1)\rho\left|\frac{\xi}{r\rho}\right|^{\frac{r}{r-1}}\sin(|S_j^\alpha|) + a\left|\frac{b}{r\rho}\right|^{\frac{1}{r-1}}\sin(S_j^\alpha)\right\} \cdot$$

$$\cdot\,\frac{(1+|\xi|)^{\frac{1}{r-1}\{2+\gamma+N-\frac{r}{2}(N+2)\}}}{t^{\frac{1+\gamma+N}{r}}(1+|b|)^{\frac{N+\lambda_N}{r-1}}}\left[\frac{1}{(1+|\xi|)^{\frac{1}{r-1}}} + \frac{1}{t^{1/r}(1+|b|)^{\frac{1+\lambda_{N+1}-\lambda_N}{r-1}}}\right].$$

Then we have the asymptotic result

$$\sup_{t\geq 1}\,\sup_{(-1)^\alpha x\geq 0}\,\frac{1}{W_N(x,t)}\left|\int_{C_j(b)}e^{ikx-i\rho k^r t}k^\gamma V(k)\,dk - A_N(x,t)\right| < \infty.$$

Proof. (Sketch) In most respects we follow the outline and method of the proof of Theorem 2.3.1. First, suppose N is even. We estimate $R_N(k;l)/(k-l)^N$ as in the proof of Lemma 1.7.3. The new feature, which appears after we change variables $\omega = kt^{1/r}$, is the factor $(\omega - l_0|\xi|^{\frac{1}{r-1}})^N = (l_0|\xi|^{\frac{1}{r-1}})^N\tilde{\omega}^N$. To estimate this we use Lemma 1.7.2(2): $|\tilde\omega| \leq C|\tilde w|$, which requires $r \geq 2$. When we express $\tilde w$ in terms of w, the correct power of $|\xi|$ appears.

When N is odd, apply the result just proved for $N+1$ and use Lemma 3.1.2 to estimate the term with $n = N$. □

The same proof actually gives us our result for the contours $C^\alpha(b)$.

Theorem 3.1.5. *Suppose $\alpha \in \{0,1\}$, $\rho > 0$ and $\gamma > -1$ are real numbers, and $J \geq 1, K \geq 0, N \geq 0$ are integers. Define the parameters $b = x/t$, $\xi = xt^{-1/r}$, $x \in \mathbb{R}$, $t \geq 1$. Define $l_0 = (-1)^\alpha i(r\rho)^{\frac{-1}{r-1}}$. Let $C_{-J}^+(b)$ and $C_K^+(b)$ denote the contours described in Lemma 1.2.2A and B, starting at the saddle point $l = l_0|b|^{\frac{1}{r-1}}$. Suppose $r = 4J - 1$, $C^\alpha(b) = C_{-J}^+(b)$ and $\tilde V = V_{-J+1}$ if $\alpha = 1$ and $r = 4K + 1$, $C^\alpha(b) = C_K^+(b)$ and $\tilde V = V_K$ if $\alpha = 0$. Suppose $V(k)$ is an entire function satisfying the bounds*

$$|V^{(N)}(k)| \leq \frac{C_N e^{a_\alpha \Im k}}{(1+|k|)^{N+\lambda_N}},$$

uniformly for k in the sector $(-1)^\alpha \tilde{V} \leq (-1)^\alpha \arg k \leq \pi/2$, *where* $C_N > 0, a_\alpha \in \mathbb{R}, \lambda_N \in \mathbb{R}$ *are constants. Define the N-term asymptotic approximation*

$$A_N(x,t) = \sum_{n=0}^{N-1} \frac{V^{(n)}(l)}{n!} \frac{A_n(\xi; r, \rho, \gamma, C^\alpha(\xi))}{t^{(1+\gamma+n)/r}},$$

where

$$A_n(\xi; r, \rho, \gamma, C^\alpha(\xi)) \overset{\text{def}}{=} \int_{C^\alpha(\xi)} e^{i\omega\xi - i\rho\omega^r} \omega^\gamma (\omega - l_0 |\xi|^{\frac{1}{r-1}})^n \, d\omega.$$

Define the weight function

$$W_N(x,t) = \exp\left\{ -(r-1)\rho \left| \frac{\xi}{r\rho} \right|^{\frac{r}{r-1}} + a_\alpha(-1)^\alpha \left| \frac{b}{r\rho} \right|^{\frac{1}{r-1}} \right\} \cdot$$
$$\cdot \frac{(1+|\xi|)^{\frac{1}{r-1}\{1+\gamma+N-\frac{r}{2}(N+1)\}}}{t^{\frac{1+\gamma+N}{r}}(1+|b|)^{\frac{N+\lambda_N}{r-1}}}.$$

Then we have the asymptotic result

$$\sup_{t \geq 1} \sup_{(-1)^\alpha x \geq 0} \frac{1}{W_N(x,t)} \left| \int_{C^\alpha(b)} e^{ikx - i\rho k^r t} k^\gamma V(k) \, dk - A_N(x,t) \right| < \infty.$$

Relation with the outer and inner expansions. Because of the previous theorem it must be the case that our uniformly valid expansion reduces in the outer and inner regions to the outer and inner expansions respectively. However, in order to see how all the expansions we have discussed tie together we will discuss this reduction. We will perform the reduction formally (without carrying the error terms) since the error has already been rigorously accounted for elsewhere.

To obtain the outer expansion from the uniform expansion, we replace $A_n(\xi; r, \rho, \gamma, \Lambda_j)$ by its asymptotic expansion from Lemma 3.1.1, and then change the order of summation. To obtain the inner expansion from the uniform expansion, we replace $A_n(\xi; r, \rho, \gamma, \Lambda_j)$ by its expression in terms of $A(\xi; r, \rho, \gamma + q, \Lambda_j)$, $q = 0, 1, \ldots, n$, and replace $V^{(n)}(l)$ by its Taylor expansion $\sum_{m=0}^{\infty} V^{(m)}(0) l^m/m!$, and interchange the order of summation, using the identity

$$\sum_{n=q}^{m} \frac{(-1)^{n-q}}{(m-n)!(n-q)!} = \begin{cases} 0 & q < m, \\ 1 & q = m. \end{cases}$$

In the overlap region, we can think about how all our expansions relate to each other as follows. Suppose in the uniform expansion we replace $A_n(\xi; r, \rho, \gamma, \Lambda_j)$

by its asymptotic expansion and $V^{(n)}(l)$ by its Taylor expansion to get the triple sum (after changing the order of summation)

$$e^{-p(l)t}\sum_{s=0}^{\infty}\sum_{n=0}^{2s}\sum_{m=n}^{\infty}\Gamma(s+\tfrac{1}{2})2n!d_n^{2s}(\gamma)\frac{V^{(m)}(0)}{m!}\binom{m}{n}\frac{l^{1+\gamma+m}}{[l^2 p_0(l)t]^{s+\frac{1}{2}}}.$$

The nth term of the uniform expansion is obtained by summing (formally) in s and m in this triple sum. The (s,n)th term of the outer expansion is obtained by summing in m. The mth term of the inner expansion is obtained by summing (formally) in s and n. The (s,m)th term of the overlap expansion is obtained by summing in n.

This explains why when $\beta = 2/(r+1)$ the uniform expansion yields a gap which is twice as large as the gap in the outer or inner expansions. The leading-order term $V(l)t^{-(1+\gamma)/r}A(\xi; r, \rho, \gamma, \Lambda_j)$ of the uniform expansion already contains in it the two terms $(s,m,n) = (1,0,0), (0,1,0)$ which would have dominated the error for the outer or inner expansions. In the overlap region the term dominating the error for the uniform expansion is $(s,m,n) = (1,1,1)$; thus the larger gap.

3.2 The Uniform Expansion of $I(x,t)$

We began chapter 1 with the goal of deriving a large time asymptotic expansion of

$$I(x,t; r, \rho, \gamma, V) = \int_{\Gamma_\theta} e^{ikx - i\rho k^r t}k^\gamma V(k)\, dk,$$

where Γ_θ is the ray $\arg(k) = \theta$, and θ is a fixed number satisfying $-\frac{\pi}{r} < \theta < 0$. As we saw in section 1.2 the contour Γ_θ can be replaced by a union of steepest descent contours; the exact nature of these contours depends however on the sign of x. We have now succeeded in deriving uniformly valid asymptotic expansions for each of the integrals comprising $I(x,t)$. Now we must assemble our results in the two cases: $x \le 0$ (case A) and $x \ge 0$ (case B). Certain assumptions and notations will be the same in both cases, so we will explain them only once, here at the beginning.

As usual, we assume $r \ge 2$, $\rho > 0$ and $\gamma > -1$ are real numbers. The special variables are $b = xt^{-1}$ and $\xi = xt^{-1/r}$, where $x \in \mathbb{R}$ and $t > 0$. $\alpha \in \{0,1\}$ is related to x by $(-1)^\alpha x \ge 0$. We define $J = [\frac{r+1}{4}]$ and $K = [\frac{r-1}{4}]$. If $-J \le j \le K$ the arguments of the saddle points are given by $S_j^\alpha = \frac{\pi}{r-1}(2j+\alpha)$; the saddle points themselves are given by $k_j(b) = |\frac{b}{r\rho}|^{\frac{1}{r-1}}e^{iS_j^\alpha}$. The contours $\tilde{C}(b)$, $C_j(b)$, $C_{-J}^+(b)$ and $C_K^+(b)$ are defined and described in detail in section 1.2. The contours $\tilde{C}(\xi)$, $C_j(\xi)$, $C_{-J}^+(\xi)$ and $C_K^+(\xi)$ are obtained from these by the scaling transformation $\omega = kt^{1/r}$ as described in section 2.1; generally $k \in C(b)$ if and only if $\omega \in C(\xi)$, where C represents any of these contours.

We assume $V(k)$ is an entire function of k whose derivatives satisfy various growth estimates in different sectors in the complex k plane. Define the argument of the asymptotic center of the jth valley to be $V_j = \frac{\pi}{r}(-\frac{1}{2} + 2j)$. Define $\epsilon_0 = 0$ if $2 \le r \le 3$. If $r > 3$ define $\tilde{\epsilon}_0 = \frac{1}{r}\sin^{-1}(\cos^r(V_1))$ and ϵ_0 to be the unique number in the interval $(0, \tilde{\epsilon}_0)$ satisfying the equation $\epsilon_0 = \frac{1}{r}\sin^{-1}(\cos^r(V_1 + \epsilon_0))$. We assume for all integers $N \ge 0$ that

$$|V^{(N)}(k)| \le \frac{C_N^0 e^{\max\{a_- \Im k, a_+ \Im k\}}}{(1 + |k|)^{N + \lambda_N^0}},$$

uniformly for $V_0 \le \arg(k) \le V_1 + \epsilon_0$ where $C_N^0 > 0, a_- \le a_+ \in \mathbb{R}, \lambda_N^0 \in \mathbb{R}$ are constants. On the sector $V_{-J} \le \arg(k) \le V_0$ (when $r \ne 4J - 1$) or $-\pi/2 \le \arg(k) \le V_0$ (when $r = 4J - 1$) we assume $V^{(N)}$ satisfies

$$|V^{(N)}(k)| \le \frac{C_N^- e^{a_- \Im k}}{(1 + |k|)^{N + \lambda_N^-}},$$

where $C_N^- > 0$ and $\lambda_N^- \in \mathbb{R}$ are constants. Also on the sector $V_1 \le \arg(k) \le V_{K+1}$ (when $r \ne 4K + 1$) or $V_1 \le \arg(k) \le \pi/2$ (when $r = 4K + 1$) we assume $V^{(N)}$ satisfies

$$|V^{(N)}(k)| \le \frac{C_N^+ e^{a_+ \Im k}}{(1 + |k|)^{N + \lambda_N^+}},$$

where $C_N^+ > 0$ and $\lambda_N^+ \in \mathbb{R}$ are constants. If λ_N represents any of $\lambda_N^0, \lambda_N^-, \lambda_N^+$, then we assume $0 \le 1 + \lambda_{N+1} - \lambda_N \le 1$ for all $N \ge 0$. The case we are most interested in is $V(k) = \int_{y_-}^{y_+} e^{-iky} v(y)\, dy$, where (as we saw in section 1.5) $\lambda_N = \min\{\lambda, \lambda' - N\}$ and $0 < \lambda \le \lambda' \le \infty$ are numbers related to the degree of smoothness of $v(y)$ at its worst singularity (for the first sector) or at the endpoints of the support of $v(y)$ (for the last two sectors); also we have $a_\pm = y_\pm$. See section 1.5 for the details.

The terms of our expansion can expressed in terms of the special functions

$$F_n(r, \delta; y) = \int_0^\infty e^{-\frac{1}{r}\sigma^r + y\sigma} \sigma^{\delta-1}(\sigma - y^{\frac{1}{r-1}})^n \, d\sigma, \quad r > 1, \delta > 0, y \in S\{y^{\frac{1}{r-1}}\}.$$

$$\mathbb{F}_n(r, \delta; y, z) = \int_0^z e^{-\frac{1}{r}\sigma^r + y\sigma} \sigma^{\delta-1}(\sigma - z)^n \, d\sigma, \quad r > 1, \delta > 0, y \in \mathbb{C},$$

$$\mathbb{F}_n^c(r, \delta; y, z) = \int_z^\infty e^{-\frac{1}{r}\sigma^r + y\sigma} \sigma^{\delta-1}(\sigma - z)^n \, d\sigma, \quad r > 1, \delta > 0, y \in \mathbb{C},$$

and $z \in S\{\ln\}$, the Riemann surface of $\ln$. See section 3.1 for more details on these functions.

Theorem 3.2.1A. *Suppose in addition to all of the above that $x \le 0$ and $N \ge 0$ is an integer. Define*

$$A_n(\xi; r, \rho, \gamma, C_j(\xi)) = \left(\frac{e^{iV_{j+1}}}{(r\rho)^{1/r}}\right)^{1+\gamma+n} F_n\left(r, 1+\gamma; \frac{|\xi| e^{i(-\frac{\pi}{2} + V_{j+1})}}{(r\rho)^{1/r}}\right)$$
$$- \left(\frac{e^{iV_j}}{(r\rho)^{1/r}}\right)^{1+\gamma+n} F_n\left(r, 1+\gamma; \frac{|\xi| e^{i(\frac{3\pi}{2} + V_j)}}{(r\rho)^{1/r}}\right),$$

for all integers $n \geq 0$ and $-J \leq j \leq -1$. When N is even and $-J \leq j \leq -1$ define

$$W_N(x,t,C_j) = \exp\left\{-(r-1)\rho\left|\frac{\xi}{r\rho}\right|^{\frac{r}{r-1}}\sin(|S_j^1|) + a_-\left|\frac{b}{r\rho}\right|^{\frac{1}{r-1}}\sin(S_j^1)\right\} \cdot$$

$$\cdot\frac{(1+|\xi|)^{\frac{1}{r-1}\{1+\gamma+N-\frac{r}{2}(N+1)\}}}{t^{\frac{1+\gamma+N}{r}}(1+|b|)^{\frac{N+\lambda_N^-}{r-1}}}.$$

When N is odd and $-J \leq j \leq -1$ define

$$W_N(x,t,C_j) = \exp\left\{-(r-1)\rho\left|\frac{\xi}{r\rho}\right|^{\frac{r}{r-1}}\sin(|S_j^1|) + a_-\left|\frac{b}{r\rho}\right|^{\frac{1}{r-1}}\sin(S_j^1)\right\} \cdot$$

$$\cdot\frac{(1+|\xi|)^{\frac{1}{r-1}\{2+\gamma+N-\frac{r}{2}(N+2)\}}}{t^{\frac{1+\gamma+N}{r}}(1+|b|)^{\frac{N+\lambda_N^-}{r-1}}}\left[\frac{1}{(1+|\xi|)^{\frac{1}{r-1}}} + \frac{1}{t^{1/r}(1+|b|)^{\frac{1+\lambda_{N+1}^- - \lambda_N^-}{r-1}}}\right].$$

(1) *Suppose $4J - 1 < r < 4J + 3$. Define*

$$A(\xi;r,\rho,\gamma+n,\tilde{C}(\xi)) = \left(\frac{e^{iV_{-J}}}{(r\rho)^{1/r}}\right)^{1+\gamma+n}F_0\left(r,1+\gamma+n;\frac{i\xi e^{iV_{-J}}}{(r\rho)^{1/r}}\right)$$

for all $n \geq 0$. Also define

$$W_N(x,t,\tilde{C}) = \frac{1}{[t^{1/r}(1+|\xi|)]^{1+\gamma+N}}.$$

Then we have the asymptotic result

$$I(x,t) = \sum_{n=0}^{N-1}\frac{V^{(n)}(0)}{n!}\frac{A(\xi;r,\rho,\gamma+n,\tilde{C}(\xi))}{t^{(1+\gamma+n)/r}} + O(W_N(x,t,\tilde{C}))$$

$$+ \sum_{j=1}^{J}\left\{\sum_{n=0}^{N-1}\frac{V^{(n)}(k_{-j}(b))}{n!}\frac{A_n(\xi;r,\rho,\gamma,C_{-j}(\xi))}{t^{(1+\gamma+n)/r}} + O(W_N(x,t,C_{-j}))\right\}.$$

The constants implicit in the error terms are independent of $x \leq 0$ and $t \geq 1$.

(2) *Suppose $r = 4J - 1$, where $J \geq 1$. Define*

$$A_n(\xi;r,\rho,\gamma,C_{-J}^+(\xi))$$

$$= \left(\frac{e^{-i\frac{\pi}{2}(r-2)}}{r\rho}\right)^{\frac{1+\gamma+n}{r}}\mathbb{F}_n^c\left(r,1+\gamma,\frac{|\xi|e^{-i\pi\frac{r-1}{r}}}{(r\rho)^{1/r}},\left[\frac{|\xi|e^{-i\pi\frac{r-1}{r}}}{(r\rho)^{1/r}}\right]^{\frac{1}{r-1}}\right),$$

for all $n \geq 0$. Define

$$W_N(x,t,C^+_{-J}) = \exp\left\{-(r-1)\rho\left|\frac{\xi}{r\rho}\right|^{\frac{r}{r-1}} - a_-\left|\frac{b}{r\rho}\right|^{\frac{1}{r-1}}\right\}.$$

$$\cdot \frac{(1+|\xi|)^{\frac{1}{r-1}\{1+\gamma+N-\frac{r}{2}(N+1)\}}}{t^{\frac{1+\gamma+N}{r}}(1+|b|)^{\frac{N+\lambda_N^-}{r-1}}}.$$

Define also

$$A(\xi; r, \rho, \gamma+n, \tilde{C}(\xi))$$

$$= \left(\frac{e^{i\frac{\pi}{2}}}{(r\rho)^{1/r}}\right)^{1+\gamma+n} \mathbb{F}_0\left(r, 1+\gamma+n; \frac{|\xi|}{(r\rho)^{1/r}}, e^{-i\pi}\left|\frac{\xi}{(r\rho)^{1/r}}\right|^{\frac{1}{r-1}}\right),$$

for all $n \geq 0$. Furthermore define

$$W_N(x,t,\tilde{C}) = \left[\frac{|\xi|^{1/(r-1)}}{t^{1/r}(1+|\xi|)^{r/(r-1)}}\right]^{1+\gamma+N}.$$

Then we have the asymptotic result

$$I(x,t) = \sum_{n=0}^{N-1} \frac{V^{(n)}(0)}{n!} \frac{A(\xi; r, \rho, \gamma+n, \tilde{C}(\xi))}{t^{(1+\gamma+n)/r}} + O(W_N(x,t,\tilde{C}))$$

$$+ \sum_{j=1}^{J-1}\left\{\sum_{n=0}^{N-1} \frac{V^{(n)}(k_{-j}(b))}{n!} \frac{A_n(\xi; r, \rho, \gamma, C_{-j}(\xi))}{t^{(1+\gamma+n)/r}} + O(W_N(x,t,C_{-j}))\right\}$$

$$+ \sum_{n=0}^{N-1} \frac{V^{(n)}(k_{-J}(b))}{n!} \frac{A_n(\xi; r, \rho, \gamma, C^+_{-J}(\xi))}{t^{(1+\gamma+n)/r}} + O(W_N(x,t,C^+_{-J})).$$

The constants implicit in the error terms are independent of $x \leq 0$ and $t \geq 1$.

Remark. We have given a separate error term for each component contour, even though they can all be incorporated into a single error term $O(W_N(x,t,\tilde{C}))$ from part (1). However, it is clear from what we have written that if for some reason the contribution of the contour $\tilde{C}$ was not present, then a much smaller (decaying exponentially as $|\xi| \to \infty$) error term is appropriate. This situation arises in certain important special cases for linear dispersive equations. See section 4.6.

Theorem 3.2.1B. *Suppose in addition to all of the assumptions listed before the previous theorem that $x \geq 0$ and $N \geq 0$ is an integer. Define*

$$A_n(\xi; r, \rho, \gamma, C_j(\xi)) = \left(\frac{e^{iV_j}}{(r\rho)^{1/r}}\right)^{1+\gamma+n} F_n\left(r, 1+\gamma; \frac{|\xi|e^{i(\frac{\pi}{2}+V_j)}}{(r\rho)^{1/r}}\right)$$

$$- \left(\frac{e^{iV_{j+1}}}{(r\rho)^{1/r}}\right)^{1+\gamma+n} F_n\left(r, 1+\gamma; \frac{|\xi|e^{i(-\frac{3\pi}{2}+V_{j+1})}}{(r\rho)^{1/r}}\right).$$

for all integers $n \geq 0$ and $0 \leq j \leq K$. Let λ_N^j denote λ_N^0 when $j = 0$ and λ_N^+ when $1 \leq j \leq K$. When N is even and $0 \leq j \leq K$ define

$$W_N(x,t,C_j) = \exp\left\{-(r-1)\rho\left|\frac{\xi}{r\rho}\right|^{\frac{r}{r-1}}\sin(|S_j^0|) + a_+\left|\frac{b}{r\rho}\right|^{\frac{1}{r-1}}\sin(S_j^0)\right\} \cdot$$

$$\cdot\frac{(1+|\xi|)^{\frac{1}{r-1}\{1+\gamma+N-\frac{r}{2}(N+1)\}}}{t^{\frac{1+\gamma+N}{r}}(1+|b|)^{\frac{N+\lambda_N^j}{r-1}}}.$$

When N is odd and $0 \leq j \leq K$ define

$$W_N(x,t,C_j) = \exp\left\{-(r-1)\rho\left|\frac{\xi}{r\rho}\right|^{\frac{r}{r-1}}\sin(|S_j^0|) + a_+\left|\frac{b}{r\rho}\right|^{\frac{1}{r-1}}\sin(S_j^0)\right\} \cdot$$

$$\cdot\frac{(1+|\xi|)^{\frac{1}{r-1}\{2+\gamma+N-\frac{r}{2}(N+2)\}}}{t^{\frac{1+\gamma+N}{r}}(1+|b|)^{\frac{N+\lambda_N^j}{r-1}}}\left[\frac{1}{(1+|\xi|)^{\frac{1}{r-1}}} + \frac{1}{t^{1/r}(1+|b|)^{\frac{1+\lambda_{N+1}^j-\lambda_N^j}{r-1}}}\right].$$

(1) *Suppose $4K+1 < r < 4K+5$. Define*

$$A(\xi;r,\rho,\gamma+n,\tilde{C}(\xi)) = \left(\frac{e^{iV_{K+1}}}{(r\rho)^{1/r}}\right)^{1+\gamma+n} F_0\left(r,1+\gamma+n;\frac{i\xi e^{iV_{K+1}}}{(r\rho)^{1/r}}\right)$$

for all $n \geq 0$. Also define

$$W_N(x,t,\tilde{C}) = \frac{1}{[t^{1/r}(1+|\xi|)]^{1+\gamma+N}}.$$

Then we have the asymptotic result

$$I(x,t) = \sum_{n=0}^{N-1}\frac{V^{(n)}(0)}{n!}\frac{A(\xi;r,\rho,\gamma+n,\tilde{C}(\xi))}{t^{(1+\gamma+n)/r}} + O(W_N(x,t,\tilde{C}))$$

$$+ \sum_{j=0}^{K}\left\{\sum_{n=0}^{N-1}\frac{V^{(n)}(k_j(b))}{n!}\frac{A_n(\xi;r,\rho,\gamma,C_j(\xi))}{t^{(1+\gamma+n)/r}} + O(W_N(x,t,C_j))\right\}.$$

The constants implicit in the error terms are independent of $x \geq 0$ and $t \geq 1$.

(2) *Suppose $r = 4K+1$, where $K \geq 1$. Define*

$$A_n(\xi;r,\rho,\gamma,C_K^+(\xi))$$

$$= \left(\frac{e^{i\frac{\pi}{2}(r-2)}}{r\rho}\right)^{\frac{1+\gamma+n}{r}} \mathbb{F}_n^c\left(r,1+\gamma,\frac{|\xi|e^{i\pi\frac{r-1}{r}}}{(r\rho)^{1/r}},\left[\frac{|\xi|e^{i\pi\frac{r-1}{r}}}{(r\rho)^{1/r}}\right]^{\frac{1}{r-1}}\right),$$

for all $n \geq 0$. Define

$$W_N(x,t,C_K^+) = \exp\left\{-(r-1)\rho\left|\frac{\xi}{r\rho}\right|^{\frac{r}{r-1}} + a_+\left|\frac{b}{r\rho}\right|^{\frac{1}{r-1}}\right\} \cdot$$

$$\cdot \frac{(1+|\xi|)^{\frac{1}{r-1}\{1+\gamma+N-\frac{r}{2}(N+1)\}}}{t^{\frac{1+\gamma+N}{r}}(1+|b|)^{\frac{N+\lambda_N^+}{r-1}}}.$$

Define also

$$A(\xi;r,\rho,\gamma+n,\tilde{C}(\xi))$$

$$= \left(\frac{e^{-i\frac{\pi}{2}}}{(r\rho)^{1/r}}\right)^{1+\gamma+n} \mathbb{F}_0\left(r,1+\gamma+n; \frac{|\xi|}{(r\rho)^{1/r}}, e^{i\pi}\left|\frac{\xi}{(r\rho)^{1/r}}\right|^{\frac{1}{r-1}}\right),$$

for all $n \geq 0$. Furthermore define

$$W_N(x,t,\tilde{C}) = \left[\frac{|\xi|^{1/(r-1)}}{t^{1/r}(1+|\xi|)^{r/(r-1)}}\right]^{1+\gamma+N}.$$

Then we have the asymptotic result

$$I(x,t) = \sum_{n=0}^{N-1} \frac{V^{(n)}(0)}{n!}\frac{A(\xi;r,\rho,\gamma+n,\tilde{C}(\xi))}{t^{(1+\gamma+n)/r}} + O(W_N(x,t,\tilde{C}))$$

$$+ \sum_{j=0}^{K-1}\left\{\sum_{n=0}^{N-1} \frac{V^{(n)}(k_j(b))}{n!}\frac{A_n(\xi;r,\rho,\gamma,C_j(\xi))}{t^{(1+\gamma+n)/r}} + O(W_N(x,t,C_j))\right\}$$

$$+ \sum_{n=0}^{N-1} \frac{V^{(n)}(k_K(b))}{n!}\frac{A_n(\xi;r,\rho,\gamma,C_K^+(\xi))}{t^{(1+\gamma+n)/r}} + O(W_N(x,t,C_K^+)).$$

The constants implicit in the error terms are independent of $x \geq 0$ and $t \geq 1$.

Remark. All the error terms can be incorporated into a pair of error terms: $O(W_N(x,t,\tilde{C}))$ from part (1) and $O(W_N(x,t,C_0))$. As can be seen from the error terms, some terms are exponentially smaller (as $|\xi| \to \infty$) than others. We have retained such exponentially small terms because of their importance when $|\xi|$ is not too large. It is known that retaining such terms can yield significant improvements in numerical accuracy in practical applications, such as when computing $I(x,t)$ via its asymptotic expansion.

Even though it is nice to be able to write down these neat expansions which are uniformly valid for all $(-1)^{\alpha}x \geq 0$ as $t \to \infty$, we pay the price of needing to know a lot about the special functions involved. These expansions "simplify" considerably in the outer regions (roughly where $|\xi| \to \infty$), in that special functions of ξ can be replaced by elementary functions. We will not bother however to write down the full asymptotic representation of $I(x,t)$ in the outer region. It can be easily extracted from our results in chapter one. (See also Lemma 2.1.1 and Lemma 3.1.1.)

The uniform expansions for $x \leq 0$ and $x \geq 0$ look different from each other. This raises the question of how they match with each other at $x = 0$. As we noted in section 3.1, the uniform expansion over the contours C_j, C_{-j}^+, and C_K^+ becomes the inner expansion over those contours in the inner region (roughly where $b \to 0$). Over the contour $\tilde{C}$ the uniform and inner expansions are the same. Since we saw in section 2.4 that the inner expansion over the original contour of integration is the same, regardless of whether $x \geq 0$ or $x \leq 0$, the two uniform expansions evaluated at $x = 0$ agree with each other, and with the inner expansion. This matching is essentially simpler than the matching we studied (in section 2.4) between the outer and inner expansions, where there was a breakdown in the asymptotic ordering of the expansions near the boundary of their regions of validity. In the present case, the difference of appearance between the uniform expansions for $x \leq 0$ and $x \geq 0$ does not reflect any misbehavior as $x \to 0$.

CHAPTER 4

SPECIAL RESULTS FOR SPECIAL CASES

4.1 The Case $r = 2$

In this section we will express the terms of the uniform expansions of $I(x, t; \tilde{C})$ and $I(x, t, C_0)$ in the case $r = 2$ in terms of special functions. Besides facilitating certain desired applications, this will also help us to understand the super-anomalous cases, namely $r = 2$, $\gamma \geq 0$ an integer.

Our first task is to realize the special function

$$F_n(2, 1 + \gamma; y) = \int_0^\infty e^{-\frac{1}{2}\sigma^2 + y\sigma} \sigma^\gamma (\sigma - y)^n \, d\sigma,$$

introduced in sections 2.1 and 3.1, in terms of well-known and studied special functions. This function was defined and studied briefly on page 346 of Olver [O1]. When $n = 0$ we have the following representation in terms of the Parabolic Cylinder function (Exercise 6.1, page 208 of Olver [O1], or 19.5.3, page 687, [AS])

$$F_0(2, 1 + \gamma; y) = \Gamma(1 + \gamma)e^{y^2/4}U(\gamma + \tfrac{1}{2}, -y).$$

Then using 19.12.4, page 691 [AS] we can express $e^{y^2/4}U(\gamma + \tfrac{1}{2}, -y)$ in terms of the multi-valued Confluent Hypergeometric function $U(a, b, z)$ to yield

$$F_0(2, 1 + \gamma; y) = \frac{\Gamma(1 + \gamma)}{2^{\frac{1+\gamma}{2}}}U(\tfrac{1+\gamma}{2}, \tfrac{1}{2}, \tfrac{1}{2}(-y)^2)$$

$$= \frac{\Gamma(1 + \gamma)(-y)}{2^{1+\frac{\gamma}{2}}}U(1 + \tfrac{\gamma}{2}, \tfrac{3}{2}, \tfrac{1}{2}(-y)^2).$$

In the above, $U(a, \tfrac{1}{2}, z)$ and $U(a, \tfrac{3}{2}, z)$ are defined for z in the Riemann surface $S\{z^{1/2}\} = \{(w^2, w) \in \mathbb{C}^2 \mid w \in \mathbb{C}\}$; the expression $\tfrac{1}{2}(-y)^2$ refers to the function $\mathbb{C} \to S\{z^{1/2}\}: y \mapsto (\tfrac{1}{2}y^2, -\tfrac{y}{\sqrt{2}})$. This expression is useful when dealing with the contour $\tilde{C}$.

On the other hand, using 19.12.3, page 691 [AS] we can express $e^{y^2/4}U(\gamma + \tfrac{1}{2}, -y)$ in terms of the entire Confluent Hypergeometric function $M(a, b, z)$

$$e^{y^2/4}U(\gamma + \tfrac{1}{2}, -y) = \frac{\sqrt{\pi}}{2^{\frac{1+\gamma}{2}}\Gamma(\tfrac{\gamma}{2} + 1)}M(\tfrac{1+\gamma}{2}, \tfrac{1}{2}, \tfrac{1}{2}y^2) + \frac{\sqrt{\pi}y}{2^{\frac{\gamma}{2}}\Gamma(\tfrac{1+\gamma}{2})}M(\tfrac{\gamma}{2} + 1, \tfrac{3}{2}, \tfrac{1}{2}y^2).$$

If we use the duplication formula $\Gamma(2z) = 2^{2z-1/2}\Gamma(z)\Gamma(z+1/2)/\sqrt{2\pi}$ and the Kummer transformation $M(a, b, z) = e^z M(b - a, b, -z)$ we get the formula

$$F_0(2, 1 + \gamma; y) = e^{y^2/2}\left[2^{\frac{\gamma-1}{2}}\Gamma(\tfrac{1+\gamma}{2})M(-\tfrac{\gamma}{2}, \tfrac{1}{2}, -\tfrac{1}{2}y^2)\right.$$
$$\left. + 2^{\frac{\gamma}{2}}\Gamma(1 + \tfrac{\gamma}{2})yM(\tfrac{1-\gamma}{2}, \tfrac{3}{2}, -\tfrac{1}{2}y^2)\right].$$

This expression is useful when dealing with the contour C_0.

These formulae allow us to prove the following.

Lemma 4.1.1. *Suppose* $r = 2$, $\rho > 0$, *and* $\gamma > -1$.

(1) *Suppose* $\xi \leq 0$. *Then*

$$A(\xi; 2, \rho, \gamma, \tilde{C}) = \Gamma(1 + \gamma)\left(\frac{e^{-i\pi/2}}{4\rho}\right)^{\frac{1+\gamma}{2}} U(\tfrac{1+\gamma}{2}, \tfrac{1}{2}, \tfrac{|\xi|^2}{4\rho}e^{i\pi/2}).$$

(2) *Suppose* $\xi \geq 0$. *Then*

$$A(\xi; 2, \rho, \gamma, \tilde{C}) = \Gamma(1 + \gamma)\left(\frac{e^{i3\pi/2}}{4\rho}\right)^{\frac{1+\gamma}{2}} U(\tfrac{1+\gamma}{2}, \tfrac{1}{2}, \tfrac{|\xi|^2}{4\rho}e^{i\pi/2}).$$

(3) *Suppose* $\xi \geq 0$. *Then*

$$A(\xi; 2, \rho, \gamma, C_0) = e^{\frac{i\xi^2}{4\rho}}\left[\left(\frac{1 + e^{i\pi\gamma}}{2}\right)\frac{\Gamma(\tfrac{1+\gamma}{2})}{(\rho i)^{\frac{1+\gamma}{2}}}M(-\tfrac{\gamma}{2}, \tfrac{1}{2}, -\tfrac{i\xi^2}{4\rho})\right.$$
$$\left. + \left(\frac{1 - e^{i\pi\gamma}}{2}\right)\frac{\Gamma(1 + \tfrac{\gamma}{2})}{(\rho i)^{1+\frac{\gamma}{2}}}i\xi M(\tfrac{1-\gamma}{2}, \tfrac{3}{2}, -\tfrac{i\xi^2}{4\rho})\right].$$

where $\arg(\rho i) = \pi/2$.

Further simplification of these formulae is possible when $\gamma \geq 0$ is an integer. For example, when $\gamma \geq 0$ is an integer,

$$U(\tfrac{1+\gamma}{2}, \tfrac{1}{2}, z^2) = \sqrt{\pi}2^{\gamma}e^{z^2}i^{\gamma}\text{erfc}(z),$$

where $i^{\gamma}\text{erfc}(z)$ is the γth repeated integral of the error function (page 299, [AS]). In particular, when $\gamma = 0$ we get

$$A(\xi; 2, \rho, 0, \tilde{C}) = -(-1)^{\alpha}\left(\frac{\pi}{4\rho i}\right)^{1/2}w\left(\frac{|\xi|e^{i3\pi/4}}{\sqrt{4\rho}}\right),$$

where $w(z) = e^{-z^2}\text{erfc}(-iz)$, $(-1)^{\alpha}\xi \geq 0$, and $\arg(\rho i) = \pi/2$. When $\gamma = 2m$ is an even integer,

$$M(-\tfrac{\gamma}{2}, \tfrac{1}{2}, -\tfrac{i\xi^2}{4\rho}) = \frac{m!}{(2m)!}(-2)^m\text{He}_{2m}(\tfrac{\xi}{\sqrt{2\rho i}}),$$

where $\mathrm{He}_\gamma(z)$ is the γth Hermite polynomial (page 778, [AS]). When $\gamma = 2m+1$ is an odd integer,

$$M(\tfrac{1-\gamma}{2}, \tfrac{3}{2}, -\tfrac{i\xi^2}{4\rho}) = \frac{m!}{(2m+1)!}(-2)^m(\tfrac{\xi}{\sqrt{2\rho i}})^{-1}\mathrm{He}_{2m+1}(\tfrac{\xi}{\sqrt{2\rho i}}).$$

Thus when $\gamma \geq 0$ is an integer $A(\xi; 2, \rho, \gamma, C_0)$ can be written as an oscillating exponential times a polynomial function of $\xi \geq 0$.

A similar result holds for $A_n(\xi; 2, \rho, \gamma, C_0)$ for all $n \geq 0$ when $\gamma \geq 0$ is an integer. To see this we use the original definition

$$A_n(\xi; 2, \rho, \gamma, C_0) = \int_{C_0(\xi)} e^{i\omega\xi - i\rho\omega^2} \omega^\gamma (\omega - \tfrac{\xi}{2\rho})^n \, d\omega.$$

Note that $\xi \geq 0$. $C_0(\xi)$ can be parameterized by $\omega = \tfrac{\xi}{2\rho} + \tfrac{w}{\sqrt{\rho i}}$, where $\arg(\rho i) = \pi/2$, and $w \in \mathbb{R}$. Using this change of variables in the defining integral yields

$$A_n(\xi; 2, \rho, \gamma, C_0) = (\rho i)^{-(n+1)/2} e^{\frac{i\xi^2}{4\rho}} \int_{-\infty}^{\infty} e^{-w^2} (\tfrac{\xi}{2\rho} + \tfrac{w}{\sqrt{\rho i}})^\gamma w^n \, dw.$$

When γ is an integer we can use the Binomial Theorem to obtain the desired result.

Lemma 4.1.2. *Suppose $r = 2$, $\rho > 0$, and $n, \gamma \geq 0$ are integers. If x is a real number, let $[x]$ denote the largest integer less than or equal to x, and let $((x))$ denote the smallest integer greater than or equal to x. Then*

$$A_n(\xi; 2, \rho, \gamma, C_0) = e^{\frac{i\xi^2}{4\rho}} \sum_{s=((\frac{n}{2}))}^{[\frac{\gamma+n}{2}]} \binom{\gamma}{2s-n} \left(\frac{\xi}{2\rho}\right)^{\gamma+n-2s} \frac{\Gamma(s + \frac{1}{2})}{(\rho i)^{s+\frac{1}{2}}}.$$

Notice that in the above expression the integer index s of the summation always obeys $0 \leq 2s - n \leq \gamma$. This lemma covers the super-anomalous cases. It shows (with a bit more calculation) in particular that when $\gamma \geq 0$ is an integer the uniformly valid expansion

$$\sum_{n=0}^{\infty} \frac{V^{(n)}(\frac{x}{2\rho t})}{n!} t^{-(1+\gamma+n)/2} A_n(\tfrac{x}{\sqrt{t}}; 2, \rho, \gamma, C_0)$$

coincides with the outer expansion (see Corollary 1.7.6)

$$e^{\frac{ix^2}{4\rho t}} \sum_{s=0}^{\infty} \frac{\Gamma(s + \frac{1}{2})}{(i\rho t)^{s+\frac{1}{2}}} \left(\frac{x}{2\rho t}\right)^{\gamma-2s} \sum_{n=\max\{0,((2s-\gamma))\}}^{2s} \frac{V^{(n)}(\frac{x}{2\rho t})}{n!} \binom{\gamma}{2s-n} \left(\frac{x}{2\rho t}\right)^n.$$

Hence in the super-anomalous cases there is no need for the inner expansion over the contour C_0. One further interesting consequence of the above result is that when $\gamma = 0$ and n is odd then $A_n(\xi; 2, \rho, 0, C_0) = 0$. When $\gamma = 0$ and $n = 2s$ is even we have

$$A_n(\xi; 2, \rho, 0, C_0) = \frac{\Gamma(s + \frac{1}{2})}{(\rho i)^{s + \frac{1}{2}}} e^{\frac{i\xi^2}{4\rho}}.$$

For the sake of desired applications, we want to make sure that we have proved the uniform expansion to be valid in these super-anomalous cases. In section 1.6 where we studied the terms of the outer expansion, we based our weight functions on the size of the terms in the "generic" case, i.e. ignoring anomalies. Hence the weight functions in Corollary 1.7.6 do not vanish as $b = x/t \to 0^+$ rapidly enough (when $2M > \gamma$) to give a sharp result even in the outer region. On the other hand, in section 3.1 where we studied the uniform expansion, we based our weight functions on the size of $A_n(\xi; 2, \rho, \gamma, \Lambda_0)$ as given in Lemma 3.1.2. When $r = 2$ the only situation where Lemma 3.1.2 does not apply is $\gamma = 0$, $n \geq 1$ odd; this is the situation where, as we saw above, $A_n(\xi; 2, \rho, \gamma, \Lambda_0)$ vanishes identically. Thus Lemma 3.1.2 gives a sharp estimate as $\xi \to \infty$ for all the other super-anomalous cases. We saw in general that the term with n odd is so small that it does not form a good basis for assessing the size of the error in the uniform expansion (this is especially true when $r = 2, \gamma = 0$). Hence we stated our error estimates in Theorem 3.1.4 using the even term as the basis for the size of the error. Hence, in regard to behavior as $\xi \to \infty$, the result stated in Theorem 3.1.4 is sharp for the super-anomalous cases as well as the generic cases. However, Lemma 3.1.2 is not a sharp estimate as $\xi \to 0^+$ in some of the super-anomalous cases. For example, when $\gamma = 1$ we have

$$A_n(\xi; 2, \rho, 1, \Lambda_0) = \begin{cases} \frac{\Gamma(\frac{1+n}{2})}{(\rho i)^{\frac{1+n}{2}}} \left(\frac{\xi}{2\rho}\right) e^{\frac{i\xi^2}{4\rho}} & n \text{ even}, \\ \frac{\Gamma(1 + \frac{n}{2})}{(\rho i)^{1 + \frac{n}{2}}} e^{\frac{i\xi^2}{4\rho}} & n \text{ odd}. \end{cases}$$

Also, when $\gamma = 2$ we have

$$A_n(\xi; 2, \rho, 2, \Lambda_0) = \begin{cases} \frac{\Gamma(\frac{1+n}{2})}{(\rho i)^{\frac{1+n}{2}}} \left[\left(\frac{\xi}{2\rho}\right)^2 + \frac{1+n}{2\rho i}\right] e^{\frac{i\xi^2}{4\rho}} & n \text{ even}, \\ 2 \frac{\Gamma(1 + \frac{n}{2})}{(\rho i)^{1 + \frac{n}{2}}} \left(\frac{\xi}{2\rho}\right) e^{\frac{i\xi^2}{4\rho}} & n \text{ odd}. \end{cases}$$

The principle illustrated by these two examples is that $A_n(\xi; 2, \rho, \gamma, \Lambda_0)$ might vanish at $\xi = 0$ for n even, and not vanish at $\xi = 0$ for n odd, or *vice versa*. Thus it will make no sense to try to incorporate this vanishing into the weight functions in the error estimates; indeed if the weight function vanished at $\xi = 0$ then the desired weighted error estimate would not hold uniformly as $\xi \to 0^+$. Hence Theorem 3.1.4 is the optimal result, even in the super-anomalous cases, for all $\xi \geq 0$. The possible vanishing of $A_n(\xi; 2, \rho, \gamma, \Lambda_0)$ at $\xi = 0$ is treated exactly like any other possible zero of that function. It is a place where the error decays faster than usual. The exact decay rate near such zeros can be predicted using Theorem 3.1.4.

When γ is not an integer, it is more difficult to express $A_n(\xi; 2, \rho, \gamma, \Lambda_0)$ in terms of special functions for all $n \geq 1$ as we have done when $n = 0$. In section 3.1 we mentioned Bleistein's technique for generating an asymptotic expansion, and that our expansion and Bleistein's could be related by using recursion relations satisfied by $F_n(r, 1 + \gamma; y)$. We will now show how this can be done when $r = 2$, and in the process obtain expressions for $A_n(\xi; 2, \rho, \gamma, \Lambda_0)$, $n = 1, 2, 3$ in terms of the special functions appearing in a Bleistein-type expansion.

Let $\delta = 1 + \gamma$, and abbreviate $F_n = F_n(2, \delta; y)$ and $F_0' = F_0(2, 1 + \delta; y)$. For each $n \geq 0$ we will find polynomials $p_n(y; \delta), q_n(y; \delta)$ such that $F_n = p_n F_0 + q_n F_0'$. We have the following table.

n	p_n	q_n
0	1	0
1	$-y$	1
2	$\delta + y^2$	$-y$
3	$-2\delta y - y^3$	$1 + \delta + y^2$

The values of (p_n, q_n) for $n = 0, 1$ are obvious. For $n \geq 2$ they can be determined from the recursion relations (c.f. exercise 9.1, page 346, Olver [O1])

$$p_n = -yp_{n-1} + (n - 2 + \delta)p_{n-2} + (n - 2)yp_{n-3},$$

$$q_n = -yq_{n-1} + (n - 2 + \delta)q_{n-2} + (n - 2)yq_{n-3}.$$

By Lemma 3.1.3 we have for $\xi \geq 0$ that

$$A_n(\xi; 2, \rho, \gamma, \Lambda_0) = \frac{1}{(2\rho i)^{\frac{\delta+n}{2}}} \left[F_n(2, \delta; y) - e^{i\pi(\delta+n)} F_n(2, \delta; -y) \right],$$

where $y = \frac{\xi e^{i\frac{\pi}{4}}}{(2\rho)^{1/2}}$ and $\arg(\rho i) = \pi/2$. By induction we have $(-1)^n p_n(-y) = p_n(y)$ and $(-1)^{n-1} q_n(-y) = q_n(y)$. Thus we obtain the following expression

$$A_n(\xi; 2, \rho, \gamma, \Lambda_0) = \frac{p_n(y; \delta)}{(2\rho i)^{n/2}} A_0(\xi; 2, \rho, \gamma, \Lambda_0) + \frac{q_n(y; \delta)}{(2\rho i)^{(n-1)/2}} A_0(\xi; 2, \rho, \gamma + 1, \Lambda_0).$$

We have proved the following.

Lemma 4.1.3. *If $\xi \geq 0$ we have*

$$A_1(\xi; 2, \rho, \gamma, \Lambda_0) = -\left(\frac{\xi}{2\rho}\right) A_0(\xi; 2, \rho, \gamma, \Lambda_0) + A_0(\xi; 2, \rho, \gamma + 1, \Lambda_0),$$

$$A_2(\xi; 2, \rho, \gamma, \Lambda_0) = \left[\frac{\delta}{2\rho i} + \left(\frac{\xi}{2\rho}\right)^2\right] A_0(\xi; 2, \rho, \gamma, \Lambda_0)$$

$$- \left(\frac{\xi}{2\rho}\right) A_0(\xi; 2, \rho, \gamma + 1, \Lambda_0),$$

$$A_3(\xi; 2, \rho, \gamma, \Lambda_0) = -\left[\frac{\delta}{\rho i}\left(\frac{\xi}{2\rho}\right) + \left(\frac{\xi}{2\rho}\right)^3\right] A_0(\xi; 2, \rho, \gamma, \Lambda_0)$$

$$+ \left[\frac{1+\delta}{2\rho i} + \left(\frac{\xi}{2\rho}\right)^2\right] A_0(\xi; 2, \rho, \gamma + 1, \Lambda_0).$$

Now we can understand how our expansion relates to the one obtained by Bleistein's procedure (see section 3.1)

$$
\int_{\Lambda_0} e^{ikx - i\rho k^2 t} k^\gamma V(k)\, dk = \frac{A(\xi; 2, \rho, \gamma, \Lambda_0)}{t^{(1+\gamma)/2}} \left[a_0^0(b) + \frac{a_0^1(b)}{t} \right]
$$

$$
+ \frac{A(\xi; 2, \rho, \gamma + 1, \Lambda_0)}{t^{(2+\gamma)/2}} \left[a_1^0(b) + \frac{a_1^1(b)}{t} \right]
$$

$$
+ \frac{1}{t^2} \int_{\Lambda_0} e^{ikx - i\rho k^2 t} k^\gamma V_2(k; b)\, dk.
$$

The coefficients $a_h^m(b)$ are given by

$$
a_0^0(b) = V(0), \qquad a_1^0(b) = \frac{V(l) - V(0)}{l},
$$

$$
W_0(k; b) = \frac{1}{2\rho i} \frac{V(k) - a_0^0(b) - a_1^0(b)k}{k(k-l)}, \quad V_1(k; b) = \delta W_0(k; b) + k \frac{\partial}{\partial k} W_0(k; b),
$$

$$
a_0^1(b) = V_1(0; b), \qquad a_1^1(b) = \frac{V_1(l; b) - V_1(0; b)}{l},
$$

$$
W_1(k; b) = \frac{1}{2\rho i} \frac{V_1(k; b) - a_0^1(b) - a_1^1(b)k}{k(k-l)}, \quad V_2(k; b) = \delta W_1(k; b) + k \frac{\partial}{\partial k} W_1(k; b),
$$

where $l = \frac{b}{2\rho}$. If we expand $V(k)$ and $V(0)$ in a Taylor series about the point $k = l$ then we get the following:

$$
a_0^0(b) = V(l) - V^{(1)}(l)l + \frac{V^{(2)}(l)}{2} l^2 - \frac{V^{(3)}(l)}{6} l^3 + O(l^4),
$$

$$
a_1^0(b) = V^{(1)}(l) - \frac{V^{(2)}(l)}{2} l + \frac{V^{(3)}(l)}{6} l^2 - O(l^3),
$$

$$
a_0^1(b) = \frac{\delta}{2\rho i} \left[\frac{V^{(2)}(l)}{2} - \frac{V^{(3)}(l)}{6} l + O(l^2) \right],
$$

$$
a_1^1(b) = \frac{1}{2\rho i} \left[\frac{V^{(3)}(l)}{6} (1 + \delta) + O(l) \right].
$$

If we use these expansions in Bleistein's expansion, collect terms involving a given derivative of V at l, and compare with Lemma 4.1.3, we obtain the first four terms of our uniform expansion.

One consequence of this comparison is the fact that uniformly valid expansions, the terms of which are of the form $a(b)A(\xi)t^{-c}$, are not uniquely determined. The first term of the Bleistein-style expansion is the same as the first term of our inner expansion. Our uniform expansion has the benefit that the first term is the correct leading-order term for all $\xi \geq 0$, whereas, as we have seen in section 3.1, the first term of the Bleistein-style expansion is only a valid leading-order term in the inner region. We will not attempt to study the asymptotic character of the Bleistein expansion when $r = 2$ and l is not tending to

0; when $r = 3$ see section 4.5. We are content with our expansion, and its nice behavior even as $l \to \infty$ (as long as $t \to \infty$). However, as we will see in section 4.5, the Bleistein-style expansion does simplify (in some cases) the matching between uniform expansions for $\xi \geq 0$ and for $\xi \leq 0$.

4.2 Large x asymptotics when $r = 2$

One of the main features of the case $r = 2$ which separates it from the cases $r > 2$ is that successive terms of the expansion of $I(x, t; C_0)$ do not give asymptotic improvement in the limit $x \to \infty$ ($t \geq 1$ fixed). From our previous results we can infer that $I(x, t; C_0)$ satisfies certain decay estimates as $x \to \infty$, but we do not yet know if those estimates are sharp; we have not yet shown that it is impossible for some cancellation to take place between the "leading-term" and the error term so that $I(x, t; C_0)$ would decay faster as $x \to \infty$ than the "leading-order term" does. Actually, it turns out that the leading-order term as $t \to \infty$ and $b \to \infty$ is correct as $b \to \infty$ and $t \geq 1$ except for a t-dependent phase shift which vanishes as $t \to \infty$. Thus the decay rate as $b \to \infty$ predicted by the leading-order term we already have is in fact correct. The situation is similar to what we have already encountered with the outer and inner regions: there are two "leading-order terms", each valid in its own region, but the regions overlap and the two terms can be matched to each other in this overlap region. In this section we will derive the leading-order term for $I(x, t; C_0)$ when $b \to \infty$, prove that it is valid, and investigate the extent to which this leading-order term remains valid as $t \to \infty$.

In section 1.5 we assumed $V(k) = \int_{y_-}^{y_+} e^{-iky} v(y) \, dy$, and under various regularity assumptions on the function v we derived the leading-order asymptotics of $V(k)$ as $k \to \infty$ in various sectors in the complex k plane. In this section we will assume the function V satisfies

$$V(k) = \frac{ae^{-iy_0 k}}{k^\lambda} + o\left(\frac{e^{\max\{y_- \Im k, y_+ \Im k\}}}{|k|^\lambda}\right)$$

as $k \to \infty$, $-\pi/4 \leq \arg k \leq 3\pi/4$. In the above $a \in \mathbb{C}$, $\lambda > 0$ is a real number, $y_- \leq y_0 \leq y_+$, and the implied estimate holds independently of $\arg k$ within the stated bounds. The reader may consult section 1.5, and in particular the discussion surrounding Lemma 1.5.2, for an understanding of conditions on a function $v(y)$, supported on the interval $[y_-, y_+]$, which imply that $V(k)$ satisfies the above asymptotic condition. One should note that the most important aspect is the decay rate in $|k|$ when k is real; the above assumption gives no leading-order information if $|\Im k| \to \infty$. It is quite convenient for us to ignore in this section the details of how the number a in the above condition depends on $v(y)$ and λ. It is also possible to assume V satisfies a more general condition where finitely many terms of the form $ae^{-iy_0 k} k^{-\lambda}$ are included with different values of a and y_0 but with the same value of λ. This immediate modification is left to the interested reader.

Theorem 4.2.1. *Suppose $V(k)$ satisfies the above stated asymptotic condition. Then*

$$\int_{C_0(b)} e^{ikx - i\rho k^2 t} k^\gamma V(k)\, dk = \frac{a\sqrt{\pi}}{\sqrt{i\rho t}} \left(\frac{x - y_0}{2\rho t}\right)^{\gamma - \lambda} \exp\left[\frac{i(x - y_0)^2}{4\rho t}\right] + o\left(\frac{b^{\gamma - \lambda}}{\sqrt{t}}\right)$$

as $b \to \infty$, uniformly for $t \geq 1$.

Proof. First we will show that

$$\int_{C_0(b)} e^{ikx - i\rho k^2 t} k^\gamma V(k)\, dk = a \int_{C_0(b)} e^{ik(x - y_0) - i\rho k^2 t} k^{\gamma - \lambda}\, dk + o\left(\frac{b^{\gamma - \lambda}}{\sqrt{t}}\right)$$

as $b \to \infty$, uniformly for $t \geq 1$. Define $\tilde{V}(k) = V(k) - ae^{-iy_0 k} k^{-\lambda}$. Then we must show that

$$\lim_{b \to \infty} \sup_{t \geq 1} \sqrt{t} b^{\lambda - \gamma} \left| \int_{C_0(b)} e^{ikx - i\rho k^2 t} k^\gamma \tilde{V}(k)\, dk \right| = 0.$$

The contour $C_0(b)$ can be parameterized by $k = \frac{b}{2\rho} + \frac{w}{\sqrt{i\rho t}}$, $w \in \mathbb{R}$, $\arg(i\rho) = \pi/2$. Changing variables in the integral to be estimated we obtain

$$\frac{1}{\sqrt{i\rho t}} e^{\frac{i\xi^2}{4\rho}} \int_{-\infty}^{\infty} e^{-w^2} \left[\frac{b}{2\rho} + \frac{w}{\sqrt{i\rho t}}\right]^\gamma \tilde{V}\left(\frac{b}{2\rho} + \frac{w}{\sqrt{i\rho t}}\right) dw,$$

where, as usual, $\xi = xt^{-1/2}$. Since $\Im(\frac{b}{2\rho} + \frac{w}{\sqrt{i\rho t}}) = \frac{w}{\sqrt{2\rho t}}$, and $|\frac{b}{2\rho} + \frac{w}{\sqrt{i\rho t}}|^{-\lambda} \leq (\frac{b}{2\sqrt{2\rho}})^{-\lambda}$ for all $b > 0$ and all $w \in \mathbb{R}, t > 0$, we have by assumption

$$\left|\tilde{V}\left(\frac{b}{2\rho} + \frac{w}{\sqrt{i\rho t}}\right)\right| = \left|\frac{b}{2\rho} + \frac{w}{\sqrt{i\rho t}}\right|^{-\lambda} e^{\frac{1}{\sqrt{2\rho t}} \max\{y_- w, y_+ w\}} o(1)$$

$$\leq \left(\frac{b}{2\sqrt{2\rho}}\right)^{-\lambda} e^{\frac{1}{\sqrt{2\rho}} \max\{y_- w, y_+ w\}} o(1)$$

as $b \to \infty$ uniformly for $w \in \mathbb{R}$ and $t \geq 1$. Thus we have

$$\sup_{t \geq 1} \sqrt{t} b^{\lambda - \gamma} \left| \int_{C_0(b)} e^{ikx - i\rho k^2 t} k^\gamma \tilde{V}(k)\, dk \right|$$

$$\leq \frac{1}{\sqrt{\rho}} \int_{-\infty}^{\infty} e^{-w^2} \left[\frac{1}{2\rho} + \frac{|w|}{b\sqrt{\rho}}\right]^\gamma \left(\frac{1}{2\sqrt{2\rho}}\right)^{-\lambda} e^{\frac{1}{\sqrt{2\rho}} \max\{y_- w, y_+ w\}} o(1)\, dw$$

as $b \to \infty$. By the Dominated Convergence Theorem this quantity vanishes in the limit $b \to \infty$.

As the second and final step we must show that

$$a \int_{C_0(b)} e^{ik(x-y_0)-i\rho k^2 t} k^{\gamma-\lambda} \, dk = \frac{a\sqrt{\pi}}{\sqrt{i\rho t}} \left(\frac{x-y_0}{2\rho t}\right)^{\gamma-\lambda} \exp\left[\frac{i(x-y_0)^2}{4\rho t}\right]$$

$$+ o\left(\frac{b^{\gamma-\lambda}}{\sqrt{t}}\right)$$

as $b \to \infty$, uniformly for $t \geq 1$. Define $\tilde{x} = x - y_0$, $\tilde{b} = \tilde{x}/t$, and $\tilde{\xi} = \tilde{x} t^{-1/2}$. For $\tilde{b}$ sufficiently large we can change contours from $C_0(b)$ to $C_0(\tilde{b})$ and then introduce the change of variable $k = \eta\tilde{b}$. This yields

$$a \int_{C_0(b)} e^{ik(x-y_0)-i\rho k^2 t} k^{\gamma-\lambda} \, dk = a\tilde{b}^{1+\gamma-\lambda} \int_{C_0(1)} e^{i(\eta-\rho\eta^2)\tilde{\xi}^2} \eta^{\gamma-\lambda} \, d\eta.$$

This last integral is a standard Laplace contour integral over a fixed contour for which a full asymptotic expansion as $\tilde{\xi} \to \infty$ can be written down using the standard method summarized in section 1.1. Thus in particular we have

$$\int_{C_0(1)} e^{i(\eta-\rho\eta^2)\tilde{\xi}^2} \eta^{\gamma-\lambda} \, d\eta = e^{\frac{i\tilde{\xi}^2}{4\rho}} \Gamma(\tfrac{1}{2}) \frac{1}{(2\rho)^{\gamma-\lambda}} \frac{1}{(i\rho\tilde{b}^2 t)^{1/2}} + O\left(\frac{1}{[\tilde{b}^2 t]^{3/2}}\right)$$

as $\tilde{\xi}, \tilde{b} \to \infty$, where $\arg(i\rho t) = \pi/2$. Multiplying both sides by $a\tilde{b}^{1+\gamma-\lambda}$ yields the desired result. $\square$

It is interesting to compare the leading-order term given in the previous theorem with the leading-order term of the outer expansion as $t \to \infty$ (see Corollary 1.7.6):

$$\int_{C_0(b)} e^{ikx-i\rho k^2 t} k^{\gamma} V(k) \, dk \sim \frac{\sqrt{\pi}}{\sqrt{i\rho t}} \left(\frac{x}{2\rho t}\right)^{\gamma} \exp\left[\frac{ix^2}{4\rho t}\right] V\left(\frac{x}{2\rho t}\right)$$

as $t \to \infty$ and in particular when $b = x/t \geq \epsilon > 0$. This has been proved under assumptions on $V(k)$ which are consistent with the asymptotic condition we are assuming in this section. This asymptotic condition on $V(k)$ becomes relevant if $b \to \infty$ and $t \to \infty$. In this case the leading-order behavior is

$$\int_{C_0(b)} e^{ikx-i\rho k^2 t} k^{\gamma} V(k) \, dk \sim \frac{a\sqrt{\pi}}{\sqrt{i\rho t}} \left(\frac{x}{2\rho t}\right)^{\gamma-\lambda} \exp\left[\frac{ix^2}{4\rho t} - \frac{iy_0 x}{2\rho t}\right].$$

If we cause $t \to \infty$ in the leading-order term in Theorem 4.2.1 we find essential agreement, i.e. we have matching. The main difference between the two leading-order terms is a phase shift of magnitude $\frac{y_0^2}{4\rho t}$, which becomes insignificant for large t. Given the size of the error term in Theorem 4.2.1, it is clear that it is immaterial whether we write $(\frac{x-y_0}{2\rho t})^{\gamma-\lambda}$ or $(\frac{x}{2\rho t})^{\gamma-\lambda}$. Thus, under these assumptions on $V(k)$, the decay estimate (see section 3.1)

$$\left| \int_{C_0(b)} e^{ikx-i\rho k^2 t} k^{\gamma} V(k) \, dk \right| \leq C \frac{(1+|\xi|)^{\gamma}}{t^{(1+\gamma)/2}(1+|b|)^{\lambda}},$$

where $C > 0$ is a constant independent of $x \geq 0, t \geq 1$, is sharp.

4.3 The Linear Schrödinger Equation

Now we will show how our general results specialize in the case of solutions of the linear Schrödinger equation $u_t - i\rho u_{xx} = 0$. We assume $u(x,0) = u_0(x)$, where the initial condition u_0 is of the form $u_0 = |D|^\gamma(v_0 + \mathcal{H}v_1)$, where v_0 and v_1 are (at least integrable) functions with compact support. In terms of the integral we have studied

$$I(x,t;r,\rho,\gamma,V) = \int_0^\infty e^{ikx - i\rho k^r t} k^\gamma V(k)\, dk$$

we can express the solution $u(x,t)$ as

$$u(x,t) = \frac{1}{2\pi}\left[I(x,t;2,\rho,\gamma,\hat{v}_0 - i\hat{v}_1) + I(-x,t;2,\rho,\gamma,\hat{v}_0^{\checkmark} + i\hat{v}_1^{\checkmark})\right]$$

where $\phi^{\checkmark}(k) = \phi(-k)$. We need to assume that both of the functions $\hat{v}_0(k) - i\hat{v}_1(k)$ and $\hat{v}_0(-k) + i\hat{v}_1(-k)$ satisfy the hypotheses on $V(k)$ enumerated in section 3.2. However, in the present context it is more natural to impose conditions directly on the functions $v_+(x) = v_0(x) - iv_1(x)$ and $v_-(x) = v_0(-x) + iv_1(-x)$. Even though we sacrifice a bit of generality by doing this, we will gain a degree of concreteness, and we will be able to assert that our results are sharp. Thus we will make the standard assumptions on $v_\pm(x) = v_0(\pm x) \mp iv_1(\pm x)$ (see the end of section 1.5 where these standard assumptions are carefully stated). As a consequence of this assumption we have for both choices of the function v the existence of a nonzero value of the limit

$$\lim_{l\to\infty} l^{n+\lambda_n^0(v)}|V^{(n)}(l)|, \qquad V(k) = \hat{v}(k), \qquad l \in \mathbb{R},$$

for all integers $n \geq 0$, where the decay parameter $\lambda_n^0(v)$ also satisfies the relation $0 \leq 1 + \lambda_{n+1}^0(v) - \lambda_n^0(v) \leq 1$ for all $n \geq 0$.

Applying Theorem 3.2.1A(1) and Theorem 3.2.1B(1) we obtain the following.

Theorem 4.3.1. *Suppose $\rho > 0$ and $\gamma > -1$. Suppose $v_\pm(x) = v_0(\pm x) \mp iv_1(\pm x)$ satisfy the standard assumptions, and $\lambda_n^0(v_\pm)$ are the associated decay parameters, as defined above. Define $v = v_0 + \mathcal{H}v_1$, $b = x/t$, and $\xi = xt^{-1/2}$. Suppose $A(\xi; 2, \rho, \gamma + n, \tilde{C}(\xi))$ and $A_n(\xi; 2, \rho, \gamma, C_0(\xi))$ are defined as in sections 2.1 and 3.1. (See section 4.1 for expressions for these in terms of more familiar special functions.) If $N \geq 0$ is an even integer and $\pm x \geq 0$ then define*

$$W_N(x,t) = \frac{(1 + |\xi|)^\gamma}{t^{\frac{1+\gamma+N}{2}}(1 + |b|)^{N+\lambda_N^0(v_\pm)}}.$$

If $N \geq 1$ is an odd integer and $\pm x \geq 0$ then define

$$W_N(x,t) = \frac{(1 + |\xi|)^\gamma}{t^{\frac{1+\gamma+N}{2}}(1 + |b|)^{N+\lambda_N^0(v_\pm)}}.$$
$$\cdot \left[\frac{|\gamma|}{(1 + |\xi|)} + \frac{1}{t^{\frac{1}{2}}(1 + |b|)^{1+\lambda_{N+1}^0(v_\pm) - \lambda_N^0(v_\pm)}}\right].$$

(1) *For all $\pm x \geq 0$ and $t \geq 1$ and all integers $N \geq 0$ we have*

$$u(x,t) = \frac{1}{2\pi} \sum_{n=0}^{N-1} \frac{1}{n! t^{(1+\gamma+n)/2}} \left\{ (\pm 1)^n \hat{v}^{(n)} \left(\frac{x}{2\rho t} \right) A_n(\pm \xi; 2, \rho, \gamma, C_0(\pm \xi)) \right.$$

$$\left. + [\hat{v}^{(n)}(0^+) - e^{\mp i \pi \gamma} \hat{v}^{(n)}(0^-)] A(\xi; 2, \rho, \gamma + n, \tilde{C}(\xi)) \right\}$$

$$+ O\left([t^{1/2}(1 + |\xi|)]^{-(1+\gamma+N)} \right) + O\left(W_N(x,t) \right).$$

(2) *If $\sin(\frac{\pi}{2}\gamma)\hat{v}_0(k) \mp \cos(\frac{\pi}{2}\gamma)\hat{v}_1(k) = 0$ for all k then for all $\pm x \geq 0$ and $t \geq 1$ and all integers $N \geq 0$ we have*

$$u(x,t) = \frac{1}{2\pi} \sum_{n=0}^{N-1} \frac{(\pm 1)^n}{n! t^{(1+\gamma+n)/2}} \hat{v}^{(n)} \left(\frac{x}{2\rho t} \right) A_n(\pm \xi; 2, \rho, \gamma, C_0(\pm \xi))$$

$$+ O\left(W_N(x,t) \right).$$

Remarks. (1) If $\gamma = 0$ then, as we noted in section 4.1, $A_n(\xi; 2, \rho, \gamma, C_0(\xi)) = 0$ for all $\xi \geq 0$ and $n \geq 1$ odd. This vanishing is reflected in the form of the error term $W_N(x,t)$ when N is odd (notice the $|\gamma|$ in the numerator of the first term).
(2) Another way to write $\hat{v}^{(n)}(0^+) - e^{\mp i \pi \gamma} \hat{v}^{(n)}(0^-)$ is

$$[1 - e^{\mp i \pi \gamma}] \hat{v}_0^{(n)}(0) - i[1 + e^{\mp i \pi \gamma}] \hat{v}_1^{(n)}(0)$$

$$= \pm 2i e^{\mp i \frac{\pi}{2} \gamma} \left[\sin(\tfrac{\pi}{2}\gamma) \hat{v}_0^{(n)}(0) \mp \cos(\tfrac{\pi}{2}\gamma) \hat{v}_1^{(n)}(0) \right].$$

This quantity will vanish under the conditions of part (2).

From the point of view of applications the greatest interest is focussed on the leading-order term. This term can already be expressed in terms of special functions, as we have shown in section 4.1. But we will introduce one more special function which is related to the fundamental solution of the linear Schrödinger equation $u_t - \frac{i}{2} u_{xx} = 0$, $u(x,t) = t^{-1/2} G(\xi)$, $G(\xi) = \frac{1}{\sqrt{2\pi i}} e^{i \xi^2 / 2}$. Since $\hat{G}(k) = e^{-ik^2/2}$ we have that

$$[|D|^\gamma P^\pm G](\xi) = \frac{1}{2\pi} \int_0^\infty e^{\pm i \omega \xi - i \omega^2 / 2} \omega^\gamma \, d\omega = \frac{e^{-i \frac{\pi}{4}(1+\gamma)}}{2\pi} F_0(2, 1 + \gamma; \pm i \xi e^{-i\pi/4}),$$

where $P^\pm = \frac{1}{2}(1 \pm i \mathcal{H})$. Thus $[|D|^\gamma P^- G](\xi) = [|D|^\gamma P^+ G](-\xi)$. We can express our usual special functions in terms of this new function

$$A(\xi; 2, \rho, \gamma, \tilde{C}(\xi)) = \frac{2\pi}{(2\rho)^{\frac{1+\gamma}{2}}} [|D|^\gamma P^+ G] \left(\frac{\xi}{\sqrt{2\rho}} \right), \qquad \xi \leq 0,$$

$$A(\xi; 2, \rho, \gamma, \tilde{C}(\xi)) = \frac{2\pi e^{i\pi(1+\gamma)}}{(2\rho)^{\frac{1+\gamma}{2}}} [|D|^\gamma P^- G] \left(\frac{\xi}{\sqrt{2\rho}} \right), \qquad \xi \geq 0,$$

$$A(\xi; 2, \rho, \gamma, C_0(\xi)) = \frac{2\pi}{(2\rho)^{\frac{1+\gamma}{2}}} \left\{ [|D|^\gamma P^+ G] \left(\frac{\xi}{\sqrt{2\rho}} \right) + e^{i\pi\gamma} [|D|^\gamma P^- G] \left(\frac{\xi}{\sqrt{2\rho}} \right) \right\},$$

the last equation holding when $\xi \geq 0$. From this we can extract the leading-order asymptotics of this function.

$$[|D|^\gamma P^+ G](\xi) = \frac{\Gamma(1+\gamma)}{2\pi}\left(\frac{i}{\xi}\right)^{1+\gamma}[1+O(\xi^{-2})], \qquad \xi \to -\infty,$$

$$[|D|^\gamma P^+ G](\xi) = \frac{\xi^\gamma}{\sqrt{2\pi i}}e^{i\xi^2/2}[1+O(\xi^{-2})] + \frac{\Gamma(1+\gamma)}{2\pi}\left(\frac{i}{\xi}\right)^{1+\gamma}[1+O(\xi^{-2})],$$

the last equation holding when $\xi \to \infty$. Using this function we can rewrite the leading-order term of the expansion of $u(x,t)$:

$$u(x,t) = \left\{\hat{v}\left(\frac{x_+}{2\rho t}\right) + e^{i\pi\gamma}\left[\hat{v}\left(\frac{x_-}{2\rho t}\right) - \hat{v}(0^-)\right]\right\}\frac{1}{(2\rho t)^{\frac{1+\gamma}{2}}}[|D|^\gamma P^+ G]\left(\frac{\xi}{\sqrt{2\rho}}\right)$$

$$+ \left\{\hat{v}\left(\frac{x_-}{2\rho t}\right) + e^{i\pi\gamma}\left[\hat{v}\left(\frac{x_+}{2\rho t}\right) - \hat{v}(0^+)\right]\right\}\frac{1}{(2\rho t)^{\frac{1+\gamma}{2}}}[|D|^\gamma P^- G]\left(\frac{\xi}{\sqrt{2\rho}}\right)$$

$$+ O(W_1(x,t)) + O([t^{1/2}(1+|\xi|)]^{-(2+\gamma)}),$$

where $x_+ = \max\{0,x\}$, $x_- = \min\{0,x\}$, and $W_1(x,t)$ is defined in the statement of Theorem 4.3.1. In the above we assume $\hat{v}(l_+) = \hat{v}(0^+)$ when $l \leq 0$ and $\hat{v}(l_-) = \hat{v}(0^-)$ when $l \geq 0$. Since $G = P^+G + P^-G$ and $\mathcal{H}G = -iP^+G + iP^-G$ we can also rewrite this leading-order term in the following manner.

$$u(x,t) = \frac{1}{(2\rho t)^{\frac{1+\gamma}{2}}}\left\{\hat{v}_0\left(\frac{x}{2\rho t}\right)[|D|^\gamma G]\left(\frac{\xi}{\sqrt{2\rho}}\right) + \hat{v}_1\left(\frac{x}{2\rho t}\right)[|D|^\gamma \mathcal{H}G]\left(\frac{\xi}{\sqrt{2\rho}}\right)\right.$$

$$+ \left[(1 - e^{i\pi\gamma})\left[\hat{v}_0(0) - \hat{v}_0\left(\frac{x}{2\rho t}\right)\right]\right.$$

$$\left.+ i\,\mathrm{sgn}(x)(1 + e^{i\pi\gamma})\left[\hat{v}_1(0) - \hat{v}_1\left(\frac{x}{2\rho t}\right)\right]\right][|D|^\gamma P^- G]\left(\frac{|\xi|}{\sqrt{2\rho}}\right)\right\}$$

$$+ O(W_1(x,t)) + O([t^{1/2}(1+|\xi|)]^{-(2+\gamma)}).$$

Under the assumptions of part (2) of Theorem 4.3.1, only the first line survives (and the first error term). When $\gamma = 0$ and $v_1 \equiv 0$ this agrees with the leading-term given by Ablowitz-Segur on page 360 of [AblSe2]. The first line of the leading-order term arises immediately from a naive application of the method of stationary phase to the integral defining $u(x,t)$. The rest of the leading term (lines two and three) may or may not be significant, depending on the relative size of γ, $\lambda_0^0(v_\pm)$, and on the asymptotic region (how x varies in relation to t).

Even though this leading-order term is uniformly valid for all $x \in \mathbb{R}$ as $t \to \infty$, it is nevertheless fairly complex and involves special functions whose asymptotics are not very familiar. Thus we will supplement this uniformly valid leading-term with simpler expressions valid in the inner and outer regions. Suppose $\zeta = xt^{-2/3}$, and $0 < \epsilon < 1$. The inner region is defined by $|\zeta| \leq \epsilon^{-1}$. In the inner region we have

$$
u(x,t) = \frac{1}{(2\rho t)^{\frac{1+\gamma}{2}}} \left\{ \hat{v}(0^+)[|D|^\gamma P^+ G]\left(\frac{\xi}{\sqrt{2\rho}}\right) + \hat{v}(0^-)[|D|^\gamma P^- G]\left(\frac{\xi}{\sqrt{2\rho}}\right) \right\}
$$
$$
+ O\left(\frac{(1+|\xi|)^{1+\gamma}}{t^{\frac{2+\gamma}{2}}}\right) + O\left(\frac{1}{t^{\frac{2+\gamma}{2}}(1+|\xi|)^{2+\gamma}}\right)
$$
$$
= \frac{1}{(2\rho t)^{\frac{1+\gamma}{2}}} \left\{ \hat{v}_0(0)[|D|^\gamma G]\left(\frac{\xi}{\sqrt{2\rho}}\right) + \hat{v}_1(0)[|D|^\gamma \mathcal{H} G]\left(\frac{\xi}{\sqrt{2\rho}}\right) \right\}
$$
$$
+ O\left(\frac{(1+|\xi|)^{1+\gamma}}{t^{\frac{2+\gamma}{2}}}\right).
$$

In the above we used Theorem 2.2.1, and Corollary 2.3.2.

The outer regions $R_\pm$ are defined by $|\zeta| \geq \epsilon$, $\pm x \geq 0$. If $(x,t) \in R_\pm$ then we have

$$
u(x,t) = [\hat{v}(0^+) - e^{\mp i\pi\gamma}\hat{v}(0^-)]\frac{\Gamma(1+\gamma)}{2\pi}\left(\frac{i}{x}\right)^{1+\gamma} + O\left(\frac{1+|b|^{-1}}{|x|^{2+\gamma}}\right)
$$
$$
+ \frac{1}{\sqrt{4\pi i \rho t}}\left(\frac{\pm x}{2\rho t}\right)^\gamma \exp\left(\frac{ix^2}{4\rho t}\right)\hat{v}\left(\frac{x}{2\rho t}\right) + O\left(\frac{|b|^{\gamma-2}}{t^{3/2}(1+|b|)^{\lambda_2^0(v_\pm)}}\right).
$$

The first line vanishes (including the error term) under the conditions of Theorem 4.3.1(2). In order for the above to provide a "leading-order term", we must either have $t \to \infty$, or $2 + \lambda_2^0(v_\pm) > \lambda_0^0(v_\pm)$, $t \geq 1$ and $|x| \to \infty$.

In the common case $2 + \lambda_2^0(v_\pm) = \lambda_0^0(v_\pm)$ we can use Theorem 4.2.1 to understand the behavior of $u(x,t)$ as $t \geq 1$ and $|x| \to \infty$. Recall we are making the standard assumptions on both v_+ and v_-. Using the notation established in section 1.5 (adapted to section 3.2) we are assuming that

$$
v_\pm(y) - [a_0(v_\pm) + a_1(v_\pm)\,\mathrm{sgn}(y - y_0(v_\pm))]|y - y_0(v_\pm)|^{\lambda_0^0(v_\pm)}
$$

is sufficiently smooth in a neighborhood of $y = y_0(v_\pm)$. Hence, applying Lemma 1.5.2 we have

$$
\hat{v}_\pm(l) \sim \frac{a(v_\pm)e^{-ily_0(v_\pm)}}{l^{\lambda_0^0(v_\pm)}}, \qquad l \to \infty,
$$

where the coefficient $a(v_\pm)$ is given by

$$
a(v_\pm) = 2\Gamma(\lambda_0^0(v_\pm))[a_0(v_\pm)\cos(\tfrac{\pi}{2}\lambda_0^0(v_\pm)) - ia_1(v_\pm)\sin(\tfrac{\pi}{2}\lambda_0^0(v_\pm))].
$$

Therefore for all $t \geq 1$ we have

$$
u(x,t) = [\hat{v}(0^+) - e^{\mp i\pi\gamma}\hat{v}(0^-)]\frac{\Gamma(1+\gamma)}{2\pi}\left(\frac{i}{x}\right)^{1+\gamma} + O\left(\frac{1}{|x|^{2+\gamma}}\right)
$$
$$
+ \frac{a(v_\pm)}{\sqrt{4\pi i \rho t}}\left(\frac{\pm x - y_0(v_\pm)}{2\rho t}\right)^{\gamma - \lambda_0^0(v_\pm)} \exp\left[\frac{i(\pm x - y_0(v_\pm))^2}{4\rho t}\right]
$$
$$
+ o\left(\frac{|b|^{\gamma - \lambda_0^0(v_\pm)}}{\sqrt{t}}\right)
$$

as $\pm b \to \infty$.

There must be hundreds of papers on various sorts of Schrödinger equations, containing various asymptotic expansions for the linear constant coefficient case. See, for example Cazenave [Caz] and Constantin [Con]. But as far as we know, the above expansions are new in the generality considered here.

4.4 The Linearized Benjamin-Ono Equation

Another commonly studied example of a linear dispersive equation with $r = 2$ is the Linearized Benjamin-Ono equation, $u_t - \rho \mathcal{H} u_{xx} = 0$, where $\mathcal{H}$ is the Hilbert transform, and $\rho > 0$. As usual we assume the initial data $u(x, 0) = u_0(x)$ is of the form $u_0 = |D|^\gamma (v_0 + \mathcal{H} v_1)$ where v_0, v_1 are integrable functions with compact support. As we have seen, the solution $u(x, t)$ can be written as

$$u(x, t) = \frac{1}{2\pi} \left[\overline{I(-x, t; 2, \rho, \gamma, (\hat{v}_0 - i\hat{v}_1)^\sim)} + I(-x, t; 2, \rho, \gamma, \hat{v}_0^\vee + i\hat{v}_1^\vee) \right]$$

where $\phi^\sim(k) = \overline{\phi(\bar{k})}$ and $\phi^\vee(k) = \phi(-k)$. Note that $(\hat{v}_0 - i\hat{v}_1)^\sim = (v_0^\vee - iv_1^\vee)^\frown$ and $\hat{v}_0^\vee + i\hat{v}_1^\vee = (v_0^\vee + iv_1^\vee)^\frown$. We will therefore make standard assumptions on $\bar{v}_-$ and v_+, where $v_\pm = v_0^\vee \pm iv_1^\vee$. Our first result gives a full uniformly-valid expansion of $u(x, t)$ as $t \to \infty$.

Theorem 4.4.1. *Suppose $\rho > 0$ and $\gamma > -1$. Suppose $\bar{v}_-$ and v_+ satisfy the standard assumptions, where $\lambda_n^0(\bar{v}_-), \lambda_n^0(v_+)$ are the associated decay parameters (as in Theorem 4.3.1). Define $\lambda_N^0 = \min\{\lambda_N^0(v_+), \lambda_N^0(\bar{v}_-)\}$. Define $v = v_0 + \mathcal{H} v_1$, $b = x/t$, and $\xi = xt^{-1/2}$. Suppose $A(\xi; 2, \rho, \gamma, \tilde{C}(\xi))$ and $A_n(\xi; 2, \rho, \gamma, C_0(\xi))$ are defined as in sections 2.1 and 3.1.*

(1) *If $N \geq 0$ is an integer and $x \leq 0$ then define*

$$W_N(x, t) \overset{N \text{ even}}{=} \frac{(1 + |\xi|)^\gamma}{t^{\frac{1+\gamma+N}{2}} (1 + |b|)^{N + \lambda_N^0}},$$

$$W_N(x, t) \overset{N \text{ odd}}{=} \frac{(1 + |\xi|)^\gamma}{t^{\frac{1+\gamma+N}{2}} (1 + |b|)^{N + \lambda_N^0}} \left[\frac{|\gamma|}{(1 + |\xi|)} + \frac{1}{t^{1/2}(1 + |b|)^{1 + \lambda_{N+1}^0 - \lambda_N^0}} \right].$$

Then for all integers $N \geq 0$ and all $x \leq 0$, $t \geq 1$ we have

$$u(x, t) = \frac{1}{2\pi} \sum_{n=0}^{N-1} \frac{1}{n! t^{(1+\gamma+n)/2}} \left\{ \hat{v}^{(n)}(0^+) \overline{A(-\xi; 2, \rho, \gamma + n, \tilde{C}(-\xi))} \right.$$

$$+ (-1)^n \hat{v}^{(n)}(0^-) A(-\xi; 2, \rho, \gamma + n, \tilde{C}(-\xi))$$

$$+ \hat{v}^{(n)} \left(\frac{-x}{2\rho t} \right) \overline{A_n(-\xi; 2, \rho, \gamma, C_0(-\xi))}$$

$$\left. + (-1)^n \hat{v}^{(n)} \left(\frac{x}{2\rho t} \right) A_n(-\xi; 2, \rho, \gamma, C_0(-\xi)) \right\}$$

$$+ O\left([t^{1/2}(1 + |\xi|)]^{-(1+\gamma+N)} \right) + O(W_N(x, t)).$$

(2) *For all integers $N \geq 0$ and all $x \geq 0$, $t \geq 1$ we have*

$$u(x,t) = \frac{1}{2\pi} \sum_{n=0}^{N-1} \frac{1}{n! t^{(1+\gamma+n)/2}} \left\{ \hat{v}^{(n)}(0^+)\overline{A(-\xi; 2, \rho, \gamma + n, \tilde{C}(-\xi))} \right.$$

$$\left. + (-1)^n \hat{v}^{(n)}(0^-) A(-\xi; 2, \rho, \gamma + n, \tilde{C}(-\xi)) \right\}$$

$$+ O\left([t^{1/2}(1+|\xi|)]^{-(1+\gamma+N)}\right)$$

Remark. We can express $A(\xi; 2, \rho, \gamma, \tilde{C}(\xi))$ and $A_n(\xi; 2, \rho, \gamma, C_0(\xi))$ in terms of other (more well-known) special functions; c.f. section 4.1.

The expansion described in the previous theorem has the drawback that it is not manifestly real-valued when v_0 and v_1 are real-valued. Our next task will be to rearrange each term of our expansion so that it has this desirable property. For this purpose it will be helpful to introduce another special function, the fundamental solution $u(x,t) = t^{-1/2} G(\xi)$ of the Linearized Benjamin-Ono equation: $u_t - \frac{1}{2} \mathcal{H} u_{xx} = 0$. The G in this section differs from the G introduced in section 4.3. $\hat{G}(k) = e^{\frac{i}{2} k |k|}$. G and its Hilbert transform $\mathcal{H}G$ are given by

$$G(\xi) = \frac{1}{\sqrt{2\pi}} g\left(\frac{\xi}{\sqrt{2\pi}}\right), \qquad \mathcal{H}G(\xi) = \frac{1}{\sqrt{2\pi}} f\left(\frac{\xi}{\sqrt{2\pi}}\right),$$

in terms of the auxiliary functions g, f associated to the Fresnel integral in [AS]. Clearly

$$2[|D|^\gamma P^\pm G](\xi) = [|D|^\gamma G](\xi) \pm i[|D|^\gamma \mathcal{H}G](\xi) = \frac{e^{\pm i \frac{\pi}{4}(1+\gamma)}}{\pi} F_0(2, 1 + \gamma; \xi e^{\pm i 3\pi/4}),$$

where $P^\pm = \frac{1}{2}(1 \pm i\mathcal{H})$. Thus $[|D|^\gamma P^- G](\xi) = \overline{[|D|^\gamma P^+ G](\xi)}$. (Notice that $|D|^\gamma P^- G$ agrees with the function of the same name from section 4.3.) Using Lemma 2.1.2 we can express our usual special functions in terms of this new function

$$A(\xi; 2, \rho, \gamma, \tilde{C}(\xi)) = \frac{2\pi}{(2\rho)^{\frac{1+\gamma}{2}}} [|D|^\gamma P^- G]\left(\frac{-\xi}{\sqrt{2\rho}}\right), \qquad \xi \leq 0,$$

$$A(\xi; 2, \rho, \gamma, \tilde{C}(\xi)) = \frac{2\pi e^{i\pi(1+\gamma)}}{(2\rho)^{\frac{1+\gamma}{2}}} [|D|^\gamma P^- G]\left(\frac{\xi}{\sqrt{2\rho}}\right), \qquad \xi \geq 0,$$

$$A(\xi; 2, \rho, \gamma, C_0(\xi)) = \frac{2\pi}{(2\rho)^{\frac{1+\gamma}{2}}} \left\{ [|D|^\gamma P^- G]\left(\frac{-\xi}{\sqrt{2\rho}}\right) + e^{i\pi\gamma} [|D|^\gamma P^- G]\left(\frac{\xi}{\sqrt{2\rho}}\right) \right\},$$

the last equation holding when $\xi \geq 0$. From these formulae and Lemma 2.1.1 we can extract the asymptotics of these functions.

$$[|D|^\gamma P^+ G](\xi) = \frac{\Gamma(1+\gamma)}{2\pi} \left(\frac{i}{\xi}\right)^{1+\gamma} [1 + O(\tfrac{1}{\xi^2})], \qquad \xi \to \infty,$$

$$[|D|^\gamma P^+ G](\xi) = \frac{(-\xi)^\gamma}{\sqrt{-2\pi i}} e^{-\frac{i}{2}\xi^2} [1 + O(\tfrac{1}{\xi^2})] + \frac{\Gamma(1+\gamma)}{2\pi} \left(\frac{i}{\xi}\right)^{1+\gamma} [1 + O(\tfrac{1}{\xi^2})],$$

the last equation holding as $\xi \to \infty$. (Here we have given only the leading-order terms.) Combining these results we get the following.

Lemma 4.4.2.

(1) *As* $\xi \to \infty$ *we have*

$$[|D|^\gamma G](\xi) \sim \frac{1}{\pi} \sum_{s=0}^{\infty} \frac{\Gamma(1+\gamma+2s)}{2^s s!} \frac{\cos[\frac{\pi}{2}(1+\gamma+3s)]}{\xi^{1+\gamma+2s}}.$$

(2) *As* $\xi \to -\infty$ *we have*

$$[|D|^\gamma G](\xi) \sim \frac{1}{\pi} \left\{ \sum_{s=0}^{\infty} \Gamma(s+\tfrac{1}{2}) \binom{\gamma}{2s} 2^{s+1/2}(-\xi)^{\gamma-2s} \cos[\tfrac{1}{2}\xi^2 - \tfrac{\pi}{2}(s+\tfrac{1}{2})] \right.$$
$$\left. + \sum_{s=0}^{\infty} \frac{\Gamma(1+\gamma+2s)}{2^s s!} \frac{\cos[\frac{\pi}{2}(1+\gamma+s)]}{(-\xi)^{1+\gamma+2s}} \right\}.$$

(3) *As* $\xi \to \infty$ *we have*

$$[|D|^\gamma \mathcal{H}G](\xi) \sim \frac{1}{\pi} \sum_{s=0}^{\infty} \frac{\Gamma(1+\gamma+2s)}{2^s s!} \frac{\cos[\frac{\pi}{2}(2+\gamma+s)]}{\xi^{1+\gamma+2s}}.$$

(4) *As* $\xi \to -\infty$ *we have*

$$[|D|^\gamma \mathcal{H}G](\xi) \sim \frac{1}{\pi} \left\{ \sum_{s=0}^{\infty} \Gamma(s+\tfrac{1}{2}) \binom{\gamma}{2s} 2^{s+1/2}(-\xi)^{\gamma-2s} \cos[\tfrac{1}{2}\xi^2 - \tfrac{\pi}{2}(s-\tfrac{1}{2})] \right.$$
$$\left. + \sum_{s=0}^{\infty} \frac{\Gamma(1+\gamma+2s)}{2^s s!} \frac{\cos[\frac{\pi}{2}(2+\gamma+s)]}{(-\xi)^{1+\gamma+2s}} \right\}.$$

In order to express $A_n(\xi; 2, \rho, \gamma, C_0(\xi))$ in terms of real-valued functions we define

$$G_n(y; \gamma) \pm iH_n(y; \gamma) = \frac{e^{\pm i\frac{\pi}{4}(1+\gamma-n)}}{\pi} F_n\left(2, 1+\gamma; ye^{\pm i\frac{3\pi}{4}}\right)$$

for all $y \in \mathbb{R}$ and $n \geq 0$. Notice that $G_0(y; \gamma) = [|D|^\gamma G](y)$ and $H_0(y; \gamma) = [|D|^\gamma \mathcal{H}G](y)$. To understand how G_n and H_n are generalizations of these functions we write down the (formal) Fourier integrals (which follow from the definitions)

$$G_n(y; \gamma) = \frac{(-i)^n}{2\pi} \int_{-\infty}^{\infty} e^{iky+\frac{i}{2}k|k|} |k|^\gamma [k + \mathrm{sgn}(k)y]^n \, dk$$

$$H_n(y; \gamma) = \frac{(-i)^{n+1}}{2\pi} \int_{-\infty}^{\infty} e^{iky+\frac{i}{2}k|k|} |k|^\gamma \, \mathrm{sgn}(k)[k + \mathrm{sgn}(k)y]^n \, dk$$

for $y \in \mathbb{R}$. G_n and H_n can also be expressed in terms of fractional derivatives of G and $\mathcal{H}G$

$$G_n(y; \gamma) = \sum_{q=0}^{n} \binom{n}{q} y^{n-q} [|D|^{\gamma+q} \mathcal{H}^n G](y),$$

$$H_n(y; \gamma) = \sum_{q=0}^{n} \binom{n}{q} y^{n-q} [|D|^{\gamma+q} \mathcal{H}^{n+1} G](y).$$

When v_0 and v_1 are real-valued, so is $v = v_0 + \mathcal{H}v_1$. In that case $\overline{\hat{v}(k)} = \hat{v}(-k)$ and the combinations

$$\mu_n(k) = i^n \frac{\hat{v}^{(n)}(k) + \hat{v}^{(n)}(-k)}{2}, \qquad \eta_n(k) = i^{n+1} \frac{\hat{v}^{(n)}(k) - \hat{v}^{(n)}(-k)}{2},$$

are also real-valued when $k > 0$. Even if v_0 and v_1 are not real-valued we define $\mu_n(k)$ and $\eta_n(k)$ by the above formulae when $k > 0$, so that $i^n \hat{v}^{(n)}(\pm k) = \mu_n(k) \mp i\eta_n(k)$. In terms of v_0 and v_1 we can write

$$\mu_n(k) = \int_{-\infty}^{\infty} [\cos(kx)v_0(x) - \sin(kx)v_1(x)]x^n \, dx$$

$$\eta_n(k) = \int_{-\infty}^{\infty} [\sin(kx)v_0(x) + \cos(kx)v_1(x)]x^n \, dx.$$

Note that $\mu_n(0^+) = i^n \hat{v}_0^{(n)}(0)$ and $\eta_n(0^+) = i^n \hat{v}_1^{(n)}(0)$.

We will also need rotated versions of these quantities. Suppose $\delta \geq 0$ is a real number. Define

$$M_n(k; \delta) = \mu_n(k)\cos(\pi\delta) - \eta_n(k)\sin(\pi\delta)$$
$$= \int_{-\infty}^{\infty} [\cos(kx + \pi\delta)v_0(x) - \sin(kx + \pi\delta)v_1(x)]x^n \, dx$$

$$N_n(k; \delta) = \mu_n(k)\sin(\pi\delta) + \eta_n(k)\cos(\pi\delta)$$
$$= \int_{-\infty}^{\infty} [\sin(kx + \pi\delta)v_0(x) + \cos(kx + \pi\delta)v_1(x)]x^n \, dx.$$

Thus $M_n(k; \delta) \pm iN_n(k; \delta) = [\mu_n(k) \pm i\eta_n(k)]e^{\pm i\pi\delta}$. These quantities are clearly real-valued when v_0 and v_1 are. We will also use these quantities in section 4.6.

Now we can write down the rearranged expansion.

Theorem 4.4.3. *Assume the hypotheses of Theorem 4.4.1. Adopt the notation just described. Define $\delta = 1 + \gamma$.*

(1) *If $x \leq 0$ and $t \geq 1$ then for all integers $N \geq 0$ we have*

$$u(x,t) = \sum_{n=0}^{N-1} \frac{1}{n!(2\rho t)^{\frac{1+\gamma+n}{2}}} \left\{ \mu_n\left(\tfrac{-b}{2\rho}\right) G_n\left(\tfrac{\xi}{\sqrt{2\rho}}; \gamma\right) + \eta_n\left(\tfrac{-b}{2\rho}\right) H_n\left(\tfrac{\xi}{\sqrt{2\rho}}; \gamma\right) \right.$$

$$+ (-1)^n \left[M_n(0^+; \delta)[|D|^{\gamma+n}\mathcal{H}^n G]\left(\tfrac{-\xi}{\sqrt{2\rho}}\right) - M_n\left(\tfrac{-b}{2\rho}; \delta\right) G_n\left(\tfrac{-\xi}{\sqrt{2\rho}}; \gamma\right) \right.$$

$$\left. + N_n(0^+; \delta)[|D|^{\gamma+n}\mathcal{H}^{n+1}G]\left(\tfrac{-\xi}{\sqrt{2\rho}}\right) - N_n\left(\tfrac{-b}{2\rho}; \delta\right) H_n\left(\tfrac{-\xi}{\sqrt{2\rho}}; \gamma\right) \right] \right\}$$

$$+ O\left([t^{1/2}(1 + |\xi|)]^{-(1+\gamma+N)}\right) + O(W_N(x,t)).$$

(2) *If $x \geq 0$ and $t \geq 1$ then for all integers $N \geq 0$ we have*

$$u(x,t) = \sum_{n=0}^{N-1} \frac{1}{n!(2\rho t)^{\frac{1+\gamma+n}{2}}} \left\{ \mu_n(0^+)[|D|^{\gamma+n}\mathcal{H}^n G]\left(\frac{\xi}{\sqrt{2\rho}}\right)\right.$$

$$+ \eta_n(0^+)[|D|^{\gamma+n}\mathcal{H}^{n+1}G]\left(\frac{\xi}{\sqrt{2\rho}}\right)\Big\}$$

$$+ O\left([t^{1/2}(1+|\xi|)]^{-(1+\gamma+N)}\right).$$

Proof. (1) In terms of the special functions we have introduced we have by Lemma 3.1.3

$$i^n A_n(-\xi; 2, \rho, \gamma, C_0(-\xi)) = \frac{\pi}{(2\rho)^{\frac{1+\gamma+n}{2}}}\left\{ G_n\left(\frac{\xi}{\sqrt{2\rho}};\gamma\right) - iH_n\left(\frac{\xi}{\sqrt{2\rho}};\gamma\right)\right.$$

$$- (-1)^n e^{i\pi(1+\gamma)}\left[G_n\left(\frac{-\xi}{\sqrt{2\rho}};\gamma\right) - iH_n\left(\frac{-\xi}{\sqrt{2\rho}};\gamma\right)\right]\Big\}.$$

Since $l = \frac{-x}{2\rho t} > 0$ we have

$$\hat{v}^{(n)}(l)\overline{A_n(-\xi; 2, \rho, \gamma, C_0(-\xi))} + (-1)^n \hat{v}^{(n)}(-l)A_n(-\xi; 2, \rho, \gamma, C_0(-\xi))$$

$$= i^n \hat{v}^{(n)}(l) \cdot \overline{i^n A_n(-\xi; 2, \rho, \gamma, C_0(-\xi))} + i^n \hat{v}^{(n)}(-l) \cdot i^n A_n(-\xi; 2, \rho, \gamma, C_0(-\xi))$$

$$= \frac{\pi}{(2\rho)^{\frac{1+\gamma+n}{2}}}\left\{ [\mu_n(l) - i\eta_n(l)]\left[G_n\left(\frac{\xi}{\sqrt{2\rho}};\gamma\right) + iH_n\left(\frac{\xi}{\sqrt{2\rho}};\gamma\right)\right]\right.$$

$$+ [\mu_n(l) + i\eta_n(l)]\left[G_n\left(\frac{\xi}{\sqrt{2\rho}};\gamma\right) - iH_n\left(\frac{\xi}{\sqrt{2\rho}};\gamma\right)\right]$$

$$- (-1)^n[\mu_n(l) - i\eta_n(l)]e^{-i\pi(1+\gamma)}\left[G_n\left(\frac{-\xi}{\sqrt{2\rho}};\gamma\right) + iH_n\left(\frac{-\xi}{\sqrt{2\rho}};\gamma\right)\right]$$

$$- (-1)^n[\mu_n(l) + i\eta_n(l)]e^{i\pi(1+\gamma)}\left[G_n\left(\frac{-\xi}{\sqrt{2\rho}};\gamma\right) - iH_n\left(\frac{-\xi}{\sqrt{2\rho}};\gamma\right)\right]\Big\}$$

$$= \frac{2\pi}{(2\rho)^{\frac{1+\gamma+n}{2}}}\left\{ \mu_n(l)G_n\left(\frac{\xi}{\sqrt{2\rho}};\gamma\right) + \eta_n(l)H_n\left(\frac{\xi}{\sqrt{2\rho}};\gamma\right)\right.$$

$$- (-1)^n\left[M_n(l; 1+\gamma)G_n\left(\frac{-\xi}{\sqrt{2\rho}};\gamma\right) + N_n(l; 1+\gamma)H_n\left(\frac{-\xi}{\sqrt{2\rho}};\gamma\right)\right]\Big\}.$$

Likewise, we saw earlier in this section how to write $A(-\xi; 2, \rho, \gamma+n, \tilde{C}(-\xi))$ in terms of $|D|^{\gamma+n}P^-G$. Since $i^n P^- = \frac{1}{2}(\mathcal{H}^n - i\mathcal{H}^{n+1})$ we have

$$i^n A(-\xi; 2, \rho, \gamma+n, \tilde{C}(-\xi))$$

$$= \frac{\pi e^{i\pi(1+\gamma+n)}}{(2\rho)^{\frac{1+\gamma+n}{2}}}\left[[|D|^{\gamma+n}\mathcal{H}^n G]\left(\frac{-\xi}{\sqrt{2\rho}}\right) - i[|D|^{\gamma+n}\mathcal{H}^{n+1}G]\left(\frac{-\xi}{\sqrt{2\rho}}\right)\right].$$

So we have

$$\hat{v}^{(n)}(0^+)\overline{A(-\xi;2,\rho,\gamma+n,\tilde{C}(-\xi))} + (-1)^n\hat{v}^{(n)}(0^-)A(-\xi;2,\rho,\gamma+n,\tilde{C}(-\xi))$$

$$= i^n\hat{v}^{(n)}(0^+) \cdot \overline{i^n A(-\xi;2,\rho,\gamma+n,\tilde{C}(-\xi))}$$

$$+ i^n\hat{v}^{(n)}(0^-) \cdot i^n A(-\xi;2,\rho,\gamma+n,\tilde{C}(-\xi))$$

$$= \frac{(-1)^n\pi}{(2\rho)^{\frac{1+\gamma+n}{2}}}\left\{[\mu_n(0^+) - i\eta_n(0^+)]e^{-i\pi(1+\gamma)}.\right.$$

$$\cdot\left[[|D|^{\gamma+n}\mathcal{H}^nG]\left(\frac{-\xi}{\sqrt{2\rho}}\right) + i[|D|^{\gamma+n}\mathcal{H}^{n+1}G]\left(\frac{-\xi}{\sqrt{2\rho}}\right)\right]$$

$$+ [\mu_n(0^+) + i\eta_n(0^+)]e^{i\pi(1+\gamma)}.$$

$$\left.\cdot\left[[|D|^{\gamma+n}\mathcal{H}^nG]\left(\frac{-\xi}{\sqrt{2\rho}}\right) - i[|D|^{\gamma+n}\mathcal{H}^{n+1}G]\left(\frac{-\xi}{\sqrt{2\rho}}\right)\right]\right\}$$

$$= \frac{(-1)^n2\pi}{(2\rho)^{\frac{1+\gamma+n}{2}}}\left\{M_n(0^+;\delta)[|D|^{\gamma+n}\mathcal{H}^nG]\left(\frac{-\xi}{\sqrt{2\rho}}\right)\right.$$

$$\left. + N_n(0^+;\delta)[|D|^{\gamma+n}\mathcal{H}^{n+1}G]\left(\frac{-\xi}{\sqrt{2\rho}}\right)\right\}.$$

Combining these expressions in Theorem 4.4.1(1) yields the result.

(2) Similar to the proof of part (1) (in fact much easier).   □

It is not difficult to write down a single expression for the leading-order term of this expansion—an expression which works for both $x \leq 0$ and $x \geq 0$. Suppose $x_- = \min\{x,0\}$ and $\delta = 1+\gamma$. Then for all $x \in \mathbb{R}$ and all $t \geq 1$ we have

$$u(x,t) = \frac{1}{(2\rho t)^{\frac{1+\gamma}{2}}}\left\{\mu_0\left(\frac{-b_-}{2\rho}\right)[|D|^\gamma G]\left(\frac{\xi}{\sqrt{2\rho}}\right) + \eta_0\left(\frac{-b_-}{2\rho}\right)[|D|^\gamma\mathcal{H}G]\left(\frac{\xi}{\sqrt{2\rho}}\right)\right.$$

$$+ \left[M_0(0^+;\delta) - M_0\left(\frac{-b_-}{2\rho};\delta\right)\right][|D|^\gamma G]\left(\frac{-\xi}{\sqrt{2\rho}}\right)$$

$$\left. + \left[N_0(0^+;\delta) - N_0\left(\frac{-b_-}{2\rho};\delta\right)\right][|D|^\gamma\mathcal{H}G]\left(\frac{-\xi}{\sqrt{2\rho}}\right)\right\}$$

$$+ O\left([t^{1/2}(1+|\xi|)]^{-(2+\gamma)}\right) + O(W_1(x,t)[1 - \text{sgn}(x)]),$$

where

$$W_1(x,t) = \frac{(1+|\xi|)^\gamma}{t^{\frac{2+\gamma}{2}}(1+|b|)^{1+\lambda_1^0}}\left[\frac{|\gamma|}{(1+|\xi|)} + \frac{1}{t^{1/2}(1+|b|)^{1+\lambda_2^0-\lambda_1^0}}\right].$$

The second and third lines of the expansion vanish when $x \geq 0$ since we make the convention that $M_0\left(\frac{-b_-}{2\rho};\delta\right) = M_0(0^+;\delta)$ when $b \geq 0$, and similarly for N_0. Even when $x < 0$ these two lines may or may not be significant, depending on γ and the degree of smoothness of v. Certainly in the inner region the second and third lines do not contribute to the leading-order term, since those lines clearly vanish in the limit as $b \to 0^-$.

We will supplement the above uniformly valid leading-order term with simplified terms valid in the inner and outer regions. Define $\zeta = xt^{-2/3}$, and suppose $0 < \epsilon < 1$. The inner region is defined by the conditions $|\zeta| \leq \epsilon^{-1}, t \geq 1$. In this region we have (using Theorem 2.2.1, Corollary 2.3.2) the estimate

$$
u(x,t) = \frac{1}{(2\rho t)^{\frac{1+\gamma}{2}}} \left\{ \hat{v}_0(0)[|D|^\gamma G]\left(\frac{\xi}{\sqrt{2\rho}}\right) + \hat{v}_1(0)[|D|^\gamma \mathcal{H}G]\left(\frac{\xi}{\sqrt{2\rho}}\right) \right\}
$$
$$
+ O\left([t^{1/2}(1+|\xi|)]^{-(2+\gamma)}\right) + O\left(\frac{(1+|\xi|)^{1+\gamma}}{t^{\frac{2+\gamma}{2}}}[1-\text{sgn}(x)]\right).
$$

By Theorem 4.4.3(2) this same estimate holds in the outer region $\zeta \geq \epsilon$. However, a term involving simpler functions can be used when $\zeta \geq \epsilon$ (but not throughout the inner region):

$$
u(x,t) = \left\{ \frac{\Gamma(1+\gamma)}{\pi x^{1+\gamma}} - \frac{\Gamma(5+\gamma)\rho^2 t^2}{2\pi x^{5+\gamma}} \right\} \left[-\sin(\tfrac{\pi}{2}\gamma)\hat{v}_0(0) + \cos(\tfrac{\pi}{2}\gamma)\hat{v}_1(0) \right]
$$
$$
+ \frac{\Gamma(2+\gamma)}{\pi x^{2+\gamma}} \left[-\sin(\tfrac{\pi}{2}\gamma)i\hat{v}_0'(0) + \cos(\tfrac{\pi}{2}\gamma)i\hat{v}_1'(0) \right]
$$
$$
+ \frac{\Gamma(3+\gamma)\rho t}{\pi x^{3+\gamma}} \left[\cos(\tfrac{\pi}{2}\gamma)\hat{v}_0(0) + \sin(\tfrac{\pi}{2}\gamma)\hat{v}_1(0) \right]
$$
$$
- \frac{\Gamma(3+\gamma)}{2\pi x^{3+\gamma}} \left[-\sin(\tfrac{\pi}{2}\gamma)\hat{v}_0^{(2)}(0) + \cos(\tfrac{\pi}{2}\gamma)\hat{v}_1^{(2)}(0) \right] + O\left(\frac{1+t+|b|^{-3}}{|x|^{4+\gamma}}\right).
$$

We have included so many terms to insure that we have given the leading-order term as $t \to \infty$ throughout the outer region even in certain anomalous cases where the first line (and possibly the second as well) vanishes. The error is always bounded as we have claimed, although in certain anomalous situations it could be asymptotically smaller than we have indicated. The above result follows from Theorem 1.4.9.

In the outer region $\zeta \leq -\epsilon$, $t \to \infty$, we have a similar result:

$$
u(x,t) = \left\{ -\frac{\Gamma(1+\gamma)}{\pi(-x)^{1+\gamma}} + \frac{\Gamma(5+\gamma)\rho^2 t^2}{2\pi(-x)^{5+\gamma}} \right\} \left[\sin(\tfrac{\pi}{2}\gamma)\hat{v}_0(0) + \cos(\tfrac{\pi}{2}\gamma)\hat{v}_1(0) \right]
$$
$$
+ \frac{\Gamma(2+\gamma)}{\pi(-x)^{2+\gamma}} \left[\sin(\tfrac{\pi}{2}\gamma)i\hat{v}_0'(0) + \cos(\tfrac{\pi}{2}\gamma)i\hat{v}_1'(0) \right]
$$
$$
+ \frac{\Gamma(3+\gamma)\rho t}{\pi(-x)^{3+\gamma}} \left[-\cos(\tfrac{\pi}{2}\gamma)\hat{v}_0(0) + \sin(\tfrac{\pi}{2}\gamma)\hat{v}_1(0) \right]
$$
$$
\frac{\Gamma(3+\gamma)}{2\pi(-x)^{3+\gamma}} \left[\sin(\tfrac{\pi}{2}\gamma)\hat{v}_0^{(2)}(0) + \cos(\tfrac{\pi}{2}\gamma)\hat{v}_1^{(2)}(0) \right] + O\left(\frac{1+t+|b|^{-3}}{|x|^{4+\gamma}}\right)
$$
$$
\frac{1}{\sqrt{\pi\rho t}} \left(\frac{-x}{2\rho t}\right)^\gamma \left\{ \mu_0\left(\frac{-x}{2\rho t}\right)\cos\left(\frac{\xi^2}{4\rho} - \frac{\pi}{4}\right) - \eta_0\left(\frac{-x}{2\rho t}\right)\sin\left(\frac{\xi^2}{4\rho} - \frac{\pi}{4}\right) \right\}
$$
$$
+ O\left(\frac{|b|^{\gamma-2}}{t^{3/2}(1+|b|)^{\lambda_2^0}}\right).
$$

The expression on the fifth line could fail to decay as $b \to -\infty$, $t \geq 1$, more rapidly than the error term on the sixth line. If so then we must use Theorem 4.2.1 to obtain the true leading-order term in this limit. Recall we are assuming $\bar{v}_-$ and v_+ satisfy the standard assumptions, where $v_\pm = v_0^{\vee} \pm iv_1^{\vee}$. This means among other things that the functions

$$\overline{v_-(y)} - [a_0(\bar{v}_-) + a_1(\bar{v}_-)\operatorname{sgn}(y - y_0(\bar{v}_-))]|y - y_0(\bar{v}_-)|^{\lambda_0^0(\bar{v}_-)}$$
$$v_+(y) - [a_0(v_+) + a_1(v_+)\operatorname{sgn}(y - y_0(v_+))]|y - y_0(v_+)|^{\lambda_0^0(v_+)}$$

are sufficiently smooth in a neighborhood of $y = y_0(\bar{v}_-)$ and $y = y_0(v_+)$ respectively. Define $y_0(v_-) = y_0(\bar{v}_-)$, $\lambda_0^0(v_-) = \lambda_0^0(\bar{v}_-)$, $a_j(v_-) = \overline{a_j(\bar{v}_-)}$, $j = 0, 1$, and

$$a(v_\pm) = 2\Gamma(\lambda_0^0(v_\pm))\left[a_0(v_\pm)\cos(\tfrac{\pi}{2}\lambda_0^0(v_\pm)) \mp ia_1(v_\pm)\sin(\tfrac{\pi}{2}\lambda_0^0(v_\pm))\right].$$

The non-degeneracy conditions inherent in our standard assumptions imply that $a(v_-) \neq 0$ and $a(v_+) \neq 0$. Then by Theorem 4.2.1 the fifth and sixth lines of the above expression for $u(x,t)$ should be replaced by

$$\frac{a(v_-)}{\sqrt{-4\pi i\rho t}}\left(\frac{-x - y_0(v_-)}{2\rho t}\right)^{\gamma - \lambda_0^0(v_-)} \exp\left[\frac{-i(-x - y_0(v_-))^2}{4\rho t}\right]$$
$$+ \frac{a(v_+)}{\sqrt{4\pi i\rho t}}\left(\frac{-x - y_0(v_+)}{2\rho t}\right)^{\gamma - \lambda_0^0(v_+)} \exp\left[\frac{i(-x - y_0(v_+))^2}{4\rho t}\right] + o\left(\frac{|b|^{\gamma - \lambda_0^0}}{\sqrt{t}}\right),$$

where $\lambda_0^0 = \min\{\lambda_0^0(v_-), \lambda_0^0(v_+)\}$, so that this modified expression for $u(x,t)$ will be the leading-order term in the limit $b \to -\infty$, $t \geq 1$. These terms cannot cancel. Notice that when v_0 and v_1 are real-valued, we have $v_- = \overline{v_+}$, and hence $y_0(v_+) = y_0(v_-)$, $\lambda_0^0(v_+) = \lambda_0^0(v_-)$, and $a(v_-) = \overline{a(v_+)}$. In that case the above two lines describe a real-valued quantity. It can be rewritten in the form:

$$\frac{2\Gamma(\lambda_0^0)}{\sqrt{\pi\rho t}}\left(\frac{-x - y_0}{2\rho t}\right)^{\gamma - \lambda_0^0} \cdot$$

$$\cdot \left\{ [a_0(v_0^{\vee})\cos(\tfrac{\pi}{2}\lambda_0^0) + a_1(v_1^{\vee})\sin(\tfrac{\pi}{2}\lambda_0^0)]\cos\left[\frac{(-x - y_0)^2}{4\rho t} - \frac{\pi}{4}\right] \right.$$

$$+ \left. [-a_0(v_1^{\vee})\cos(\tfrac{\pi}{2}\lambda_0^0) + a_1(v_0^{\vee})\sin(\tfrac{\pi}{2}\lambda_0^0)]\sin\left[\frac{(-x - y_0)^2}{4\rho t} - \frac{\pi}{4}\right] \right\}$$

$$+ o\left(\frac{|b|^{\gamma - \lambda_0^0}}{\sqrt{t}}\right)$$

where $y_0 = y_0(v_+) = y_0(v_-)$ and $a_j(v_\pm) = a_j(v_0^{\vee}) \pm ia_j(v_1^{\vee})$, $j = 0, 1$.

4.5 The Case $r = 3$

In this section we will study the special functions appearing in our uniform expansion of $I(x,t)$ in the case $r = 3$. In particular our goal will be first of all to establish the relation between our foundational special functions $F_n, \mathbb{F}_n, \mathbb{F}_n^c$ and previously studied real-valued special functions such as Airy functions and their relatives. Then we wish to re-express the special functions A_n appearing in our asymptotic expansions in terms of those real-valued functions, or generalizations of them. We will also study in this section the relation between our uniform expansion of $I(x,t)$, c.f. Theorem 3.2.1A(2) and Theorem 3.2.1B(1), and another different uniformly valid expansion which can in this case be derived by a method discussed by Bleistein (see section 3.1). It was the desire to derive an analog of Bleistein's expansion for general r which motivated much of our work on uniformly-valid expansions, but we have not succeeded in generalizing the integration by parts approach to non-integral r. However, we have produced an uniformly valid expansion for general r by a different method, and so it seems desirable to compare our expansion to the extensively studied Bleistein expansion, whenever that expansion is known to exist; in particular in the case $r = 3$. For this purpose it will be necessary to derive recursions satisfied by $A(\xi; 3, \rho, \gamma + n, \hat{C})$ and $A_n(\xi; 3, \rho, \gamma, C_j)$ in the variable n.

First we note that $F_0(3, 1; y) = \pi \mathrm{Hi}(y) = \int_0^\infty e^{-\frac{1}{3}\sigma^3 + y\sigma} \, d\sigma$, where Hi is *Scorer's function*, c.f. page 332, [O1]. Hence when $\gamma \geq 0$ is an integer, we have $F_0(3, 1 + \gamma; y) = \pi \mathrm{Hi}^{(\gamma)}(y)$ for all $y \in \mathbb{R}$. Thus for general real $\gamma > -1$ and real y we interpret $F_0(3, 1 + \gamma; y)$ as some sort of fractional derivative (or integral) of $\mathrm{Hi}(y)$; in that case when we write $\mathrm{Hi}^{(\gamma)}(y)$, we mean simply $\frac{1}{\pi} F_0(3, 1 + \gamma; y)$. Since $\mathrm{Hi}(y) \sim \frac{1}{\sqrt{\pi} y^{1/4}} e^{\frac{2}{3} y^{3/2}}$ as $y \to \infty$ we will not attempt to interpret this type of fractional derivative as a Fourier multiplier operator applied to Hi. Thus in Theorem 3.2.1B(1) we have

$$A(\xi; 3, \rho, \gamma + n, \hat{C}(\xi)) = \left(\frac{e^{i\frac{\pi}{2}}}{(3\rho)^{1/3}} \right)^{1+\gamma+n} \pi \mathrm{Hi}^{(\gamma+n)} \left(\frac{-\xi}{(3\rho)^{1/3}} \right)$$

for all $\xi \geq 0$.

This function satisfies recursion relations in the argument $n \geq 0$. A simple integration by parts shows that

$$F_0(3, 1 + \gamma + n; y) = y F_0(3, 1 + \gamma + n - 2; y) + (\gamma + n - 2) F_0(3, 1 + \gamma + n - 3; y)$$

when $n \geq 3$. This relationship is equivalent to the third order differential equation satisfied by $\mathrm{Hi}^{(\gamma)}(y)$:

$$\left(\frac{d}{dy} \right)^3 \mathrm{Hi}^{(\gamma)}(y) = \frac{d}{dy} \left(y \mathrm{Hi}^{(\gamma)}(y) \right) + \gamma \mathrm{Hi}^{(\gamma)}(y).$$

Thus we can write $F_0(3, 1 + \gamma + n; y) = \sum_{j=0}^2 \tilde{p}_n^j(y; \gamma) F_0(3, 1 + \gamma + j; y)$, where the functions $\tilde{p}_n^j(y; \gamma)$ satisfy the recursion

$$\tilde{p}_n^j(y; \gamma) = y \tilde{p}_{n-2}^j(y; \gamma) + (\gamma + n - 2) \tilde{p}_{n-3}^j(y; \gamma)$$

for all $n \geq 3$ and $j = 0, 1, 2$. The following is a table of the first few of these functions.

n	$\tilde{p}_n^0(y; \gamma)$	$\tilde{p}_n^1(y; \gamma)$	$\tilde{p}_n^2(y; \gamma)$	$\tilde{q}_n(y, z; \gamma)$
0	1	0	0	0
1	0	1	0	0
2	0	0	1	0
3	$1 + \gamma$	y	0	1
4	0	$2 + \gamma$	y	z
5	$(1 + \gamma)y$	y^2	$3 + \gamma$	$y + z^2$

Thus we obtain the reduction formula

$$A(\xi; 3, \rho, \gamma + n, \tilde{C}(\xi)) = \sum_{j=0}^{2} \left(\frac{e^{i\frac{\pi}{2}}}{(3\rho)^{1/3}} \right)^{n-j} \tilde{p}_n^j \left(\frac{-\xi}{(3\rho)^{1/3}}; \gamma \right) A(\xi; 3, \rho, \gamma + j, \tilde{C}(\xi)),$$

for all integers $n \geq 0$ and $\xi > 0$. When $\gamma = 0$ this picture simplifies a bit further, since Hi satisfies the inhomogeneous second order differential equation $\mathrm{Hi}^{(2)}(y) = y\mathrm{Hi}(y) + \frac{1}{\pi}$.

It is clear that $\mathbb{F}_0(3, 1; y, z) = \int_0^z e^{-\frac{1}{3}\sigma^3 + y\sigma} \, d\sigma$ should be called an *incomplete Scorer's function*. $\mathbb{F}_0(3, 1 + \gamma; y, z) = \int_0^z e^{-\frac{1}{3}\sigma^3 + y\sigma} \sigma^\gamma \, d\sigma$ is an incomplete form of the integral defining $\pi\mathrm{Hi}^{(\gamma)}(y)$, but not the γth derivative of the incomplete Scorer's function. We will not introduce a new special notation for this function; $\mathbb{F}_0(3, 1 + \gamma; y, z)$ will be a sufficiently good symbol for our purposes. By Theorem 3.2.1A(2) we have

$$A(\xi; 3, \rho, \gamma + n, \tilde{C}(\xi))$$
$$= \frac{e^{i\frac{\pi}{2}(1 - \gamma + n)}}{(3\rho)^{\frac{1 + \gamma + n}{3}}} e^{i\pi\gamma} \mathbb{F}_0 \left(3, 1 + \gamma + n; \frac{-\xi}{(3\rho)^{1/3}}, e^{-i\pi} \left| \frac{\xi}{(3\rho)^{1/3}} \right|^{\frac{1}{2}} \right),$$

for all $\xi < 0$. If we agree that $\arg(\xi) = -\pi$ and $-\xi$ has argument -2π, then the z argument in the above is the square root of the y argument. However, when γ is not an integer it will not be the case that $\mathbb{F}_0(3, 1 + \gamma; y, z)$ is real-valued, since $\arg z = -\pi$. But the combination $e^{i\pi\gamma} \mathbb{F}_0(3, 1 + \gamma; y, z)$ will be real-valued, so we have expressed things in terms of this combination.

As before it is possible to obtain a reduction formula. The recursion relations for $\mathbb{F}_0(3, 1 + \gamma + n; y, z)$ are inhomogeneous. Integration by parts yields

$$\mathbb{F}_0(3, 1 + \gamma + n; y, z)$$
$$= y\mathbb{F}_0(3, 1 + \gamma + n - 2; y, z) + (\gamma + n - 2)\mathbb{F}_0(3, 1 + \gamma + n - 3; y, z)$$
$$- z^{\gamma + n - 2} e^{-\frac{1}{3}z^3 + zy},$$

where $n \geq 3$. As before we write

$$\mathbb{F}_0(3, 1 + \gamma + n; y, z) = \sum_{j=0}^{2} \tilde{p}_n^j(y; \gamma) \mathbb{F}_0(3, 1 + \gamma + j; y, z) - \tilde{q}_n(y, z; \gamma) z^{1 + \gamma} e^{-\frac{1}{3}z^3 + zy}.$$

where the functions $\tilde{p}_n^j(y; \gamma)$ were defined above, and the new functions $\tilde{q}_n(y, z; \gamma)$ satisfy the inhomogeneous recursion relations

$$\tilde{q}_n(y, z; \gamma) = y\tilde{q}_{n-2}(y, z; \gamma) + (\gamma + n - 2)\tilde{q}_{n-3}(y, z; \gamma) + z^{n-3}$$

for $n \geq 3$. The first several instances of this function are in the last column of the previous table. We will only need the case where $\arg y = -2\pi$ and $z = y^{1/2} = \sqrt{|y|}e^{-i\pi}$. These relations can be used to express $A(\xi; 3, \rho, \gamma+n, \tilde{C}(\xi))$ when $\xi < 0$:

$$A(\xi; 3, \rho, \gamma + n, \tilde{C}(\xi))$$

$$= \sum_{j=0}^{2} \left(\frac{e^{i\frac{\pi}{2}}}{(3\rho)^{1/3}} \right)^{n-j} \tilde{p}_n^j \left(\frac{-\xi}{(3\rho)^{1/3}}; \gamma \right) A(\xi; 3, \rho, \gamma + j, \tilde{C}(\xi))$$

$$+ \frac{e^{i\frac{\pi}{2}(1-\gamma+n)}}{(3\rho)^{n/3}} \tilde{q}_n \left(\frac{-\xi}{(3\rho)^{1/3}}, -\left| \frac{\xi}{(3\rho)^{1/3}} \right|^{\frac{1}{2}}; \gamma \right) \left| \frac{\xi}{3\rho} \right|^{\frac{1+\gamma}{2}} \exp \left[-2\rho \left| \frac{\xi}{3\rho} \right|^{3/2} \right].$$

In the above we considered $F_0(3, 1 + \gamma; y)$ or $F_0(3, 1 + \gamma; y, z)$ where y and z were real. However, it will be helpful to express $F_0(3, 1 + \gamma; y)$ for y lying on the rays in the complex plane with $\arg y = \pm\pi/3, \pm 2\pi/3$ in terms of fractional derivatives (integrals) of the familiar Airy function

$$\text{Ai}(\xi) = \frac{1}{\pi} \int_0^{\infty} \cos(\tfrac{1}{3}\sigma^3 + \xi\sigma) \, d\sigma = \frac{1}{2\pi} \int_{-\infty}^{\infty} e^{iw\xi} e^{\frac{i}{3}w^3} \, dw$$

and its (possibly less familiar) Hilbert transform (another of Scorer's functions)

$$\text{Gi}(\xi) = \frac{1}{\pi} \int_0^{\infty} \sin(\tfrac{1}{3}\sigma^3 + \xi\sigma) \, d\sigma = \frac{1}{2\pi} \int_{-\infty}^{\infty} e^{iw\xi} (-i) \, \text{sgn}(w) e^{\frac{i}{3}w^3} \, dw.$$

It is easy enough to see that $\text{Gi} = \mathcal{H}\text{Ai}$ since $\mathcal{H} = -i \, \text{sgn}(D)$, where $D = -i\partial_x$. We denote by $P^{\pm}$ the projections onto the subspace of functions whose Fourier transform is supported on the interval from 0 to $\pm\infty$. Hence $P^{\pm} = \frac{1}{2}(1 \pm i\mathcal{H})$. The fractional differentiation (when $\gamma > 0$) or integration (when $-1 < \gamma < 0$) operator is $|D|^{\gamma}$, and is an ordinary Fourier multiplier operator, like $\mathcal{H}$ or $P^{\pm}$. We have the following result.

Lemma 4.5.1. *Suppose $\gamma > -1$ is a real number. Then for all $y \in \mathbb{R}$ we have*

$$[|D|^{\gamma} P^{\pm} \text{Ai}] \, (y) = \frac{e^{\pm i \frac{\pi}{6}(1+\gamma)}}{2\pi} F_0(3, 1 + \gamma; y e^{\pm i \frac{2\pi}{3}}),$$

$$[|D|^{\gamma} \text{Ai}] \, (y) \pm i \, [|D|^{\gamma} \text{Gi}] \, (y) = \frac{e^{\pm i \frac{\pi}{6}(1+\gamma)}}{\pi} F_0(3, 1 + \gamma; y e^{\pm i \frac{2\pi}{3}}).$$

In particular, we note that $|D|^{\gamma} P^{\pm} \text{Ai}$ are complex conjugates of one another.

Proof. Change the contour in the Fourier integrals defining the functions on the left to the ray passing out into the center of the valley. $\square$

We will also write $G_0(y;\gamma) = [|D|^\gamma\mathrm{Ai}](y)$ and $H_0(y;\gamma) = [|D|^\gamma\mathrm{Gi}](y)$ for all $y \in \mathbb{R}$. Later in this section we will define $G_n(y;\gamma)$ and $H_n(y;\gamma)$ for all $n \geq 1$.

On the other hand if $y > 0$, $\arg y = 0$, then a rotation of the contour of integration shows that

$$e^{\pm i\frac{\pi}{6}(1+\gamma)}\mathbb{F}_0^c\left(3, 1+\gamma; ye^{\pm i\frac{2\pi}{3}}, y^{1/2}e^{\pm i\frac{\pi}{3}}\right)$$

$$= \int_{\pm iy^{1/2}}^\infty \cos(\tfrac{1}{3}\sigma^3 + \sigma y)\sigma^\gamma\, d\sigma \pm i\int_{\pm iy^{1/2}}^\infty \sin(\tfrac{1}{3}\sigma^3 + \sigma y)\sigma^\gamma\, d\sigma.$$

Thus this function is closely related to *incomplete Airy functions*, which are discussed on pages 108–112 of [AM]. The functions $\int_z^\infty \cos(\tfrac{1}{3}\sigma^3 + \sigma y)\sigma^\gamma\, d\sigma$ and $\int_z^\infty \sin(\tfrac{1}{3}\sigma^3 + \sigma y)\sigma^\gamma\, d\sigma$ are real-valued when y and $z \geq 0$ are real, but not in general when they are complex. In this case we are more interested in dealing with real-valued functions than with previously studied ones. So we define $G_0^c(y;\gamma)$ and $H_0^c(y;\gamma)$ to be real-valued such that

$$G_0^c(y;\gamma) \pm iH_0^c(y;\gamma) = \frac{e^{\pm i\frac{\pi}{6}(1+\gamma)}}{\pi}\mathbb{F}_0^c\left(3, 1+\gamma; ye^{\pm i\frac{2\pi}{3}}, y^{1/2}e^{\pm i\frac{\pi}{3}}\right)$$

for $y > 0$. Later, in this section we will define $G_n^c(y;\gamma)$ and $H_n^c(y;\gamma)$ for all $n \geq 1$ and $y > 0$.

Thus in light of Lemma 2.1.2, Lemma 3.1.3 ($n = 0$), and all these definitions we have the following.

Lemma 4.5.2. *Suppose $r = 3$, $\rho > 0$ and $\gamma > -1$ are real numbers.*

(1) *If $\xi \in \mathbb{R}$ then*

$$A(\xi; 3, \rho, \gamma, \Gamma_{V_0}) = \frac{2\pi}{(3\rho)^{\frac{1+\gamma}{3}}}\left[|D|^\gamma P^-\mathrm{Ai}\right]\left(\frac{-\xi}{(3\rho)^{\frac{1}{3}}}\right).$$

(2) *If $\xi > 0$ then*

$$A(\xi; 3, \rho, \gamma, C_0(\xi)) = \frac{2\pi}{(3\rho)^{\frac{1+\gamma}{3}}}\left[|D|^\gamma P^-\mathrm{Ai}\right]\left(\frac{-\xi}{(3\rho)^{\frac{1}{3}}}\right) - \frac{e^{i\frac{\pi}{2}(1+\gamma)}}{(3\rho)^{\frac{1+\gamma}{3}}}\pi\mathrm{Hi}^{(\gamma)}\left(\frac{-\xi}{(3\rho)^{\frac{1}{3}}}\right).$$

(3) *If $\xi < 0$ then*

$$A(\xi; 3, \rho, \gamma, C_{-1}^+(\xi)) = \frac{\pi}{(3\rho)^{\frac{1+\gamma}{3}}}\left[G_0^c\left(\frac{-\xi}{(3\rho)^{1/3}};\gamma\right) - iH_0^c\left(\frac{-\xi}{(3\rho)^{1/3}};\gamma\right)\right].$$

Remark. The functions $G_0^c(y;\gamma)$ and $H_0^c(y;\gamma)$ decay super-exponentially as $y \to \infty$, as can be seen from Lemma 3.1.2 and part (3) of the above. However, for some purposes it will be useful to write them as the difference of two more slowly decaying functions. From the relation

$$F_0(3, 1+\gamma; y) = \mathbb{F}_0(3, 1+\gamma; y, y^{1/2}) + \mathbb{F}_0^c(3, 1+\gamma; y, y^{1/2})$$

mentioned in section 3.1, and a change of contour in the integral $\mathbb{F}_0(3, 1 + \gamma; y, y^{1/2})$ we have

$$G_0^c(y; \gamma) = [|D|^\gamma \mathrm{Ai}](y) - \sin(\tfrac{\pi}{2}\gamma)\frac{e^{i\pi\gamma}}{\pi}\mathbb{F}_0(3, 1 + \gamma; y, y^{1/2}e^{-i\pi}),$$

$$H_0^c(y; \gamma) = [|D|^\gamma \mathrm{Gi}](y) + \cos(\tfrac{\pi}{2}\gamma)\frac{e^{i\pi\gamma}}{\pi}\mathbb{F}_0(3, 1 + \gamma; y, y^{1/2}e^{-i\pi}),$$

for all $y > 0$.

The first result of this last Lemma is useful in representing all of the terms of the inner expansion. The second and third results concern the leading term of the uniform expansion over the contours $C_0(b)$ and $C_{-1}^+(b)$. To describe the higher order terms we must use Lemma 3.1.3, which in the case $\xi > 0$ becomes

$$A_n(\xi; 3, \rho, \gamma, C_0(\xi)) = \frac{1}{(3\rho)^{\frac{1+\gamma+n}{3}}}\left[e^{-i\frac{\pi}{6}(1+\gamma+n)}F_n\left(3, 1 + \gamma; \frac{(\xi e^{i\pi})e^{-i\frac{2\pi}{3}}}{(3\rho)^{1/3}}\right)\right.$$
$$\left. - e^{i\frac{\pi}{2}(1+\gamma+n)}F_n\left(3, 1 + \gamma; \frac{\xi e^{-i\pi}}{(3\rho)^{1/3}}\right)\right].$$

In the above, the positive number ξ has argument 0. We write $\xi e^{\pm i\pi}$ instead of $-\xi$ because of the fractional powers which are involved in F_n for $n \geq 1$. It is convenient and natural to define for all $y > 0$ and all $n \geq 0$ the real-valued functions G_n and H_n such that

$$G_n(y; \gamma) \pm iH_n(y; \gamma) \overset{\mathrm{def}}{=} \frac{e^{\pm i\frac{\pi}{6}(1+\gamma-2n)}}{\pi}F_n\left(3, 1 + \gamma; ye^{\pm i\frac{2\pi}{3}}\right),$$

$$G_n(-y; \gamma) \pm iH_n(-y; \gamma) \overset{\mathrm{def}}{=} \frac{e^{\pm i\frac{\pi}{6}(1+\gamma-2n)}}{\pi}F_n\left(3, 1 + \gamma; (ye^{\mp i\pi})e^{\pm i\frac{2\pi}{3}}\right).$$

Notice that when $n = 0$ in the above, we obtain the definitions $G_0(y; \gamma) = [|D|^\gamma \mathrm{Ai}](y)$ and $H_0(y; \gamma) = [|D|^\gamma \mathrm{Gi}](y)$ for all $y \in \mathbb{R}$ we made earlier. These functions G_n and H_n obviously differ from the functions with the same names discussed in section 4.4. To see that these definitions are natural generalizations of the case $n = 0$ it is helpful to write down the (formal) Fourier integrals

$$G_n(y; \gamma) = \frac{(-i)^n}{2\pi}\int_{-\infty}^{\infty}e^{iky+\frac{i}{3}k^3}|k|^\gamma(k - iy^{1/2})^n\,dk$$

$$H_n(y; \gamma) = \frac{(-i)^{n+1}}{2\pi}\int_{-\infty}^{\infty}e^{iky+\frac{i}{3}k^3}|k|^\gamma\,\mathrm{sgn}(k)(k - iy^{1/2})^n\,dk$$

when $y > 0$ and

$$G_n(y; \gamma) = \frac{(-i)^n}{2\pi}\int_{-\infty}^{\infty}e^{iky+\frac{i}{3}k^3}|k|^\gamma(k - \mathrm{sgn}(k)|y|^{1/2})^n\,dk$$

$$H_n(y; \gamma) = \frac{(-i)^{n+1}}{2\pi}\int_{-\infty}^{\infty}e^{iky+\frac{i}{3}k^3}|k|^\gamma\,\mathrm{sgn}(k)(k - \mathrm{sgn}(k)|y|^{1/2})^n\,dk$$

when $y < 0$. These can be obtained from the definitions by changing contours. Increasing n increases the degree of vanishing in the vicinity of the saddle point(s), as might be expected. When $y < 0$ the steepest descent contour passes through both saddle points, and both feature prominently in these integrals. When $y > 0$ there are still two saddle points, but the steepest descent contour passes through only the one in the upper-half plane; it is exactly this saddle point which appears in these integrals. The factors of the powers of $-i$ are needed to cause the functions to be real-valued. The case $y < 0$ applies directly to the formula for A_n we wrote above. We will apply the case $y > 0$ later in this section.

When $y > 0$ and $n \geq 1$ we have

$$F_n(3, 1+\gamma; ye^{\pm i\pi}) = \int_0^\infty e^{-\frac{1}{3}\sigma^3 - y\sigma} \sigma^\gamma (\sigma \mp iy^{1/2})^n \, d\sigma$$

$$= \sum_{q=0}^n \binom{n}{q} (\mp iy^{1/2})^q F_0(3, 1+\gamma+n-q; -y).$$

Since $F_0(3, 1+\gamma+n-q; -y) = \pi \mathrm{Hi}^{(\gamma+n-q)}(-y)$ is real-valued, we see that each of the terms of this sum with q even are real-valued, and the other terms are purely imaginary. The term with $q = 0$ is $\pi \mathrm{Hi}^{(\gamma+n)}(-y)$, so we will think of both the real and imaginary parts of $F_n(3, 1+\gamma; ye^{\pm i\pi})$ as being generalized Scorer's functions. We will not introduce any new notation for these functions beyond $\Re F_n(3, 1+\gamma; ye^{i\pi})$ and $\Im F_n(3, 1+\gamma; ye^{i\pi})$. Thus

$$F_n(3, 1+\gamma; ye^{\pm i\pi}) = \Re F_n(3, 1+\gamma; ye^{i\pi}) \pm i\Im F_n(3, 1+\gamma; ye^{i\pi}).$$

These functions, together with $G_n(y; \gamma)$ and $H_n(y; \gamma)$ can be used to express $A_n(\xi; 3, \rho, \gamma, C_0(\xi))$ for $\xi > 0$ in terms of real-valued functions as follows.

$$i^n A_n(\xi; 3, \rho, \gamma, C_0(\xi)) = \frac{\pi}{(3\rho)^{\frac{1+\gamma+n}{3}}} \Big\{ G_n(-y; \gamma) - iH_n(-y; \gamma)$$

$$- \frac{(-1)^n}{\pi} e^{i\frac{\pi}{2}(1+\gamma)} \big[\Re F_n(3, 1+\gamma; ye^{i\pi}) - i\Im F_n(3, 1+\gamma; ye^{i\pi}) \big] \Big\},$$

where $y = \frac{\xi}{(3\rho)^{1/3}}$.

$F_n(3, 1+\gamma; y)$ satisfies recursion relations in the argument n; and such can be used to express similar relations for $A_n(\xi; 3, \rho, \gamma, C_0(\xi))$. First we will derive recursions for the function

$$F_n(3, 1+\gamma; y, z) \stackrel{\mathrm{def}}{=} \int_0^\infty e^{-\frac{1}{3}\sigma^3 + y\sigma} \sigma^\gamma (\sigma - z)^n \, d\sigma.$$

It is an exercise in integration by parts to show that

$$F_n(3, 1+\gamma; y, z) = -3z F_{n-1}(3, 1+\gamma; y, z) + (y - 3z^2) F_{n-2}(3, 1+\gamma; y, z)$$

$$+ [\gamma + n - 2 + z(y - z^2)] F_{n-3}(3, 1+\gamma; y, z) + z(n - 3) F_{n-4}(3, 1+\gamma; y, z)$$

for all $n \geq 3$. We write

$$F_n(3, 1 + \gamma; y, z) = \sum_{j=0}^{2} p_n^j(y, z; \gamma) F_0(3, 1 + \gamma + j; y).$$

Each of the functions $p_n^j(y, z; \gamma)$ satisfies the same recursion relation in n as $F_n(3, 1 + \gamma; y, z)$ does. The following table gives the first several values of this function.

n	$p_n^0(y, z; \gamma)$	$p_n^1(y, z; \gamma)$	$p_n^2(y, z; \gamma)$	$q_n(y, z; \gamma)$
0	1	0	0	0
1	$-z$	1	0	0
2	z^2	$-2z$	1	0
3	$1 + \gamma - z^3$	$3z^2 + y$	$-3z$	1
4	$-4z(1 + \gamma)$ $+z^4$	$-4z(z^2 + y)$ $+2 + \gamma$	$6z^2 + y$	$-3z$
5	$(10z^2 + y)(1 + \gamma)$ $-z^5$	$5z^4 + 10z^2 y + y^2$ $-5z(\gamma + 2)$	$-5z(2z^2 + y)$ $+3 + \gamma$	$y + 6z^2$

Define $p_n^j(y; \gamma) = p_n^j(y, y^{1/2}; \gamma)$. It is not difficult to check by induction that the functions $p_n^j(y; \gamma)$, $j = 0, 1, 2$, $n \geq 0$, possess the symmetry relation

$$p_n^j(y e^{i\frac{4\pi}{3}}) = e^{i\frac{2\pi}{3}(n-j)} p_n^j(y).$$

Thus we obtain the following reduction formula

$$A_n(\xi; 3, \rho, \gamma, C_0(\xi)) = \sum_{j=0}^{2} \left(\frac{e^{-i\frac{\pi}{2}}}{3\rho} \right)^{\frac{n-j}{3}} p_n^j \left(\frac{\xi e^{i\pi/3}}{(3\rho)^{1/3}}; \gamma \right) A(\xi; 3, \rho, \gamma + j, C_0(\xi)),$$

valid for all $\xi > 0$. This formula simplifies somewhat when $\gamma = 0$ because then $\text{Ai}''(y) = y\text{Ai}(y)$ and $\text{Gi}''(y) = y\text{Gi}(y) - \frac{1}{\pi}$.

When $\xi < 0$ the higher order terms of the uniform expansion for the contour $C_{-1}^+(b)$ are associated with the function $A_n(\xi; 3, \rho, \gamma, C_{-1}^+(\xi))$, which according to Lemma 3.1.3 is given by

$$A_n(\xi; 3, \rho, \gamma, C_{-1}^+(\xi)) = \left(\frac{e^{-i\frac{\pi}{2}}}{3\rho} \right)^{\frac{1+\gamma+n}{3}} \mathbb{F}_n \left(3, 1 + \gamma, \frac{|\xi| e^{-i\frac{2\pi}{3}}}{(3\rho)^{1/3}}, \left[\frac{|\xi| e^{-i\frac{2\pi}{3}}}{(3\rho)^{1/3}} \right]^{\frac{1}{2}} \right)$$

This can be expressed in terms of the real-valued functions $G_n^c(y; \gamma)$, $H_n^c(y; \gamma)$, which are defined for $y > 0$ and $n \geq 0$ such that

$$G_n^c(y; \gamma) \pm i H_n^c(y; \gamma) \overset{\text{def}}{=} \frac{e^{\pm i\frac{\pi}{6}(1+\gamma-2n)}}{\pi} \mathbb{F}_n \left(3, 1 + \gamma, y e^{\pm i\frac{2\pi}{3}}, y^{1/2} e^{\pm i\frac{\pi}{3}} \right).$$

Notice that when $n = 0$ this agrees with what we defined earlier. Thus for $\zeta < 0$ we have

$$i^n A_n(\xi; 3, \rho, \gamma, C^+_{-1}(\xi)) = \frac{\pi}{(3\rho)^{\frac{1+\gamma+n}{3}}} \left[G^c_n \left(\frac{|\xi|}{(3\rho)^{1/3}} \right) - i H^c_n \left(\frac{|\xi|}{(3\rho)^{1/3}} \right) \right].$$

As in that case, we can express these two super-exponentially decaying functions as a combination of more slowly decaying functions:

$$G^c_n(y; \gamma) = G_n(y; \gamma) - (-1)^n \sin(\tfrac{\pi}{2}\gamma) \frac{e^{i\pi\gamma}}{\pi} \mathbb{F}_n(3, 1 + \gamma; y, y^{1/2} e^{-i\pi}),$$

$$H^c_n(y; \gamma) = H_n(y; \gamma) + (-1)^n \cos(\tfrac{\pi}{2}\gamma) \frac{e^{i\pi\gamma}}{\pi} \mathbb{F}_n(3, 1 + \gamma; y, y^{1/2} e^{-i\pi}),$$

where $y > 0$ and $n \geq 0$. We consider the real-valued combination

$$\frac{e^{i\pi\gamma}}{\pi} \mathbb{F}_n(3, 1 + \gamma; y, y^{1/2} e^{-i\pi})$$

to be a generalized incomplete Scorer's function.

As in the complete case integration by parts yields the recursion relation

$$\mathbb{F}_n(3, 1 + \gamma; y, z) = -3z\mathbb{F}^c_{n-1}(3, 1 + \gamma; y, z) + (y - 3z^2)\mathbb{F}^c_{n-2}(3, 1 + \gamma; y, z)$$

$$+ [\gamma + n - 2 + z(y - z^2)]\mathbb{F}^c_{n-3}(3, 1 + \gamma; y, z) + z(n - 3)\mathbb{F}^c_{n-4}(3, 1 + \gamma; y, z)$$

$$+ \begin{cases} e^{-\frac{1}{3}z^3 + zy} z^{1+\gamma} & n = 3 \\ 0 & n \geq 4 \end{cases}$$

for all $n \geq 3$. We write

$$\mathbb{F}^c_n(3, 1 + \gamma; y, z) = \sum_{j=0}^{2} p^j_n(y, z; \gamma) \mathbb{F}^c_0(3, 1 + \gamma + j; y) + q_n(y, z; \gamma) e^{-\frac{1}{3}z^3 + zy} z^{1+\gamma}.$$

The functions $p^j_n(y, z; \gamma)$ are the same as in the complete case, and the function $q_n(y, z; \gamma)$ satisfies for $n \geq 4$ the same recursion in n that the $p^j_n(y, z; \gamma)$ do. Clearly $q_3(y, z; \gamma) = 1$. The first few of the functions $q_n(y, z; \gamma)$ are listed in the last column of the previous table. Using this, we can derive the following reduction formula

$$A_n(\xi; 3, \rho, \gamma, C^+_{-1}(\xi))$$

$$= \sum_{j=0}^{2} \left(\frac{e^{-i\frac{\pi}{2}}}{3\rho} \right)^{\frac{n-j}{3}} p^j_n \left(\frac{|\xi| e^{-i\frac{2\pi}{3}}}{(3\rho)^{1/3}}; \gamma \right) A(\xi; 3, \rho, \gamma + j, C^+_{-1}(\xi))$$

$$+ \left(\frac{e^{-i\frac{\pi}{2}}}{3\rho} \right)^{\frac{n}{3}} q_n \left(\frac{|\xi| e^{-i\frac{2\pi}{3}}}{(3\rho)^{1/3}}, \left[\frac{|\xi| e^{-i\frac{2\pi}{3}}}{(3\rho)^{1/3}} \right]^{\frac{1}{2}}; \gamma \right) \left[\frac{|\xi| e^{-i\pi}}{3\rho} \right]^{\frac{1+\gamma}{2}} \exp \left[-2\rho \left| \frac{\xi}{3\rho} \right|^{3/2} \right].$$

which is valid for $\xi < 0$ and $n \geq 0$.

Reduction formulae can be derived for the real-valued functions we have discussed along similar lines, although the results are more complex. Asymptotics for these real-valued functions can also be extracted from our results, but will will not need them.

Relation of our uniform expansion to a Bleistein-style expansion. It is well-known in general that a Bleistein-style uniform expansion reduces to the steepest descent expansion (our outer expansion) in the region where the saddle points are fixed and separated (i.e. $b \neq 0$ is fixed). It is also not difficult to see that when ξ is constant (our inner region) that the Bleistein-style uniform expansion reduces to our inner expansion. Since our uniform expansion also has these properties, it might seem unnecessary to compare directly our expansion to the Bleistein-style expansion. However, this comparison is not that difficult to do, at least at low orders, and the comparison will clarify the relative merits of these two uniform expansions.

We will compare the first six terms of each of these two expansions of the integral $I(x, t; 3, \rho, \gamma, V)$. This is the appropriate thing to do when $\gamma \neq 0$. When $\gamma = 0$, then the Bleistein-style expansion simplifies somewhat, so that six terms become four. The effect of these simplifications should be evident from our discussion; we will leave the details to the interested reader.

As we saw in section 3.1 the first six terms of the Bleistein-style expansion are in the form

$$\int_{\Gamma_{V_0}} e^{ikx - i\rho k^3 t} k^\gamma V(k)\, dk \sim \sum_{h=0}^{2} \sum_{m=0}^{1} \frac{a_h^m(b)}{t^{m+(1+\gamma+h)/3}} A(\xi; 3, \rho, \gamma + h, \Gamma_{V_0}).$$

Define $l = \left| \frac{x}{3\rho t} \right|^{1/2}$ when $x > 0$ and $l = -i \left| \frac{x}{3\rho t} \right|^{1/2}$ when $x < 0$. Then the first three coefficients $a_0^0(b), a_1^0(b), a_2^0(b)$ are given by

$$a_0^0(b) = V(0), \qquad a_1^0(b) = \frac{V(l) - V(-l)}{2l}, \qquad a_2^0(b) = \frac{V(l) - 2V(0) + V(-l)}{2l^2}.$$

These are chosen so that the function

$$W_0(k; b) = \frac{V(k) - \sum_{h=0}^{2} a_h^0(b) k^h}{ik(3\rho k^2 - b)}$$

is analytic in neighborhoods of the points $k = 0, k = l, k = -l$. One then defines $V_1(k; b) = (1+\gamma)W_0(k; b) + kW_0'(k; b)$, and repeats the process with $V(k)$ replaced by $V_1(k; b)$:

$$a_0^1(b) = V_1(0; b), \qquad a_1^1(b) = \frac{V_1(l; b) - V_1(-l; b)}{2l},$$

$$a_2^1(b) = \frac{V_1(l; b) - 2V_1(0; b) + V_1(-l; b)}{2l^2}.$$

We will begin our comparison in the region $x > 0$, $l \to 0$. The first six terms of our expansion are (Theorem 3.2.1B(1))

$$\int_{\Gamma_{V_0}} e^{ikx - i\rho k^3 t} k^\gamma V(k)\, dk \sim \sum_{n=0}^{5} \frac{V^{(n)}(0)}{n!} \frac{A(\xi; 3, \rho, \gamma + n, \tilde{C}(\xi))}{t^{(1+\gamma+n)/3}}$$

$$+ \sum_{n=0}^{5} \frac{V^{(n)}(l)}{n!} \frac{A_n(\xi; 3, \rho, \gamma, C_0(\xi))}{t^{(1+\gamma+n)/3}}.$$

Suppose we use the reduction formulae we have derived in the first part of this section to express the special functions in our expansion in terms of $A(\xi; 3, \rho, \gamma + h, \tilde{C}(\xi))$ and $A(\xi; 3, \rho, \gamma + h, C_0(\xi))$, $h = 0, 1, 2$. This form of our expansion will compare directly to the Bleistein-style expansion, rewritten in the form

$$\int_{\Gamma_{V_0}} e^{ikx - i\rho k^3 t} k^\gamma V(k)\, dk \sim \sum_{h=0}^{2} \sum_{m=0}^{1} \frac{a_h^m(b)}{t^{m+(1+\gamma+h)/3}} A(\xi; 3, \rho, \gamma + h, \tilde{C}(\xi))$$

$$+ \sum_{h=0}^{2} \sum_{m=0}^{1} \frac{a_h^m(b)}{t^{m+(1+\gamma+h)/3}} A(\xi; 3, \rho, \gamma + h, C_0(\xi)).$$

In the first line of the above we expand as follows:

$$a_0^0(b) = V(0)$$
$$a_1^0(b) = V^{(1)}(0) + \tfrac{1}{6} V^{(3)}(0) l^2 + \tfrac{1}{120} V^{(5)}(0) l^4 + O(l^5)$$
$$a_2^0(b) = \tfrac{1}{2} V^{(2)}(0) + \tfrac{1}{24} V^{(4)}(0) l^2 + O(l^4)$$
$$a_0^1(b) = \frac{1+\gamma}{3\rho i} \left[\tfrac{1}{6} V^{(3)}(0) + \tfrac{1}{120} V^{(5)}(0) l^2 \right] + O(l^3)$$
$$a_1^1(b) = \frac{2+\gamma}{72\rho i} V^{(4)}(0) + O(l^2)$$
$$a_2^1(b) = \frac{3+\gamma}{360\rho i} V^{(5)}(0) + O(l).$$

Comparing terms with $V^{(n)}(0) A(\xi; 3, \rho, \gamma + h, \tilde{C}(\xi))$ in common produces agreement with the first line in our expansion. In the second line of the rewritten Bleistein-style expansion we expand in a different manner:

$$a_0^0(b) = V(l) - V'(l) l + \tfrac{1}{2} V^{(2)}(l) l^2 - \tfrac{1}{6} V^{(3)}(l) l^3 + \tfrac{1}{24} V^{(4)}(l) l^4 - \tfrac{1}{120} V^{(5)}(l) l^5$$
$$+ O(l^6)$$
$$a_1^0(b) = V'(l) - V^{(2)}(l) l + \tfrac{2}{3} V^{(3)}(l) l^2 - \tfrac{1}{3} V^{(4)}(l) l^3 + \tfrac{2}{15} V^{(5)}(l) l^4 + O(l^5)$$
$$a_2^0(b) = \tfrac{1}{2} V^{(2)}(l) - \tfrac{1}{2} V^{(3)}(l) l + \tfrac{7}{24} V^{(4)}(l) l^2 - \tfrac{1}{8} V^{(5)}(l) l^3 + O(l^4)$$
$$a_0^1(b) = \frac{1+\gamma}{3\rho i} \left[\tfrac{1}{6} V^{(3)}(l) - \tfrac{1}{6} V^{(4)}(l) l + \tfrac{11}{120} V^{(5)}(l) l^2 \right] + O(l^3)$$
$$a_1^1(b) = \frac{2+\gamma}{72\rho i} \left[V^{(4)}(l) - V^{(5)}(l) l \right] + O(l^2)$$
$$a_2^1(b) = \frac{3+\gamma}{360\rho i} V^{(5)}(l) + O(l).$$

Comparing terms with $V^{(n)}(l) A(\xi; 3, \rho, \gamma + h, C_0(\xi))$ in common produces agreement with the second line in our expansion.

A comparison under the assumption $x < 0$, $l \to 0$, can be carried out in a similar manner. The only new feature is that when one uses the reduction formulae to express our expansion in terms of $A(\xi; 3, \rho, \gamma + h, \tilde{C}(\xi))$ and

$A(\xi; 3, \rho, \gamma + h, C_{-1}^{+}(\xi))$ there are elementary functions appearing in the terms $n = 3, 4, 5$ which are not present in the Bleistein-style expansion. Fortunately, when one expands $V^{(n)}(l)$ about $l = 0$ in the second line, these extra terms cancel with the extra terms coming from the first line.

We have shown how to make the Bleistein-style expansion look like our expansion when $l \to 0$. In that case the two expansions are the same in regard to the number of terms required to achieve an error of a given size. When l does not tend to 0 this is no longer true. We will see (when $\gamma \neq 0$) that $3n$ terms of the Bleistein expansion are needed to achieve an error which our expansion can achieve with only $2n - 1$ terms. (This is the situation when $x > 0$, which we assume henceforth in our discussion. Modifications for the case $x < 0$ are left to the reader.) Hence all of the first three terms in the Bleistein-style expansion are needed in order to achieve the same accuracy as our leading-order term achieves. This can be seen by relating both expansions to the outer expansion in the region where $b \neq 0$ is a constant. Our expansion is based on the simple idea of expanding $V(k)$ in a Taylor series near $k = l$. This is not true for the Bleistein approach, but the Bleistein-style expansion can be viewed as arising from a particular expansion of $V(k)$ which is not a Taylor expansion:

$$V(k) = a_0^0(b) + a_1^0(b)k + a_2^0(b)k^2 + ik(3\rho k^2 - b)W_0(k; b).$$

Actually, this will only generate the first three terms of the Bleistein-style expansion. To generate the next three terms we must solve the differential equation $V_1(k; b) = (1 + \gamma)W_0(k; b) + kW_0'(k; b)$ for $W_0(k; b)$ in terms of $V_1(k; b)$, and continue the expansion process. Since both $W_0(k; b)$ and $V_1(k; b)$ are analytic in k, the unique solution of this equation is

$$W_0(k; b) = \frac{1}{k} \int_0^k \left(\frac{\omega}{k}\right)^\gamma V_1(\omega; b)\, d\omega,$$

where the contour of integration is the ray connecting 0 with k. If we use the expansion

$$V_1(k; b) = a_0^1(b) + a_1^1(b)k + a_2^1(b)k^2 + ik(3\rho k^2 - b)W_1(k; b)$$

in the above integral, we obtain the following expansion of $V(k)$:

$$V(k) = a_0^0(b) + a_1^0(b)k + a_2^0(b)k^2 + ik(3\rho k^2 - b)\left[\frac{a_0^1(b)}{1 + \gamma} + \frac{a_1^1(b)}{2 + \gamma}k + \frac{a_2^1(b)}{3 + \gamma}k^2\right]$$

$$- (3\rho k^2 - b)\int_0^k \left(\frac{\omega}{k}\right)^\gamma \omega(3\rho\omega^2 - b)W_1(\omega; b)\, d\omega.$$

This expansion generates the first six terms of the Bleistein-style expansion of $I(x, t)$. This odd-looking expansion approximates $V(k)$ in a very interesting manner. The first three terms are simply a polynomial in k which interpolates $V(k)$ at the three points $k = 0, k = l, k = -l$. If $E_1(k) = ik(3\rho k^2 - b)W_0(k; b)$

is the error in the approximation of $V(k)$ by this quadratic polynomial, we find that the first term in the outer expansion of $I(x, t; 3, \rho, \gamma, E_1, C_0(b))$ vanishes, as one can see by looking at Corollary 1.7.6, since $E_1(l) = 0$. So three terms of the Bleistein-style expansion are needed to achieve agreement with the first term in the outer expansion. Our expansion, being based on a Taylor expansion, only requires one term to achieve the same level of agreement with the outer expansion.

Now define $E_2(k) = -(3\rho k^2 - b) \int_0^k \left(\frac{\omega}{k}\right)^\gamma \omega(3\rho\omega^2 - b)W_1(\omega; b) \, d\omega$, which is the error upon approximating $V(k)$ by the first six terms of the above expansion. Even though $E_2(l) = 0$, it is not generally true that $E_2'(l) = 0$, as one might expect if one were approximating $V(k)$ by a Hermite interpolatory polynomial. What is true is that $(\gamma - \frac{1}{2})lE_2'(l) + \frac{1}{2}l^2 E_2''(l) = 0$, as one can check by direct differentiation of this integral. This means that the first two terms of the outer expansion of $I(x, t; 3, \rho, \gamma, E_2, C_0(b))$ vanish (see Corollary 1.7.6 for the explicit form of the second term in the outer expansion). Thus six terms of the Bleistein-style expansion are needed to achieve agreement with the first two terms of the outer expansion. Our expansion, being based on a Taylor expansion, only requires three terms to achieve the same level of agreement. This type of argument can be extended to show that $3n$ terms of the Bleistein-style expansion and $2n - 1$ terms of our expansion are needed to agree with the first n terms of the outer expansion.

This assessment of the relation between our expansion and the Bleistein-style expansion continues to remain accurate in the region $b \to \infty$. Olde Daalhuis and Temme [oDT] have proved error estimates for a Bleistein-style expansion in the case $\gamma = 0$. To compare their results to ours we make the same assumptions on $V(k)$ as in section 3.2, but furthermore we assume that for all $N \geq 0$ that $N + \lambda_N^0 = \lambda_0^0 = \lambda$. (Olde Daalhuis and Temme actually estimated a sum of two $I(x, t)$s, a sum in which the contributions of the contour $\tilde{C}$ cancels. Their expression was in the case $\gamma = 0$ where the Bleistein-style expansion simplifies considerably. Our previous discussion assumed $\gamma \neq 0$, but we claim that the same ideas apply when $\gamma = 0$.) Olde Daalhuis and Temme found that the error ε_n after $2n$ terms of the Bleistein-style expansion is bounded by

$$|\varepsilon_n| \leq \frac{C_n}{(1 + |b|)^{n/2}} \frac{1}{(1 + |b|)^\lambda t^{n + \frac{1}{3}}(1 + |\xi|)^{1/4}}.$$

(These authors suggest that one should take their parameter $\theta \leq 1$ as large as possible satisfying some other conditions. This turns out to be the wrong thing to do in our case, since it would lead to a less sharp error estimate. We have chosen their parameter θ to have the value 0, leading to the above estimate.) This is the same as the error estimate $O(W_N(x, t, C_0))$ we gave for $N = 2n - 1$ terms of our uniform expansion. Thus $2n$ terms of the Bleistein-style expansion and $2n - 1$ terms of our expansion both lead to the same size error term in the region $b \geq 1$. In the case $N + \lambda_N^0 > \lambda_0^0$ our results suggest that the error estimate of Olde Daalhuis and Temme is not sharp, owing to their estimations of the size of $V^{(N)}(l)$ in terms of the size of $V(k)$ using the Cauchy integral formula. Those estimates however are probably the best one can do without detailed knowledge of $V(k)$ such as we have in our case.

In conclusion, we see that the two uniform expansions relate differently to each other, depending on whether one is in the inner region or the outer regions. The benefit of the Bleistein-style expansion is that it has the same appearance regardless of whether $x > 0$ or $x < 0$; in that sense it is like our inner expansion. The price for this benefit is that in the outer region one needs to carry more terms (when compared to our expansion) to achieve a given degree of accuracy. These additional terms provide agreement at the other saddle point $-l$, which is not asymptotically important in the outer regions. From this point of view, our expansion is more akin to the outer expansion. However, our expansion is uniformly valid in the entire region $\pm x \geq 0$, including the case $x \to 0$, whereas the outer expansion actually becomes disordered in that limit. Furthermore, the reduction formulae proved in this section enable us to write our expansion so that in its entirety it involves exactly the same special functions which appear in the Bleistein-style expansion, and no others.

4.6 The Linearized Korteweg-de Vries Equation

Our goal in this section is to study the large time behavior of solutions of the initial-value problem for the Linearized Korteweg-de Vries Equation $u_t + \rho u_{xxx} = 0$, $u(x,0) = [||D|^\gamma (v_0 + \mathcal{H}v_1)](x)$. In terms of the integral we have studied in chapters 1-3 we have that

$$u(x,t) = \frac{1}{2\pi} \left[\overline{I(-x,t;3,\rho,\gamma,(\hat{v}_0 - i\hat{v}_1)^\sim)} + I(-x,t;3,\rho,\gamma,\hat{v}_0^\curvee + i\hat{v}_1^\curvee) \right],$$

where $\phi^\sim(k) = \overline{\phi(\bar{k})}$ and $\phi^\curvee(k) = \phi(-k)$. As in the case of the linearized Benjamin-Ono Equation, we want to write down the uniformly valid asymptotic expansion of the solution in a form which will be manifestly real-valued whenever v_0 and v_1 are real-valued. We will use the formulae involving special functions established in the previous section without comment. Thus if the reader encounters an undefined special function in this section, he should consult section 4.5 for an explanation.

Our first task will be to specialize Theorems 3.2.1A(2) and 3.2.1B(1) to the case $r = 3$, and then combine these into an expansion for $u(x,t)$.

Lemma 4.6.1. *Suppose $\rho > 0$ and $\gamma > -1$. Suppose $\bar{v}_-$ and v_+ satisfy the standard assumptions, where $v_\pm = v_0^\curvee \pm i v_1^\curvee$. Suppose $\lambda_n^0(v_\pm), \lambda_n^-(v_\pm), a_-(v_\pm)$ are the associated decay parameters; i.e. we have the existence of nonzero values of the limits*

$$\lim_{l \to \infty} l^{n + \lambda_n^0(v_\pm)} |\hat{v}_0^{(n)}(\mp l) \pm i\hat{v}_1^{(n)}(\mp l)|, \quad \lim_{l \to \infty} l^{n + \lambda_n^-(v_\pm)} e^{la_-(v_\pm)} |\hat{v}_0^{(n)}(il) \pm i\hat{v}_1^{(n)}(il)|,$$

for l real and for all integers $n \geq 0$. (Note that the decay parameters $\lambda_n^0(v_-)$, $\lambda_n^-(v_-), a_-(v_-)$ are defined to be the decay parameters for the function $\bar{v}_-$, i.e.

$\lambda_n^0(\bar{v}_-), \lambda_n^-(\bar{v}_-), a_-(\bar{v}_-)$ respectively. See sections 1.5 and 3.2 for more details.) Define $v = v_0 + \mathcal{H}v_1$, $b = x/t$, $\xi = xt^{-1/3}$ and

$$\lambda_n^0 = \min\{\lambda_n^0(v_+), \lambda_n^0(v_-)\},$$
$$a_- = \min\{a_-(v_+), a_-(v_-)\},$$
$$\lambda_n^- = \begin{cases} \lambda_n^-(v_+) & a_-(v_+) < a_-(v_-) \\ \lambda_n^-(v_-) & a_-(v_+) > a_-(v_-) \\ \min\{\lambda_n^-(v_+), \lambda_n^-(v_-)\} & a_-(v_+) = a_-(v_-). \end{cases}$$

Suppose $A(\xi; 3, \rho, \gamma, \tilde{C}(\xi))$ and $A_n(\xi; 3, \rho, \gamma, C_0(\xi))$ are defined as in sections 2.1 and 3.1.

(1) If $N \geq 0$ is an integer and $x \leq 0$ then define

$$W_N(x, t, C_0) \overset{N \text{ even}}{=} \frac{(1 + |\xi|)^{\frac{\gamma}{2} - \frac{(N+1)}{4}}}{t^{\frac{1+\gamma+N}{3}}(1 + |b|)^{\frac{N+\lambda_N^0}{2}}},$$

$$W_N(x, t, C_0) \overset{N \text{ odd}}{=} \frac{(1 + |\xi|)^{\frac{\gamma}{2} - \frac{(N+2)}{4}}}{t^{\frac{1+\gamma+N}{3}}(1 + |b|)^{\frac{N+\lambda_N^0}{2}}} \left[\frac{1}{(1 + |\xi|)^{\frac{1}{2}}} + \frac{1}{t^{1/3}(1 + |b|)^{\frac{1+\lambda_{N+1}^0 - \lambda_N^0}{2}}} \right]$$

Then for all integers $N \geq 0$ and all $x \leq 0$, $t \geq 1$ we have

$$u(x, t) = \frac{1}{2\pi} \sum_{n=0}^{N-1} \frac{1}{n! t^{(1+\gamma+n)/3}} \left\{ \hat{v}^{(n)} \left(\left| \frac{b}{3\rho} \right|^{\frac{1}{2}} \right) \overline{A_n(-\xi; 3, \rho, \gamma, C_0(-\xi))} \right.$$
$$+ (-1)^n \hat{v}^{(n)} \left(-\left| \frac{b}{3\rho} \right|^{\frac{1}{2}} \right) A_n(-\xi; 3, \rho, \gamma, C_0(-\xi))$$
$$+ (-1)^n \left[\hat{v}^{(n)}(0^-) - e^{-i\pi\gamma} \hat{v}^{(n)}(0^+) \right] A(-\xi; 3, \rho, \gamma + n, \tilde{C}(-\xi)) \right\}$$
$$+ O\left([t^{1/3}(1 + |\xi|)]^{-(1+\gamma+N)} \right) + O(W_N(x, t, C_0)).$$

If $\sin(\frac{\pi}{2}\gamma)\hat{v}_0(k) + \cos(\frac{\pi}{2}\gamma)\hat{v}_1(k) \equiv 0$ then the third line and the first error term are not present.

(2) If $N \geq 0$ is an integer and $x \geq 0$ then define

$$W_N(x, t, C_{-1}^+) = \exp\left\{ -2\rho \left| \frac{\xi}{3\rho} \right|^{\frac{3}{2}} - a_- \left| \frac{b}{3\rho} \right|^{\frac{1}{2}} \right\} \cdot \frac{(1 + |\xi|)^{\frac{\gamma}{2} - \frac{(N+1)}{4}}}{t^{\frac{1+\gamma+N}{3}}(1 + |b|)^{\frac{N+\lambda_N^0}{2}}}.$$

Then for all integers $N \geq 0$ and all $x \geq 0$, $t \geq 1$ we have

$$u(x, t) = \frac{1}{2\pi} \sum_{n=0}^{N-1} \frac{1}{n! t^{\frac{1+\gamma+n}{3}}} \left\{ \left(\hat{v}_0^{(n)} - i\hat{v}_1^{(n)} \right) \left(i \left| \frac{b}{3\rho} \right|^{\frac{1}{2}} \right) \overline{A_n(-\xi; 3, \rho, \gamma, C_{-1}^+(-\xi))} \right.$$

$$+ (-1)^n \left(\hat{v}_0^{(n)} + i\hat{v}_1^{(n)} \right) \left(i \left| \frac{b}{3\rho} \right|^{\frac{1}{2}} \right) A_n(-\xi; 3, \rho, \gamma, C_{-1}^+(-\xi))$$

$$+ (-1)^n \left[(1 - e^{i\pi\gamma})\hat{v}_0^{(n)}(0) + i(1 + e^{i\pi\gamma})\hat{v}_1^{(n)}(0) \right] A(-\xi; 3, \rho, \gamma + n, \tilde{C}(-\xi)) \Big\}$$

$$+ O\left(\left[\frac{|\xi|^{1/2}}{t^{1/3}(1 + |\xi|)^{3/2}} \right]^{1+\gamma+N} \right) + O(W_N(x, t, C_{-1}^+)).$$

If $\sin(\frac{\pi}{2}\gamma)\hat{v}_0(k) - \cos(\frac{\pi}{2}\gamma)\hat{v}_1(k) \equiv 0$ then the third line and the first error term are not present.

Proof. If $\pm x \geq 0$ then transformation of contour integrals yields the relation

$$\overline{I(-x, t; 3, \rho, \gamma, (\hat{v}_0 - i\hat{v}_1)^\sim, \tilde{C}(-b))} = -e^{\pm i\pi\gamma} I(-x, t; 3, \rho, \gamma, \hat{v}_0^{\vee} - i\hat{v}_1^{\vee}, \tilde{C}(-b)).$$

Therefore

$$\overline{I(-x, t; 3, \rho, \gamma, (\hat{v}_0 - i\hat{v}_1)^\sim, \tilde{C}(-b))} + I(-x, t; 3, \rho, \gamma, \hat{v}_0^{\vee} + i\hat{v}_1^{\vee}, \tilde{C}(-b))$$
$$= I(-x, t; 3, \rho, \gamma, [1 - e^{\pm i\pi\gamma}]\hat{v}_0^{\vee} + i[1 + e^{\pm i\pi\gamma}]\hat{v}_1^{\vee}, \tilde{C}(-b)).$$

We have $[1 - e^{\pm i\pi\gamma}]\hat{v}_0^{\vee} + i[1 + e^{\pm i\pi\gamma}]\hat{v}_1^{\vee} = 2ie^{\pm i\frac{\pi}{2}\gamma}[\mp \sin(\frac{\pi}{2}\gamma)\hat{v}_0^{\vee} + \cos(\frac{\pi}{2}\gamma)\hat{v}_1^{\vee}]$. If this quantity vanishes then the contour $\tilde{C}(-b)$ makes no contribution. Otherwise, we apply Theorem 2.2.1 and 2.2.2. For the contours $C_0(-b)$ and $C_{-1}^+(-b)$ we apply Theorems 3.1.4 and 3.1.5. $\square$

Suppose that the functions $\mu_n(k), \eta_n(k), M_n(k; \delta), N_n(k; \delta)$ for $k \geq 0, \delta \geq 0$, and $n \geq 0$ are defined as in section 4.4. These functions are real-valued when v_0 and v_1 are. Using the special functions from the previous section we can rewrite our expansion in a manifestly real-valued way.

Theorem 4.6.2. *Adopt the assumptions and notation of the previous lemma. Suppose $\delta = 1 + \gamma$.*

(1) *If $x \leq 0$ and $t \geq 1$ then for all integers $N \geq 0$ we have*

$$u(x, t) = \sum_{n=0}^{N-1} \frac{1}{n!(3\rho t)^{\frac{1+\gamma+n}{3}}} \left\{ \mu_n \left(\left| \frac{b}{3\rho} \right|^{1/2} \right) G_n \left(\frac{\xi}{(3\rho)^{1/3}}; \gamma \right) \right.$$

$$+ \eta_n \left(\left| \frac{b}{3\rho} \right|^{1/2} \right) H_n \left(\frac{\xi}{(3\rho)^{1/3}}; \gamma \right)$$

$$+ (-1)^n \left[M_n(0^+; \frac{\delta}{2})\text{Hi}^{(\gamma+n)} \left(\frac{\xi}{(3\rho)^{1/3}} \right) \right.$$

$$- M_n \left(\left| \frac{b}{3\rho} \right|^{1/2}; \frac{\delta}{2} \right) \frac{1}{\pi} \Re F_n \left(3, \delta; \frac{|\xi|e^{i\pi}}{(3\rho)^{1/3}} \right)$$

$$\left. \left. - N_n \left(\left| \frac{b}{3\rho} \right|^{1/2}; \frac{\delta}{2} \right) \frac{1}{\pi} \Im F_n \left(3, \delta; \frac{|\xi|e^{i\pi}}{(3\rho)^{1/3}} \right) \right] \right\}$$

$$+ O \left([t^{1/3}(1 + |\xi|)]^{-(1+\gamma+N)} \right) + O(W_N(x, t, C_0)).$$

$M_n(0^+; \frac{\delta}{2}) = -2i^n[\hat{v}_0^{(n)}(0)\sin(\frac{\pi}{2}\gamma) + \hat{v}_1^{(n)}(0)\cos(\frac{\pi}{2}\gamma)]$, *and if these quantities vanish for all $n \geq 0$ then the first error term is also not present.*

(2) *If $x \geq 0$ and $t \geq 1$ then for all integers $N \geq 0$ we have*

$$
u(x,t) = \sum_{n=0}^{N-1} \frac{1}{n!(3\rho t)^{\frac{1+\gamma+n}{3}}} \left\{ i^n \hat{v}_0^{(n)}\left(i\left|\frac{b}{3\rho}\right|^{\frac{1}{2}}\right) G_n^c\left(\frac{\xi}{(3\rho)^{1/3}};\gamma\right) \right.
$$
$$
+ i^n \hat{v}_1^{(n)}\left(i\left|\frac{b}{3\rho}\right|^{\frac{1}{2}}\right) H_n^c\left(\frac{\xi}{(3\rho)^{1/3}};\gamma\right)
$$
$$
+ (-1)^n \left[\sin(\tfrac{\pi}{2}\gamma)i^n\hat{v}_0^{(n)}(0) - \cos(\tfrac{\pi}{2}\gamma)i^n\hat{v}_1^{(n)}(0)\right] \cdot
$$
$$
\left. \cdot \frac{e^{i\pi\gamma}}{\pi} F_0\left(3, 1+\gamma+n; \tfrac{\xi}{(3\rho)^{1/3}}, \left[\tfrac{\xi}{(3\rho)^{1/3}}\right]^{1/2} e^{-i\pi}\right) \right\}
$$
$$
+ O\left(\left[\frac{|\xi|^{1/2}}{t^{1/3}(1+|\xi|)^{3/2}}\right]^{1+\gamma+N}\right) + O(W_N(x,t,C_{-1}^+)).
$$

If $\sin(\frac{\pi}{2}\gamma)i^n\hat{v}_0^{(n)}(0) - \cos(\frac{\pi}{2}\gamma)i^n\hat{v}_1^{(n)}(0)$ vanishes for all $n \geq 0$ then the first error term is also not present.

Remarks. (a) The resemblance between part (1) of the above and the expansion given for the Linearized Benjamin-Ono equation in Theorem 4.4.3(1) is strong. This is because the steepest descent contours and the positioning of the saddle points are similar for these two equations when $x \leq 0$. On the other hand the expansions for these two equations when $x \geq 0$ are quite different.

(b) When v is a real-valued function and l is purely imaginary then $i^n\hat{v}^{(n)}(l)$ is real-valued.

(c) Since G_n^c and H_n^c are super-exponentially decaying, and $\hat{v}_j^{(n)}(il)$, $j = 0, 1$, can grow exponentially as $l \to \infty$ (l is real), we have written the expansion for $x \geq 0$ in the above form to show how this growth and decay compensate one another as $b \to \infty$. In the region $b \to 0^+$ another form of the expansion is probably more revealing:

$$
u(x,t) = \sum_{n=0}^{N-1} \frac{1}{n!(3\rho t)^{\frac{1+\gamma+n}{3}}} \left\{ i^n\hat{v}_0^{(n)}(l)G_n(y;\gamma) + i^n\hat{v}_1^{(n)}(l)H_n(y;\gamma) \right.
$$
$$
+ (-1)^n\left[\sin(\tfrac{\pi}{2}\gamma)i^n\hat{v}_0^{(n)}(0) - \cos(\tfrac{\pi}{2}\gamma)i^n\hat{v}_1^{(n)}(0)\right]\cdot
$$
$$
\cdot \frac{e^{i\pi\gamma}}{\pi} F_0\left(3, 1+\gamma+n; y, y^{1/2}e^{-i\pi}\right)
$$
$$
- (-1)^n\left[\sin(\tfrac{\pi}{2}\gamma)i^n\hat{v}_0^{(n)}(l) - \cos(\tfrac{\pi}{2}\gamma)i^n\hat{v}_1^{(n)}(l)\right]\cdot
$$
$$
\left. \cdot \frac{e^{i\pi\gamma}}{\pi} F_n\left(3, 1+\gamma; y, y^{1/2}e^{-i\pi}\right) \right\}
$$
$$
+ O\left(\left[\frac{|\xi|^{1/2}}{t^{1/3}(1+|\xi|)^{3/2}}\right]^{1+\gamma+N}\right) + O(W_N(x,t,C_{-1}^+)),
$$

where $l = i\left|\frac{b}{3\rho}\right|^{\frac{1}{2}}$ and $y = \frac{\xi}{(3\rho)^{1/3}}$. The second through the fifth lines of the above are not important to the leading term ($n = 0$) in the region $b \to 0^+$.

The number of special functions involved in this expansion is considerable, but we have given the entire expansion explicitly. If we just write the leading-order term, then the appearance is not quite as formidable.

(1) If $x \leq 0$ and $t \geq 1$ then

$$u(x,t) = \frac{1}{(3\rho t)^{\frac{1+\gamma}{3}}}\left\{\mu_0\left(\left|\frac{b}{3\rho}\right|^{1/2}\right)[|D|^\gamma \mathrm{Ai}]\left(\frac{\xi}{(3\rho)^{1/3}}\right)\right.$$

$$+ \eta_0\left(\left|\frac{b}{3\rho}\right|^{1/2}\right)[|D|^\gamma \mathrm{Gi}]\left(\frac{\xi}{(3\rho)^{1/3}}\right)$$

$$\left.+ \left[M_0(0^+;\tfrac{\delta}{2}) - M_0\left(\left|\frac{b}{3\rho}\right|^{1/2};\tfrac{\delta}{2}\right)\right]\mathrm{Hi}^{(\gamma)}\left(\frac{\xi}{(3\rho)^{1/3}}\right)\right\}$$

$$+ O\left([t^{1/3}(1+|\xi|)]^{-(2+\gamma)}\right) + O(W_1(x,t,C_0)).$$

(2) If $x \geq 0$ and $t \geq 1$ then

$$u(x,t) = \frac{1}{(3\rho t)^{\frac{1+\gamma}{3}}}\left\{\hat{v}_0\left(i\left|\frac{b}{3\rho}\right|^{\frac{1}{2}}\right)G_0^c\left(\frac{\xi}{(3\rho)^{1/3}};\gamma\right)\right.$$

$$+ \hat{v}_1\left(i\left|\frac{b}{3\rho}\right|^{\frac{1}{2}}\right)H_0^c\left(\frac{\xi}{(3\rho)^{1/3}};\gamma\right)$$

$$+ \left[\sin(\tfrac{\pi}{2}\gamma)\hat{v}_0(0) - \cos(\tfrac{\pi}{2}\gamma)\hat{v}_1(0)\right]\cdot$$

$$\left.\cdot\frac{e^{i\pi\gamma}}{\pi}\mathbb{F}_0\left(3,1+\gamma;\frac{\xi}{(3\rho)^{1/3}},\left[\frac{\xi}{(3\rho)^{1/3}}\right]^{1/2}e^{-i\pi}\right)\right\}$$

$$+ O\left(\left[\frac{|\xi|^{1/2}}{t^{1/3}(1+|\xi|)^{3/2}}\right]^{2+\gamma}\right) + O(W_1(x,t,C_{-1}^+)).$$

An alternate expression for this case which is preferable in certain situations is given by:

$$u(x,t) = \frac{1}{(3\rho t)^{\frac{1+\gamma}{3}}}\left\{\hat{v}_0(l)[|D|^\gamma \mathrm{Ai}](y) + \hat{v}_1(l)[|D|^\gamma \mathrm{Gi}](y)\right.$$

$$+ \left[\sin(\tfrac{\pi}{2}\gamma)[\hat{v}_0(0) - \hat{v}_0(l)] - \cos(\tfrac{\pi}{2}\gamma)[\hat{v}_1(0) - \hat{v}_1(l)]\right]\cdot$$

$$\left.\cdot\frac{e^{i\pi\gamma}}{\pi}\mathbb{F}_0\left(3,1+\gamma;y,y^{1/2}e^{-i\pi}\right)\right\}$$

$$+ O\left(\left[\frac{|\xi|^{1/2}}{t^{1/3}(1+|\xi|)^{3/2}}\right]^{2+\gamma}\right) + O(W_1(x,t,C_{-1}^+)),$$

where $l = i\left|\frac{b}{3\rho}\right|^{\frac{1}{2}}$ and $y = \frac{\xi}{(3\rho)^{1/3}}$. For example, in the special case where

$$\sin(\tfrac{\pi}{2}\gamma)\hat{v}_0(k) - \cos(\tfrac{\pi}{2}\gamma)\hat{v}_1(k) \equiv 0$$

then the second and third lines vanish along with the first error term.

When $\gamma = 0$ and $v_1 \equiv 0$ we can give a single expression which is an uniformly-valid leading-order term for both $x \leq 0$ and $x \geq 0$. Similar expressions also exist when $\gamma \geq 0$ is an integer. When $v_1 \equiv 0$ we extend the functions $\mu_0(k)$ and $\eta_0(k)$ to all complex k by the rule

$$\mu_0(k) = \frac{\hat{v}_0(k) + \hat{v}_0(-\bar{k})}{2}, \qquad \eta_0(k) = i\frac{\hat{v}_0(k) - \hat{v}_0(-\bar{k})}{2}.$$

Suppose $\alpha \in \{0, 1\}$ such that $(-1)^\alpha x \leq 0$; assume $\arg(-x) = \alpha\pi$. Then for all real x and $t \geq 1$ we have

$$u(x,t) = \frac{1}{(3\rho t)^{\frac{1}{3}}}\left\{\mu_0\left(\sqrt{\frac{-b}{3\rho}}\right) \text{Ai}\left(\frac{\xi}{(3\rho)^{1/3}}\right) + \eta_0\left(\sqrt{\frac{-b}{3\rho}}\right) \text{Bi}\left(\frac{\xi}{(3\rho)^{1/3}}\right)\right\}$$
$$+ \text{ error term,}$$

where $\text{Bi}(y) \overset{\text{def}}{=} \text{Gi}(y) + \text{Hi}(y)$ for all $y \in \mathbb{R}$. The error term is $O(W_1(x, t; C_0))$ if $x \leq 0$ and $O(W_1(x, t; C_{-1}^+))$ if $x \geq 0$.

For the purposes of comparison it is also interesting to write down the first three terms of an uniformly valid expansion generated by Bleistein's integration-by-parts procedure.

$$u(x,t) = \frac{1}{(3\rho t)^{\frac{1+\gamma}{3}}}\left\{\hat{v}_0(0)[|D|^\gamma \text{Ai}](y) + \hat{v}_1(0)[|D|^\gamma \text{Gi}](y)\right\}$$
$$+ \frac{1}{(3\rho t)^{\frac{2+\gamma}{3}}}\left\{-i\left[\frac{\hat{v}_1(l) - \hat{v}_1(-l)}{2l}\right][|D|^{\gamma+1}\text{Ai}](y)\right.$$
$$\left. + i\left[\frac{\hat{v}_0(l) - \hat{v}_0(-l)}{2l}\right][|D|^{\gamma+1}\text{Gi}](y)\right\}$$
$$+ \frac{1}{(3\rho t)^{\frac{3+\gamma}{3}}}\left\{\left[\frac{\hat{v}_0(l) + \hat{v}_0(-l) - 2\hat{v}_0(0)}{2l^2}\right][|D|^{\gamma+2}\text{Ai}](y)\right.$$
$$\left. + \left[\frac{\hat{v}_1(l) + \hat{v}_1(-l) - 2\hat{v}_1(0)}{2l^2}\right][|D|^{\gamma+2}\text{Gi}](y)\right\} + \text{error term,}$$

where $y = \frac{\xi}{(3\rho)^{1/3}}$ and $l = \sqrt{\frac{-b}{3\rho}}$, with the same choice of $\arg(-b)$ as above. This expansion is as accurate as three terms of our uniform expansion in the inner region, but in the outer region it is only as accurate as one term of our expansion. We will not attempt to be more specific about the size of the error term here (see section 4.5). The quantity $i[\hat{v}_j(l) - \hat{v}_j(-l)]/(2l)$, $j = 0, 1$, is real-valued for all $x \in \mathbb{R}$ when v_0, v_1 are real-valued. An expansion similar to the above was written down in the special case $\gamma = 0$, $v_1(k) \equiv 0$, by Ablowitz Segur [AblSe2] (page 365; their expression had a misprint, which is corrected below). In that case one can use the differential equations satisfied by Ai and Gi to eliminate the second derivatives, obtaining

$$u(x,t) = \frac{1}{(3\rho t)^{1/3}}\left[\frac{\hat{v}_0(l) + \hat{v}_0(-l)}{2}\right]\text{Ai}(y) + \frac{1}{(3\rho t)^{2/3}}\left[\frac{\hat{v}_0(l) - \hat{v}_0(-l)}{2il}\right]\text{Ai}'(y)$$
$$+ \text{ error term.}$$

This can be directly compared to the leading-order term involving Bi recorded above. The expression involving Ai' is more accurate in the inner region than the expression involving Bi. The two expressions have the same level of accuracy in the outer regions.

For the linearized Korteweg-de Vries equation the inner region is defined using the optimal cross-over variable $\zeta = xt^{-1/2}$ (see section 3.1). If $0 < \epsilon < 1$ then we define the inner region to be such that $|\zeta| \leq \epsilon^{-1}$. In this region a leading-order term of relatively simple functional form can be given:

$$u(x,t) = \frac{1}{(3\rho t)^{\frac{1+\gamma}{3}}} \left\{ \hat{v}_0(0)[|D|^\gamma \text{Ai}] \left(\frac{\xi}{(3\rho)^{1/3}} \right) + \hat{v}_1(0)[|D|^\gamma \text{Gi}] \left(\frac{\xi}{(3\rho)^{1/3}} \right) \right\}$$

$$+ O\left(c_\alpha \left[\frac{(1 - \alpha + |\xi|)^{1/2}}{t^{1/3}(1 + |\xi|)^{3/2}} \right]^{2+\gamma} \right) + O\left(\exp\left\{ -2\rho\alpha \left| \frac{\xi}{3\rho} \right|^{\frac{3}{2}} \right\} \frac{(1 + |\xi|)^{\frac{1}{2}(\gamma + \frac{1}{2})}}{t^{\frac{2+\gamma}{3}}} \right).$$

In the above $\alpha \in \{0, 1\}$ depends on x such that $(-1)^\alpha x \leq 0$, and

$$c_\alpha = \begin{cases} 0 & \sin(\frac{\pi}{2}\gamma)\hat{v}_0(k) + (-1)^\alpha \cos(\frac{\pi}{2}\gamma)\hat{v}_1(k) \equiv 0, \\ 1 & \text{otherwise.} \end{cases}$$

These error terms follow from an application of Theorems 2.2.1, 2.2.1, 2.3.1, and Corollary 2.3.2. c.f. page 365 of Ablowitz and Segur [AblSe2].

In the outer region $\zeta \leq -\epsilon$ we can apply Theorem 1.4.9 and Corollary 1.7.6 to obtain

$$u(x,t) = -\frac{\Gamma(1+\gamma)}{\pi(-x)^{1+\gamma}} \left[\sin(\tfrac{\pi}{2}\gamma)\hat{v}_0(0) + \cos(\tfrac{\pi}{2}\gamma)\hat{v}_1(0) \right] + O\left(\frac{c_0}{|x|^{2+\gamma}} \right)$$

$$+ \frac{1}{\sqrt{\pi}(3\rho t)^{1/2}} \left| \frac{b}{3\rho} \right|^{\frac{7}{2}-\frac{1}{4}} \left\{ \mu_0 \left(\left| \frac{b}{3\rho} \right|^{1/2} \right) \cos\left(2\rho \left| \frac{\xi}{3\rho} \right|^{3/2} - \frac{\pi}{4} \right) \right.$$

$$\left. - \eta_0 \left(\left| \frac{b}{3\rho} \right|^{1/2} \right) \sin\left(2\rho \left| \frac{\xi}{3\rho} \right|^{3/2} - \frac{\pi}{4} \right) \right\}$$

$$+ O\left(\frac{|b|^{\frac{1}{2}(\gamma - \frac{7}{2})}}{t^{3/2}(1 + |b|)^{\lambda_2^0/2}} \right).$$

c.f. page 364 of Ablowitz and Segur [AblSe2]. The number λ_2^0 is defined in Lemma 4.6.1. The number c_0 was defined above. There is an infinite hierarchy of anomalies possible which make it cumbersome to write down a guaranteed leading-order term for the contribution of the contour $\tilde{C}(-b)$. The above expansion is appropriate when either $\sin(\frac{\pi}{2}\gamma)\hat{v}_0(0) + \cos(\frac{\pi}{2}\gamma)\hat{v}_1(0) \neq 0$ or $c_0 = 0$.

In the outer region $\zeta \geq \epsilon$ we can apply Theorem 1.4.9 and Theorem 1.7.5 to obtain the following expansion.

$$u(x,t) = -\frac{\Gamma(1+\gamma)}{\pi x^{1+\gamma}} \left[\sin(\tfrac{\pi}{2}\gamma)\hat{v}_0(0) - \cos(\tfrac{\pi}{2}\gamma)\hat{v}_1(0) \right] + O\left(\frac{c_1}{|x|^{2+\gamma}} \right)$$

$$+ \frac{B^{\frac{\gamma}{2}-\frac{1}{4}}}{2\sqrt{\pi}(3\rho t)^{1/2}} \exp\left(-2\rho\left|\frac{\xi}{3\rho}\right|^{\frac{3}{2}}\right) \left[\cos(\tfrac{\pi}{2}\gamma)\hat{v}_0(iB^{\frac{1}{2}}) + \sin(\tfrac{\pi}{2}\gamma)\hat{v}_1(iB^{\frac{1}{2}})\right]$$

$$+ \frac{B^{\frac{\gamma}{2}-1}}{2\pi(3\rho t)} \exp\left(-2\rho\left|\frac{\xi}{3\rho}\right|^{\frac{3}{2}}\right) \left\{(\gamma-\tfrac{1}{3})\left[\sin(\tfrac{\pi}{2}\gamma)\hat{v}_0(iB^{\frac{1}{2}}) - \cos(\tfrac{\pi}{2}\gamma)\hat{v}_1(iB^{\frac{1}{2}})\right]\right.$$

$$\left.+ B^{\frac{1}{2}}\left[\sin(\tfrac{\pi}{2}\gamma)i\hat{v}_0'(iB^{\frac{1}{2}}) - \cos(\tfrac{\pi}{2}\gamma)i\hat{v}_1'(iB^{\frac{1}{2}})\right]\right\}$$

$$+ O\left(\exp\left[-2\rho\left|\frac{\xi}{3\rho}\right|^{\frac{3}{2}} - a_-\left|\frac{b}{3\rho}\right|^{\frac{1}{2}}\right] \frac{|b|^{\frac{1}{2}(\gamma-\frac{7}{2})}}{t^{3/2}(1+|b|)^{\lambda_2^-/2}}\right).$$

c.f. page 365 of Ablowitz and Segur [AblSe2]. In the above, $B = \left|\frac{b}{3\rho}\right|$, and the numbers a_- and λ_2^- are defined in Lemma 4.6.1. The number c_1 is defined above. If $c_1 = 1$ then the contribution from the contour $\tilde{C}(-b)$ (first line) clearly dominates as $\xi \to \infty$. If $c_1 = 0$ then this slowly decaying contribution vanishes, along with lines three and four, and $u(x,t)$ is seen to decay exponentially to zero as $\xi \to \infty$.

CHAPTER 5

APPLICATIONS

In this chapter we will survey some of the different types of applications which motivated us to obtain the asymptotic results which we did in the first four chapters. Our treatment thus far has been fairly complete, but there can be no hope of completeness in our discussions of applications—there are simply too many different questions. Our purpose thus far has been to provide an adequate and solid foundation upon which a variety of different applications could be built. Here we will choose only a few applications and discuss them only briefly.

5.1 Self-Similar Asymptotic Approximations

This topic is motivated by investigations into the large time asymptotics of solutions of more general linear and nonlinear partial differential equations posed on the entire real line—especially those where dissipative effects are important, such as in the heat equation $u_t - \nu u_{xx} = 0$. Works which focus on the large time behavior for solutions of such equations include [ABS], [Bi], [BDP], [BL1], [BL2], [BL3], [BPW1], [BPW2], [D1], [D2], [D3], [EZ], [EVZ], [N1], [N2], [NS1], [NS2], [NS3], [Wa], [Z1], [Z2]. For example, in [D2] the following family of nonlinear dispersive-dissipative evolution equations

$$u_t + \partial_x P(u) + Q(D)u + iR(D)u = 0,$$
$$u(x,0) = [|D|^\gamma (v_0 + \mathcal{H}v_1)](x),$$

was studied. $P(u) = P(u, \bar{u})$ was assumed to be a polynomial function of the complex variables u and $\bar{u}$, all of whose terms have degree at least $p \geq 2$. $Q(D)$ was assumed to be the (pseudo) differential operator $Q(D) = \nu|D|^q$, where $\nu > 0$ is a constant and $q > 1$ is a real number. For example, when $q = 2$, $Q(D)u = -\nu u_{xx}$ is the dissipative term in the heat equation. The dispersive operator $R(D)$ was assumed to also be homogeneous, of degree $r > 1$, as discussed in the introduction (see the Introduction for some relevant examples). We will confine our discussion here to KdV-like equations, where $R(D) = -\rho|D|^r \operatorname{sgn}(D)$ and $\rho > 0$ is a constant. Rather than discussing the very weak assumptions on the initial data made in [D2], let us assume, in agreement with what we have assumed throughout this work, that v_0 and v_1 have compact support. This will be more than sufficient to allow us to apply the results in [D2]. The first main result of that work concerned the linear case $P(u) \equiv 0$. Under the assumption $r \geq q$ it was proved that there is a function $w(x)$ such that

$$u(x,t) = \frac{1}{t^{(1+\gamma)/q}} w\left(\frac{x}{t^{1/q}}\right) + o\left(\frac{1}{t^{(1+\gamma)/q}}\right) \qquad \text{as } t \to \infty$$

uniformly in x. The function $w(\xi)$ determines a *self-similar asymptotic approximation* to the solution $u(x,t)$ for large t. The function $w(\xi)$, called the self-similar *profile* of the solution, was identified in [D2] by means of its Fourier integral representation.

Having at our disposal the entire function $F_0(r,\delta;y) = \int_0^\infty e^{-\frac{1}{r}\sigma^r + y\sigma}\sigma^{\delta-1}\,d\sigma$, we can now be a bit more explicit about the profile. If $-\pi/2 \le \theta_0 \le 0$ then define the real-valued function

$$G(y;\theta_0) = G_{\theta_0}(y) = \frac{1}{\pi}\Re[e^{-i\theta_0/r}F_0(r,1;iye^{-i\theta_0/r})].$$

Clearly then we have for all $\gamma > -1$ and $y \in \mathbb{R}$

$$[|D|^\gamma G_{\theta_0}](y) + i[|D|^\gamma \mathcal{H}G_{\theta_0}](y) = \frac{e^{-i\theta_0(1+\gamma)/r}}{\pi}F_0(r,1+\gamma;iye^{-i\theta_0/r}).$$

(In the special case $r = q = 2$ the functions G_{θ_0} and $\mathcal{H}G_{\theta_0}$ have been studied in [D1]. When $r = 2$ the function $G(y;-\pi/2)$ is the same as the function $G(y)$ introduced in section 4.4. When $r = 3$ we have $G(y;-\pi/2) = \mathrm{Ai}(y)$.) Define also $\mu = \mu_0(0^+) = \hat{v}_0(0)$ and $\eta = \eta_0(0^+) = \hat{v}_1(0)$ and $\nu - i\rho = Re^{i\theta_0}$ where $R = \sqrt{\nu^2 + \rho^2}$ and $\theta_0 = -\tan^{-1}(\frac{\rho}{\nu})$. With these definitions we then have that the profile identified in [D2] is given by

$$w(\xi) = \begin{cases} (q\nu)^{-\frac{1+\gamma}{q}}\left\{\mu[|D|^\gamma G_0]\left(\frac{\xi}{(q\nu)^{1/q}}\right) + \eta[|D|^\gamma \mathcal{H}G_0]\left(\frac{\xi}{(q\nu)^{1/q}}\right)\right\} & r > q, \\ (rR)^{-\frac{1+\gamma}{r}}\left\{\mu[|D|^\gamma G_{\theta_0}]\left(\frac{\xi}{(rR)^{1/r}}\right) + \eta[|D|^\gamma \mathcal{H}G_{\theta_0}]\left(\frac{\xi}{(rR)^{1/r}}\right)\right\} & r = q. \end{cases}$$

When $r > q$ the function $w(\xi)$ is unaffected by the presence of dispersion. This phenomenon could have been predicted by a heuristic scaling argument discussed in [D2].

The same scaling argument would also predict, under the assumption that there is a self-similar asymptotic approximation to the solution for large t, that when $r < q$ the form of the profile $w(\xi)$ should be unaffected by the presence of dissipation. In particular, it should be given by the formula

$$w(\xi) = (r\rho)^{-\frac{1+\gamma}{r}}\left\{\mu[|D|^\gamma G_{-\frac{\pi}{2}}]\left(\frac{\xi}{(r\rho)^{1/r}}\right) + \eta[|D|^\gamma \mathcal{H}G_{-\frac{\pi}{2}}]\left(\frac{\xi}{(r\rho)^{1/r}}\right)\right\}.$$

Actually, for the scaling argument to work one needs to assume not only that

$$\lim_{t\to\infty} t^{(1+\gamma)/r}\left\|u(x,t) - \frac{1}{t^{(1+\gamma)/r}}w\left(\frac{x}{t^{1/r}}\right)\right\|_{L^\infty(x)} = 0,$$

but also for an appropriately large integer κ that

$$\lim_{t\to\infty} t^{(1+\gamma+\kappa)/r}\left\|\partial_x^\kappa u(x,t) - \frac{1}{t^{(1+\gamma+\kappa)/r}}w^{(\kappa)}\left(\frac{x}{t^{1/r}}\right)\right\|_{L^\infty(x)} = 0.$$

This sort of behavior is present when $r \geq q$, but when $r < q$ one worries about dispersive oscillations for large values of x in $\partial_x^\kappa u(x, t)$ which might cause such a result to fail. This motivates consideration of the question of which conditions on the initial conditions are necessary (and/or sufficient) to imply this type of asymptotically self-similar behavior of the solution $u(x, t)$ of the linearized dispersive-dissipative equation when $1 < r < q$.

Another question is to what extent this behavior is present in the nonlinear case $P(u) \not\equiv 0$. In [D2] it was proved that under the conditions $r \geq q$ and $p + (p - 1)\gamma > q$ the solution of the nonlinear equation has this asymptotically self-similar behavior (for all $\kappa \geq 0$) with exactly the same profile as in the linear case. The condition $p + (p-1)\gamma > q$ was termed *asymptotically weak nonlinearity* because it can be heuristically derived from the same kind of scaling arguments (and hence seems to be necessary), and because it has been proved to be sufficient to imply that the nonlinearity does not influence the form of the profile $w(\xi)$. As further evidence of the necessity of this condition, it has been proved in [D1] that in the case $p = q = r = 2$, $\gamma = 0$, that the profile $w(\xi)$ for the nonlinear case is different than it is for the linear case. Scaling arguments would suggest that when $r < q$ the condition of asymptotically weak nonlinearity is $p + (p-1)\gamma > r$. However, the reliability of the scaling argument in the case $r < q$ is questionable, so it is clear that we must settle the first question we have raised (concerning the linear equation) before any headway can be made on the nonlinear question.

In the linear case $P(u) \equiv 0$ when $r < q$ scaling suggests that dissipation would not be an influence on the profile $w(\xi)$. But it is not even known until now whether or not the linear equation behaves in an asymptotically self-similar way in the *purely dispersive* case, i.e. when $\nu = 0$. This is one of the questions which chapters one through four enable us to answer precisely. This answer is the application of our results which will occupy the remainder of this section.

First we will give some typically sharp pointwise estimates for the solution $u(x, t)$. By the phrase "typically sharp" we mean that the estimates are sharp if the conditions $\sin(\frac{\pi}{2}\gamma)\hat{v}_0(0) \pm \cos(\frac{\pi}{2}\gamma)\hat{v}_1(0) \neq 0$ hold.

Theorem 5.1.1. *Suppose $r \geq 2$, $\rho > 0$, $\gamma > -1$, and v_0 and v_1 are two complex-valued functions with compact support. Define $v_\pm = v_0^\vee \pm iv_1^\vee$. Suppose that $\bar{v}_-$ and v_+ satisfy the standard assumptions stated at the end of section 1.5. Let $\lambda_0^0(\bar{v}_-), \lambda_0^0(v_+) > 0$ be the associated decay parameters, such that the limits*

$$\lim_{l \to \infty} l^{\lambda_0^0(v_+)} |\hat{v}_0(-l) + i\hat{v}_1(-l)|, \qquad \lim_{l \to \infty} l^{\lambda_0^0(\bar{v}_-)} |\hat{v}_0(l) - i\hat{v}_1(l)|$$

exist and have nonzero values. Define $\lambda_0^0 = \min\{\lambda_0^0(v_+), \lambda_0^0(\bar{v}_-)\}$. Let

$$u(x, t) = \frac{1}{2\pi} \int_{\Gamma_\theta} e^{-ikx - i\rho k^r t} k^\gamma \overline{[\hat{v}_0(\bar{k}) - i\hat{v}_1(\bar{k})]} \, dk$$

$$+ \frac{1}{2\pi} \int_{\Gamma_\theta} e^{-ikx - i\rho k^r t} k^\gamma [\hat{v}_0(-k) + i\hat{v}_1(-k)] \, dk,$$

where Γ_θ is the oriented ray $\arg(k) = \theta \in (-\frac{\pi}{r}, 0)$ in the complex plane, be the solution (in a distributional sense) of the initial-value problem

$$u_t - \rho|D|^{r-1}u_x = 0, \qquad u(\cdot, t) \to |D|^\gamma(v_0 + \mathcal{H}v_1) \text{ in } \mathcal{S}'(\mathbb{R}) \text{ as } t \to 0^+.$$

Define $\xi = xt^{-1/r}$ *and* $b = xt^{-1}$. *Then there exists a positive constant* C *such that for all* $x \in \mathbb{R}$ *and* $t \geq 1$ *we have*

$$|u(x,t)| \leq \frac{C}{t^{\frac{1+\gamma}{r}}} \left[\frac{1}{(1+|\xi|)^{1+\gamma}} + [1 - \mathrm{sgn}(b)] \frac{(1+|\xi|)^{\frac{1}{r-1}(1+\gamma-\frac{r}{2})}}{(1+|b|)^{\frac{\lambda_0^0}{r-1}}} \right].$$

Remark. Ben-Artzi and Treves [bAT] have obtained the above estimate for $\gamma = 0$, $x \geq 0$, and all $t > 0$. When $r > 6$, $\gamma = \frac{r}{2} - 1$, $x \geq 0$, and all $t > 0$ they prove the estimate

$$|u(x,t)| \leq \frac{C}{t^{1/2}(1+|\xi|)^{\frac{r}{2(r-1)}}}$$

under significantly weaker assumptions on the initial data than ours. When their result is specialized to our initial data, we see that their temporal decay rate is correct, namely $O(t^{-1/2})$, but their decay estimate as $\xi \to \infty$ is not optimal.

Proof. (Sketch) We use Theorems 3.2.1A ($x \geq 0$) and B ($x \leq 0$) with $N = 0$. First combine the estimates coming from $\bar{v}_-$ and v_+ as in Lemma 4.6.1. Straightforward arguments show that when $x \leq 0$ the term $O(W_0(x,t,C_j))$ for $1 \leq j \leq K$ can be absorbed into $O(W_0(x,t,C_0))$. When $r = 4K + 1$, $O(W_0(x,t,C_K^+)) = O(W_0(x,t,C_K)$, and $W_0(x,t,\tilde{C})$ vanishes as $|\xi| \to 0^+$. However, when $|\xi| \leq 1$ the term $O(W_0(x,t,C_0))$ does not vanish, so we obtain the result we stated. The proof for $x \geq 0$ is very similar, except in that case even $O(W_0(x,t,C_{-1}))$ can be absorbed into $O(W_0(x,t,\tilde{C}))$ provided $r \neq 4J - 1$. $\square$

Now we can state the main result of this section giving the precise conditions on the initial data such that the solution possesses an explicit asymptotically self-similar approximation.

Theorem 5.1.2. *Adopt the assumptions and notation of the previous theorem, but assume that* $-1 < \gamma < \frac{r}{2} - 1$. *Let* $G(y) = G(y; -\frac{\pi}{2})$ *be as defined earlier in this section. Define*

$$w(\xi) = (r\rho)^{-\frac{1+\gamma}{r}} \left\{ \hat{v}_0(0)[|D|^\gamma G]\left(\frac{\xi}{(r\rho)^{1/r}}\right) + \hat{v}_1(0)[|D|^\gamma \mathcal{H}G]\left(\frac{\xi}{(r\rho)^{1/r}}\right) \right\}.$$

Then we have that

$$\lim_{t \to \infty} t^{(1+\gamma)/r} \left\| u(x,t) - \frac{1}{t^{(1+\gamma)/r}} w\left(\frac{x}{t^{1/r}}\right) \right\|_{L^\infty(x)} = 0.$$

Proof. (Sketch) We consider two cases, according as $|\zeta| = |xt^{-2/(r+1)}| \leq 1$ or as $|\zeta| \geq 1$. In the inner region we note that the self-similar asymptotic approximation we have written down is simply the first term of the inner expansion. The L^∞ norm over $|\zeta| \leq 1$ tends to zero as $t \to \infty$ by using the error terms for the inner expansion and noting that in the inner region $|\xi|^{1/(r-1)}t^{-1/r} = |b|^{1/(r-1)} = |\zeta|^{1/(r-1)}t^{-1/(r+1)}$, which tends to zero as $t \to \infty$.

In the outer regions $|\zeta| \geq 1$ we estimate the part involving $u(x,t)$ and the part involving w separately. Use Theorem 5.1.1 to estimate the term involving $u(x,t)$. The term involving w satisfies (by Lemma 2.1.1) the same estimates as $u(x,t)$ except that $\lambda_0^0 = 0$. Since the denominator $(1+|b|)^{\lambda_0^0/(r-1)}$ is not essential in order to control the L^∞ norm over the region $|\zeta| \geq 1$ we are done. $\square$

One can formulate natural conditions under which the analogous result holds but with L^s norms substituted for the L^∞ norm. A similar result can be proved for solutions of Schrödinger-type equations, $u_t + i\rho|D|^r u = 0$, $r \geq 2$.

The above theorem is not true if $\gamma \geq \frac{r}{2} - 1$. The problem occurs in the outer region, i.e. where $b = x/t < 0$ is constant. In that region, $\xi = bt^{1-1/r} \to -\infty$ and

$$\frac{w(\xi)}{t^{(1+\gamma)/r}} \sim \frac{\sqrt{2}}{\sqrt{\pi \rho r(r-1)t}} \left|\frac{b}{r\rho}\right|^{\frac{1}{r-1}(1+\gamma-\frac{r}{2})} \left\{ \hat{v}_0(0) \cos\left[(r-1)\rho\left|\frac{b}{r\rho}\right|^{\frac{r}{r-1}} t - \frac{\pi}{4}\right] \right.$$
$$\left. - \hat{v}_1(0) \sin\left[(r-1)\rho\left|\frac{b}{r\rho}\right|^{\frac{r}{r-1}} t - \frac{\pi}{4}\right] \right\}$$

On the other hand, in this region one can use Corollary 1.7.6 to show that the actual behavior of $u(x,t)$ as $t \to \infty$ is

$$u(x,t) \sim$$
$$\frac{\sqrt{2}}{\sqrt{\pi \rho r(r-1)t}} \left|\frac{b}{r\rho}\right|^{\frac{1}{r-1}(1+\gamma-\frac{r}{2})} \left\{ \mu_0\left(\left|\frac{b}{r\rho}\right|^{\frac{1}{r-1}}\right) \cos\left[(r-1)\rho\left|\frac{b}{r\rho}\right|^{\frac{r}{r-1}} t - \frac{\pi}{4}\right] \right.$$
$$\left. - \eta_0\left(\left|\frac{b}{r\rho}\right|^{\frac{1}{r-1}}\right) \sin\left[(r-1)\rho\left|\frac{b}{r\rho}\right|^{\frac{r}{r-1}} t - \frac{\pi}{4}\right] \right\},$$

where the functions $\mu_0(l)$ and $\eta_0(l)$ are defined as in section 4.4. Hence one must choose b such that

$$\mu_0\left(\left|\frac{b}{r\rho}\right|^{1/(r-1)}\right) \neq \hat{v}_0(0), \quad \text{and/or} \quad \eta_0\left(\left|\frac{b}{r\rho}\right|^{1/(r-1)}\right) \neq \hat{v}_1(0)$$

to see that the leading-order behavior of $u(x,t)$ is not self-similar. Thus we see that our worries about dispersive oscillations were well-founded; the full extent of what is true in the dissipative case $r \geq q$ does not generalize to the purely dispersive case.

Another observation is that what we use as an asymptotic approximation to $u(x,t)$ can depend on how we wish to measure the error. In Theorem 5.1.2 above, we measured the error in the L^∞ norm in the x variable. Thus the deviation in the outer region of $u(x,t)$ from $t^{-(1+\gamma)/r}w(\xi)$, which is a quantity of size $O(t^{-1/2})$, is relatively smaller than effects which are considered as leading-order, i.e. those of size $O(t^{-(1+\gamma)/r})$. By contrast, our results in Theorems 3.2.1A and

B give us an uniformly valid leading-order term (not in self-similar form) whose error is asymptotically smaller than that term itself. This is roughly like the difference between controlling absolute error and relative error. If one insists on controlling the relative error, one is pretty much forced into the analysis we have given in Theorems 3.2.1A and B. If one is willing to control only some sort of absolute error, then one can sometimes get by with simpler asymptotic approximations, such as in Theorem 5.1.2, and simpler proofs.

5.2 Sharp L^s Decay Estimates, Smoothing Effects

Decay Estimates. Much of the literature on the large-time behavior of solutions of nonlinear dispersive equations is preoccupied with the problem of obtaining decay estimates in t of the L^s norms in the x variable of the solution. The procedure, as exemplified by the paper of Sidi, Sulem, and Sulem [SSS], is to first estimate the L^s norm of the profile of the fundamental solution of the linearized equation; then one obtains decay estimates for general solutions of the linearized equation by Young's convolution inequality; finally one estimates decay rates of solutions of the nonlinear equation by various methods. As the technology of estimating has increased, so the scope of the final results has increased. To our knowledge, the most advanced work along these lines for the family of generalized Korteweg-de Vries (GKdV) equations $u_t + \partial_x(u^p) + \rho u_{xxx} = 0$ is by Christ and Weinstein [CW]. They obtained the decay estimate $\|u(\cdot,t)\|_{L^\infty} = O(t^{-1/3})$ for solutions with suitably small initial data in $W^{1,1} \cap H^2$ for all integers $p \geq 4$. For general nonlinear dispersive equations of GKdV type the reference is Kenig Ponce Vega [KPV]. In these works a variety of decay estimates of L^s norms of different (fractional) derivatives of the solution of both the linear and the nonlinear equation are proved. The authors make no statements about whether their estimates are sharp or not. If their decay estimates could be improved, it might be possible to strengthen their conclusions about solutions of the nonlinear equations.

The ultimate desired improvement is exemplified in the papers of Ablowitz and Segur [AblSe1], and Deift, Venakidis, and Zhou [DVZ] on the large-time behavior of solutions of the Korteweg-de Vries (KdV) equation ($p = 2$), and the paper of Deift and Zhou [DZ1] on the Modified Korteweg-de Vries (MKdV) equation ($p = 3$). In these works, more or less explicit leading-order terms are obtained in various regions in the (x, t) plane, accompanied by (in the last two) explicit error estimates. In some regions when $p = 3$ a complete asymptotic expansion of the solution is known [DZ2]. Comparable results for the GKdV equation in the case $p \geq 4$ are not known; indeed until now such results for even the linearized equation have not been documented in the literature (of the results on pages 363–366 in Ablowitz Segur [AblSe2], (A.1.59) was marred by a typo, and all were not accompanied by precise hypotheses or sharp error estimates.) Chapters one through four have been written to provide a solid foundation for such generalizations.

One step in the direction of obtaining generalizations of these detailed results for $p = 2, 3$ to the cases $p \geq 4$ are *linear scattering* results, such as Theorem 6.2 in

Ponce Vega [PV] (see also the references in that paper). In such results one finds solutions $u_\pm(x,t)$ of the linearized equations such that $u(x,t) - u_\pm(x,t)$ tends to zero in some norm (usually L^2 or H^1) as $\pm t \to \infty$. Such linear scattering results for the GKdV equation are known for $p \geq 5$ but not when $p = 4$; they are known to be not true when $p = 2, 3$. So far the proofs of these linear scattering results have been based on L^s decays estimates.

So the study of decay estimates and linear scattering results for solutions of generalized nonlinear dispersive equations will benefit from a detailed knowledge of sharp L^s decay estimates for solutions (and their fractional derivatives) of linear dispersive equations. Some L^∞ decay estimates when γ is an integer $0 \leq \gamma \leq \frac{r}{2} - 1$ are proved in Ben-Artzi Treves [bAT]. These are proved under general assumptions on the initial data u_0, such as $u_0 \in L^1$ or $u_0 \in L^2$. In this section we will prove general L^s "decay" estimates using our pointwise estimates (Theorem 5.1.1). (The L^s norm will not always decay; sometimes it will grow.) These will be established only under our considerably more restrictive assumptions on u_0. We will show that these "decay" estimates are sharp using the leading-term of the asymptotic expansion of the solution. Thus our results, while of interest in themselves, will also serve as a guide as to what might be possible to prove under more general assumptions on u_0.

Theorem 5.2.1. *Adopt the notation and assumptions of Theorem 5.1.1. Suppose either $1 \leq s < \infty$ and $\frac{1}{s} < 1 + \gamma < \frac{1}{s} + \lambda_0^0 + r(\frac{1}{2} - \frac{1}{s})$ or $s = \infty$ and $0 < 1 + \gamma \leq \lambda_0^0 + \frac{r}{2}$. Then for all $t \geq 1$ we have*

$$
\|u(\cdot, t)\|_{L^s} = \begin{cases} O(t^{-\frac{1}{2} + \frac{1}{s}}) & -\frac{1}{2} + \frac{1}{s} > -\frac{1}{r}(1+\gamma) + \frac{1}{rs}, \\ O(t^{-\frac{1}{2} + \frac{1}{s}}(\ln t)^{\frac{1}{s}}) & -\frac{1}{2} + \frac{1}{s} = -\frac{1}{r}(1+\gamma) + \frac{1}{rs}, \\ O(t^{-\frac{1}{r}(1+\gamma) + \frac{1}{rs}}) & -\frac{1}{2} + \frac{1}{s} < -\frac{1}{r}(1+\gamma) + \frac{1}{rs}. \end{cases}
$$

Remark. The condition $\frac{1}{s} < 1 + \gamma < \frac{1}{s} + \lambda_0^0 + r(\frac{1}{2} - \frac{1}{s})$ is sufficient to imply that $u(\cdot, t) \in L^s$ for $t \geq 1$, $1 \leq s < \infty$, but unless $\sin(\frac{\pi}{2}\gamma)\hat{v}_0(0) \pm \cos(\frac{\pi}{2}\gamma)\hat{v}_1(0) \neq 0$ they are not necessary. For example, if $\gamma = 0$ and $\hat{v}_1(0) = 0$, then $(1 + |\xi|)^{-1}$ in the pointwise estimates (Theorem 5.1.1) should be replaced by $(1 + |\xi|)^{-2}$, which is in L^s for all $1 \leq s \leq \infty$. Thus the conclusion of the above theorem would hold for $s = 1$.

Proof. If $1 + \gamma > \frac{1}{s}$ then $\|(1 + |\xi|)^{-(1+\gamma)}\|_{L^s(x)} = t^{\frac{1}{rs}}\|(1 + |\xi|)^{-(1+\gamma)}\|_{L^s(\xi)} = O(t^{\frac{1}{rs}})$. Thus the first term $t^{-(1+\gamma)/r}(1 + |\xi|)^{-(1+\gamma)}$ of the pointwise estimate of $|u(x,t)|$ in Theorem 5.1.1 has an L^s norm which is $O(t^{-\frac{1}{r}(1+\gamma) + \frac{1}{rs}})$. This works for all $1 \leq s \leq \infty$.

In the second term, in the case $1 \leq s < \infty$, we divide into two pieces

$$
\int_0^\infty \frac{(1 + \xi)^{\frac{s}{r-1}(1+\gamma-\frac{r}{2})}}{(1 + b)^{\frac{s\lambda_0^0}{r-1}}}\, dx = \left(\int_0^t + \int_t^\infty \right) \frac{(1 + \xi)^{\frac{s}{r-1}(1+\gamma-\frac{r}{2})}}{(1 + b)^{\frac{s\lambda_0^0}{r-1}}}\, dx
$$

$$
= I_1 + I_2.
$$

Clearly, in order for $I_2 < \infty$ we need $\frac{s}{r-1}(1+\gamma-\frac{r}{2}-\lambda_0^0) < -1$, which is true by our assumption. Changing variables $x = bt$ we have

$$I_2 = \int_1^\infty \frac{(1+bt^{\frac{r-1}{r}})^{\frac{s}{r-1}(1+\gamma-\frac{r}{2})}}{(1+b)^{\frac{s\lambda_0^0}{r-1}}} t\, db \leq t^{1+\frac{s}{r}(1+\gamma-\frac{r}{2})} \int_1^\infty \frac{(1+b)^{\frac{s}{r-1}(1+\gamma-\frac{r}{2})}}{(1+b)^{\frac{s\lambda_0^0}{r-1}}}\, db$$

$$= O(t^{\frac{s}{r}(1+\gamma)-s(\frac{1}{2}-\frac{1}{s})}).$$

In I_1 we note that $|b| \leq 1$; hence a good upper estimate is obtained by replacing the denominator by 1. Changing variables $x = \xi t^{1/r}$, we obtain

$$I_1 \leq \int_0^{t^{(r-1)/r}} (1+\xi)^{\frac{s}{r-1}(1+\gamma-\frac{r}{2})} t^{1/r}\, d\xi$$

We must consider three cases.

(1) Suppose $\frac{s}{r-1}(1+\gamma-\frac{r}{2}) > -1$, i.e. $\frac{s}{r}(1+\gamma) - s(\frac{1}{2}-\frac{1}{s}) > \frac{1}{r}$. Then

$$I_1 \leq t^{1/r} \frac{(1+\xi)^{1+\frac{s}{r-1}(1+\gamma-\frac{r}{2})}}{1+\frac{s}{r-1}(1+\gamma-\frac{r}{2})}\bigg|_{\xi=0}^{\xi=t^{(r-1)/r}} = O(t^{\frac{s}{r}(1+\gamma)-s(\frac{1}{2}-\frac{1}{s})}).$$

(2) Suppose $\frac{s}{r-1}(1+\gamma-\frac{r}{2}) = -1$, i.e. $\frac{s}{r}(1+\gamma) - s(\frac{1}{2}-\frac{1}{s}) = \frac{1}{r}$. Then

$$I_1 \leq t^{1/r}\ln(1+\xi)\bigg|_{\xi=0}^{\xi=t^{(r-1)/r}} = O(t^{\frac{1}{r}}\ln(t)).$$

(3) Suppose $\frac{s}{r-1}(1+\gamma-\frac{r}{2}) < -1$, i.e. $\frac{s}{r}(1+\gamma) - s(\frac{1}{2}-\frac{1}{s}) < \frac{1}{r}$. Then

$$I_1 \leq t^{1/r} \frac{(1+\xi)^{1+\frac{s}{r-1}(1+\gamma-\frac{r}{2})}}{1+\frac{s}{r-1}(1+\gamma-\frac{r}{2})}\bigg|_{\xi=0}^{\xi=t^{(r-1)/r}} = O(t^{\frac{1}{r}}).$$

Taking the $1/s$ power and multiplying by the factor $t^{-(1+\gamma)/r}$ we obtain our estimate of the second term. The three cases in the result stated correspond to cases (1), (2), and (3) above. Comparing this with our estimate of the first term yields the stated result, when $1 \leq s < \infty$.

In the second term of the pointwise estimate when $s = \infty$ and $1+\gamma \leq \lambda_0^0 + \frac{r}{2}$ it is elementary to show that

$$\sup_{b \leq 0} \frac{(1+|b|t^{\frac{r-1}{r}})^{\frac{1}{r-1}(1+\gamma-\frac{r}{2})}}{(1+|b|)^{\frac{\lambda_0^0}{r-1}}} = O(t^{\frac{1}{r}(1+\gamma)-\frac{1}{2}}).$$

Thus the L^∞ norm of the second term in the pointwise estimate is $O(t^{-1/2})$. Taking the larger of the first or second terms yields the stated result. $\square$

Theorem 5.2.2. *Adopt the assumptions and notation of Theorem 5.2.1.*

(1) *If* $-\frac{1}{2} + \frac{1}{s} > -\frac{1}{r}(1 + \gamma) + \frac{1}{rs}$ *then* $\liminf_{t\to\infty} t^{\frac{1}{2} - \frac{1}{s}} \|u(\cdot, t)\|_{L^s} > 0$ *unless the initial data is identically zero.*

(2) *If* $-\frac{1}{2} + \frac{1}{s} = -\frac{1}{r}(1 + \gamma) + \frac{1}{rs}$ *then* $\liminf_{t\to\infty} t^{\frac{1}{2} - \frac{1}{s}} (\ln t)^{-\frac{1}{s}} \|u(\cdot, t)\|_{L^s} > 0$ *unless* $\hat{v}_0(0) = 0$ *and* $\hat{v}_1(0) = 0.$

(3) *If* $-\frac{1}{2} + \frac{1}{s} < -\frac{1}{r}(1 + \gamma) + \frac{1}{rs}$ *then* $\liminf_{t\to\infty} t^{\frac{1}{r}(1+\gamma) - \frac{1}{rs}} \|u(\cdot, t)\|_{L^s} > 0$ *unless* $\hat{v}_0(0) = 0$ *and* $\hat{v}_1(0) = 0.$

Remark. In case (3) above we say the corresponding decay estimates are *typically sharp* in the sense that they fail to be sharp on a set of initial data with finite codimension.

Proof. (1) The L^s norm of $u(\cdot, t)$ over $\mathbb{R}$ is certainly larger than the same norm over the interval $x \le -t$, $t \ge 1$. The pairs (x, t) satisfying these conditions are in the outer region, so we write $u(x, t)$ as the sum of two integrals (as in the statement of Theorem 5.1.1), and decompose further the contour Γ_θ as the union of $C_0(-b), C_1(-b), \ldots, C_K(-b), \tilde{C}(-b)$ (when $r \ne 4K + 1$) or as the union of $C_0(-b), C_1(-b), \ldots, C_{K-1}(-b), C_K^+(-b), \tilde{C}(-b)$ (when $r = 4K + 1$). For each integral over any contour except $C_0(-b)$ apply Theorem 1.4.9, Theorem 1.7.5, or Corollary 1.7.6, all with $N = 0$ to estimate the size its contribution. For the two integrals over $C_0(-b)$ use Corollary 1.7.6 with $N = 2$.

The contributions of integrals over $\tilde{C}(-b)$ are $O(|x|^{-(1+\gamma)})$. Since we are assuming $\frac{1}{s} < 1 + \gamma$, this is in L^s. A short calculation shows that $t^{\frac{1}{2} - \frac{1}{s}}$ times the L^s norm of this function over $x \in (-\infty, -t]$ is $O(t^{\frac{1}{2} - (1+\gamma)})$. But $-\frac{1}{2} + \frac{1}{s} > -\frac{1}{r}(1 + \gamma) + \frac{1}{rs}$ implies that $\frac{1}{2} - (1 + \gamma) < -\frac{r-1}{r}(1 + \gamma - \frac{1}{s}) \le 0$. Thus $O(t^{\frac{1}{2} - (1+\gamma)}) = o(1)$ as $t \to \infty$.

The contributions of integrals over $C_j(-b)$, $1 \le j \le K$ (provided $K \ge 1$) are estimated by big O of

$$\exp\left\{-\sin(|S_j^0|)\left[(r-1)\rho t \left|\frac{b}{r\rho}\right|^{\frac{r}{r-1}} - a_+ \left|\frac{b}{r\rho}\right|^{\frac{1}{r-1}}\right]\right\} \frac{|b|^{\frac{1}{r-1}(1+\gamma - \frac{1}{2})}}{t^{\frac{1}{2}}(1 + |b|)^{\frac{\lambda_0^+}{r-1}}}.$$

The numbers a_+ and λ_0^+ are as described in section 3.2, and are associated either to $\bar{v}_-$ or to v_+. We can estimate this by an exponentially decaying function of t times an exponentially decaying function of b. The L^s norm of this function over the interval $b \le -1$ will be exponentially decaying, and therefore insignificant. The contributions of integrals over $C_K^+(-b)$ are dealt with in exactly the same manner.

The error term for the contributions from the contours $C_0(-b)$ is

$$O\left(\frac{|b|^{\frac{1}{r-1}(1+\gamma - \frac{3r}{2})}}{t^{3/2}(1 + |b|)^{\lambda_2^0/(r-1)}}\right) = O\left(t^{-3/2}|b|^{\frac{1}{r-1}(1+\gamma - \frac{3r}{2} - \lambda_2^0)}\right),$$

since $|b| \ge 1$. Since $0 \le 2 + \lambda_2^0 - \lambda_0^0 \le 2$, we have that $-\lambda_2^0 \le 2 - \lambda_0^0$, and thus the error term is

$$O\left(t^{-3/2}|b|^{\frac{1}{r-1}(1+\gamma - \frac{r}{2} - \lambda_0^0 - r + 2)}\right) = O\left(t^{-3/2}|b|^{\frac{1}{r-1}(1+\gamma - \frac{r}{2} - \lambda_0^0)}\right)$$

since $r \geq 2$. The assumption $1 + \gamma < \frac{1}{s} + \lambda_0^0 + r(\frac{1}{2} - \frac{1}{s})$ (for $1 \leq s < \infty$) or $1 + \gamma \leq \lambda_0^0 + \frac{r}{2}$ (for $s = \infty$) implies that this is in L^s for $b \leq -1$. A short calculation shows that $t^{\frac{1}{2} - \frac{1}{s}}$ times the L^s norm of this error is $O(t^{-1}) = o(1)$ as $t \to \infty$.

The leading-order contribution from the two integrals over the contour $C_0(-b)$ can be found, as we indicated above, from Corollary 1.7.6. It is given by the following expression.

$$\frac{\sqrt{2}}{\sqrt{\pi \rho r(r-1)t}} \left| \frac{b}{r\rho} \right|^{\frac{1}{r-1}(1+\gamma - \frac{r}{2})} \left\{ \mu_0 \left(\left| \frac{b}{r\rho} \right|^{\frac{1}{r-1}} \right) \cos \left[(r-1)\rho \left| \frac{b}{r\rho} \right|^{\frac{r}{r-1}} t - \frac{\pi}{4} \right] \right.$$
$$\left. - \eta_0 \left(\left| \frac{b}{r\rho} \right|^{\frac{1}{r-1}} \right) \sin \left[(r-1)\rho \left| \frac{b}{r\rho} \right|^{\frac{r}{r-1}} t - \frac{\pi}{4} \right] \right\}$$

where when $l > 0$ we define

$$\mu_0(l) = \frac{\hat{v}_0(l) - i\hat{v}_1(l) + \hat{v}_0(-l) + i\hat{v}_1(-l)}{2},$$
$$\eta_0(l) = i \frac{\hat{v}_0(l) - i\hat{v}_1(l) - \hat{v}_0(-l) - i\hat{v}_1(-l)}{2}.$$

If $\liminf_{t \to \infty} t^{\frac{1}{2} - \frac{1}{s}} \|u(\cdot, t)\|_{L^s} = 0$ then the same statement holds for the real and imaginary parts of $u(x, t)$, so without loss of generality we may assume v_0 and v_1 are real-valued. μ_0 and η_0 are entire functions, and so cannot vanish simultaneously on any set with an accumulation point without $\hat{v}_0(l) - i\hat{v}_1(l)$ and $\hat{v}_0(-l) + i\hat{v}_1(-l)$, and hence v_0 and v_1, vanishing identically. Suppose, by way of contradiction, that v_0 and v_1 do not vanish identically, and that $\liminf_{t \to \infty} t^{\frac{1}{2} - \frac{1}{s}} \|u(\cdot, t)\|_{L^s} = 0$. Define

$$M(b) = \left| \frac{b}{r\rho} \right|^{\frac{1}{r-1}(1+\gamma - \frac{r}{2})} \mu_0 \left(\left| \frac{b}{r\rho} \right|^{\frac{1}{r-1}} \right)$$
$$N(b) = \left| \frac{b}{r\rho} \right|^{\frac{1}{r-1}(1+\gamma - \frac{r}{2})} \eta_0 \left(\left| \frac{b}{r\rho} \right|^{\frac{1}{r-1}} \right)$$
$$\phi(b) = \arg[M(b) + iN(b)]$$
$$\theta(b) = (r-1)\rho \left| \frac{b}{r\rho} \right|^{\frac{r}{r-1}}.$$

With these definitions, the leading-order term can be rewritten as

$$\frac{\sqrt{2}}{\sqrt{\pi \rho r(r-1)t}} [M(b)^2 + N(b)^2]^{1/2} \cos[\phi(b) + \theta(b)t - \frac{\pi}{4}].$$

Choose $b_1 < b_2 < -1$ such that $[M(b)^2 + N(b)^2]^{1/2} \geq \delta > 0$ for all $b \in [b_1, b_2]$ and with $b_2 - b_1$ sufficiently small so that both $\phi(b)$ and $\theta(b)$ are very well approximated by their linear Taylor polynomials over the interval $[b_1, b_2]$. The assumption $\liminf_{t \to \infty} t^{\frac{1}{2} - \frac{1}{s}} \|u(\cdot, t)\|_{L^s} = 0$ implies that

$$\liminf_{t \to \infty} \left\| [M(b)^2 + N(b)^2]^{1/2} \cos[\phi(b) + \theta(b)t - \frac{\pi}{4}] \right\|_{L^s([b_1, b_2])} = 0.$$

Define S_t to be the set of all $b \in [b_1, b_2]$ such that $|\cos(\phi(b) + \theta(b)t - \frac{\pi}{4})| \geq 1/\sqrt{2}$. Then for sufficiently large t it is clear that the measure of S_t is close to $(b_2 - b_1)/2$. But the previous displayed assertion implies that

$$\liminf_{t \to \infty} \left\| [M(b)^2 + N(b)^2]^{1/2} \right\|_{L^s(S_t)} = 0,$$

in clear contradiction to the fact that $[M(b)^2 + N(b)^2]^{1/2} \geq \delta > 0$ for all $b \in [b_1, b_2]$. Thus we have proved that either v_0 and v_1 vanish identically or $\liminf_{t \to \infty} t^{\frac{1}{2} - \frac{1}{s}} \|u(\cdot, t)\|_{L^s} > 0$, as we had to show.

(2) This time we restrict attention to $-t^\beta \leq x \leq 0$, where $\frac{1}{r} < \beta < 1$, which is part of the inner region. As in (1) we can assume without loss of generality that v_0 and v_1 are real-valued. Decomposing the two integrals forming the solution $u(x, t)$ as in part (1), we apply Theorems 2.2.1, 2.2.2, 2.3.1, and Corollary 2.3.2 to write down the leading-order term and its error in that region. First we must show that under the assumptions of (2) the error terms decay in the L^s norm (over this region) like $o(t^{-\frac{1}{2} + \frac{1}{s}} (\ln t)^{\frac{1}{s}})$. First of all, consider the error term $W_1(x, t) = [t^{1/r}(1 + |\xi|)]^{-(2+\gamma)}$ associated to the contour $\tilde{C}(-b)$ (Theorem 2.2.1). A short calculation shows that $t^{\frac{1}{2} - \frac{1}{s}} (\ln t)^{-\frac{1}{s}}$ times the L^s norm of $W_1(x, t)$ over $x \in [-t^\beta, 0]$ is $O(t^{-\frac{1}{r}} (\ln t)^{-\frac{1}{s}}) = o(1)$ as $t \to \infty$. When $r = 4K + 1$ this error term changes form slightly (as in Theorem 2.2.2), but the same estimates apply.

Next consider the error term associated to the contours $C_j(-b)$, $1 \leq j \leq K$, as well as $C_K^+(-b)$ when $r = 4K + 1$, as described in Theorem 2.3.1, and Corollary 2.3.2. This error term is big O of

$$\exp\left\{ -(r-1)\rho \left| \frac{\xi}{r\rho} \right|^{\frac{r}{r-1}} \sin(|S_j^\alpha|) \right\} \frac{(1 + |\xi|)^{\frac{1}{r-1}[2+\gamma - \frac{r}{2}]}}{t^{\frac{2+\gamma}{r}}}.$$

Because of the exponential decay which is evident, this is in L^s over $x \in (-\infty, 0]$. Thus this error term is very similar to the last one, and the outcome is the same, namely $O(t^{-\frac{1}{r}} (\ln t)^{-\frac{1}{s}}) = o(1)$ as $t \to \infty$.

Finally, consider the error term corresponding to the contour $C_0(-b)$, which is big O of $t^{-\frac{2+\gamma}{r}} (1 + |\xi|)^{\frac{1}{r-1}[2+\gamma - \frac{r}{2}]}$. A short calculation, using the assumption characterizing case (2), shows that $t^{\frac{1}{2} - \frac{1}{s}} (\ln t)^{-\frac{1}{s}}$ times the L^s norm of this quantity over $x \in [-t^\beta, 0]$ is $O(t^{(\beta - 1)/(r-1)} (\ln t)^{-1/s}) = o(1)$ as $t \to \infty$.

It remains then to consider the leading-order term of the inner expansion. This is of the form

$$\frac{1}{2\pi t^{\frac{1+\gamma}{r}}} \left\{ [\hat{v}_0(0) - i\hat{v}_1(0)] \overline{A(-\xi; r, \rho, \gamma, \Gamma_{V_0})} + [\hat{v}_0(0) + i\hat{v}_1(0)] A(-\xi; r, \rho, \gamma, \Gamma_{V_0}) \right\}$$

$$= \frac{1}{(r\rho t)^{\frac{1+\gamma}{r}}} \left\{ \hat{v}_0(0) [|D|^\gamma G] \left(\frac{\xi}{(r\rho)^{1/r}} \right) + \hat{v}_1(0) [|D|^\gamma \mathcal{H}G] \left(\frac{\xi}{(r\rho)^{1/r}} \right) \right\}$$

$$= t^{-\frac{1+\gamma}{r}} w(\xi),$$

where G, $\mathcal{H}G$ and w were defined in Theorem 5.1.2. As we saw at the end of section 2.4, $A(-\xi; r, \rho, \gamma, \Gamma_{V_0})$ can be written as the sum of similar functions over the component contours. Each of these functions contributes an asymptotic expansion as $\xi \to -\infty$, which are found in Lemma 2.1.1. The part from $\tilde{C}(-\xi)$ is $O(|\xi|^{-(1+\gamma)})$. The parts from $C_j(-\xi)$, $1 \leq j \leq K$, and $C_K^+(-\xi)$ are exponentially decaying as $\xi \to -\infty$. The part from $C_0(-\xi)$ is $O(|\xi|^{\frac{1}{r-1}(1+\gamma-\frac{1}{2})})$. By the assumption $-\frac{1}{2} + \frac{1}{s} = -\frac{1}{r}(1+\gamma) + \frac{1}{rs}$ this is the same as $O(|\xi|^{-\frac{1}{s}})$. In view of the assumption $\frac{1}{s} < 1 + \gamma$, we have to the leading-order as $\xi \to -\infty$

$$w(\xi) \sim \frac{\sqrt{2}}{\sqrt{\pi\rho r(r-1)}} \left|\frac{\xi}{r\rho}\right|^{\frac{1}{r-1}(1+\gamma-\frac{r}{2})} \left\{ \hat{v}_0(0) \cos\left[(r-1)\rho\left|\frac{\xi}{r\rho}\right|^{\frac{r}{r-1}} - \frac{\pi}{4}\right] \right.$$
$$\left. - \hat{v}_1(0) \sin\left[(r-1)\rho\left|\frac{\xi}{r\rho}\right|^{\frac{r}{r-1}} - \frac{\pi}{4}\right] \right\}$$
$$= \frac{\sqrt{2}}{\sqrt{\pi\rho r(r-1)}} \left|\frac{\xi}{r\rho}\right|^{\frac{1}{r-1}(1+\gamma-\frac{r}{2})} |\hat{v}_0(0) + i\hat{v}_1(0)| \cos\left[(r-1)\rho\left|\frac{\xi}{r\rho}\right|^{\frac{r}{r-1}} - \frac{\pi}{4} + \phi\right]$$

where $\phi = \arg[\hat{v}_0(0) + i\hat{v}_1(0)]$. By choosing $M > 0$ sufficiently large we can insure that the L^s norm of $w(\xi)$ over the x interval $[-t^\beta, -Mt^{1/r}]$ for large t is equivalent to the L^s norm of its leading-order asymptotic expansion. So if we assume $\liminf_{t\to\infty} t^{\frac{1}{r}(1+\gamma)-\frac{1}{rs}}(\ln t)^{-1/s}\|u(\cdot,t)\|_{L^s} = 0$ then as a consequence we have that

$$\liminf_{t\to\infty} \frac{|\hat{v}_0(0) + i\hat{v}_1(0)|}{(\ln t)^{1/s}} \left\| |\xi|^{-1/s} \cos\left[(r-1)\rho\left|\frac{\xi}{r\rho}\right|^{\frac{r}{r-1}} - \frac{\pi}{4} + \phi\right] \right\|_{L^s((-t^{\beta-1/r}, -M])}$$
$$= 0.$$

When $s = \infty$ this statement implies that $\hat{v}_0(0) = 0$ and $\hat{v}_1(0) = 0$. When $s < \infty$, we can restate this as

$$|\hat{v}_0(0) + i\hat{v}_1(0)|^s \cdot \liminf_{\tau\to\infty} \frac{1}{\ln \tau} \int_M^\tau \left|\cos\left[(r-1)\rho\left|\frac{\xi}{r\rho}\right|^{\frac{r}{r-1}} - \frac{\pi}{4} + \phi\right]\right|^s \frac{d\xi}{\xi} = 0.$$

But this again implies $|\hat{v}_0(0) + i\hat{v}_1(0)| = 0$, and hence $\hat{v}_0(0) = 0$ and $\hat{v}_1(0) = 0$.

(3) Again one restricts to $-t^\beta \leq x \leq 0$, where $\frac{1}{r} < \beta < 1$. The result follows in a similar manner as (2). □

As an example of an application of our estimates, we will evaluate the degree of sharpness of the decay rates of various (semi)norms of the solution which were employed by Christ and Weinstein [CW] in their proof of the decay estimate $\|u(\cdot,t)\|_{L^\infty} = O(t^{-1/3})$ for small solutions $u(x,t)$ of the GKdV equation with nonlinearity $u^3 u_x$ (i.e. $p = 4$). We assume $r = 3$, $v_1 \equiv 0$, and $0 \leq \gamma_0 < \lambda_0^0$, so that we expect $u_0 \in L^1$ (see below). We will list first the decay estimate they proved, and then describe the sharp (or typically sharp) decay estimate for the solution of the linearized KdV equation. Case (1), (2), or (3) refers to Theorem 5.2.1.

(1) $\|u(\cdot,t)\|_{L^4} = O(t^{-1/4+\delta})$, where $0 < \delta$ is small. The sharp estimate for the linearized equation is:

$$\|u(\cdot,t)\|_{L^4} = \begin{cases} O(t^{-1/4}) & 0 < \gamma_0 < \lambda_0^0, \quad \text{Case (1),} \\ O(t^{-1/4}(\ln t)^{1/4}) & \gamma_0 = 0 \quad \text{Case (2).} \end{cases}$$

(2) $\|u(\cdot,t)\|_{L^\infty} = O(t^{-1/3})$. The sharp estimate for the linearized equation is:

$$\|u(\cdot,t)\|_{L^\infty} = \begin{cases} O(t^{-1/2}) & \frac{1}{2} \le \gamma_0 < \lambda_0^0, & \text{Cases (1) and (2),} \\ O(t^{-\frac{1}{3}(1+\gamma_0)}) & 0 \le \gamma_0 < \frac{1}{2} & \text{Case (3).} \end{cases}$$

(3) $\|\partial_x u(\cdot,t)\|_{L^\infty} = O(t^{-1/3-\epsilon})$, where $\epsilon > 0$ is small. In this case we need $\gamma_0 + \frac{1}{2} \le \lambda_0^0$. The the sharp estimate for the linearized equation is

$$\|\partial_x u(\cdot,t)\|_{L^\infty} = O(t^{-1/2}), \qquad \text{Case (1).}$$

(4) $\| |D|^{1/2} u(\cdot,t)\|_{L^4} = O(t^{-1/4})$. This estimate is already sharp, case (1).
(5) $\| |D|^{1/2} u(\cdot,t)\|_{L^\infty} = O(t^{-1/3-\epsilon})$, where $\epsilon > 0$ is small. The the sharp estimate for the linearized equation is

$$\| |D|^{1/2} u(\cdot,t)\|_{L^\infty} = O(t^{-1/2}), \qquad \text{Case (1).}$$

(6) $\| |D|^2 u(\cdot,t)\|_{L^2} = O(1)$. This estimate is already sharp, case (1).

When $\gamma_0 = 0$, (2), (4), and (6) are already sharp, (1) is almost sharp, and (3) and (5) would be sharp if $\epsilon = 1/6$. Estimate (2) illustrates the fact that the assumption $u_0 \in L^1$ does not determine the decay rate in general unless one makes the additional stipulation that $\hat{u}_0(0) \ne 0$, which forces $\gamma = 0$. If the assumption were $u_0 \in L^2$ the situation would be even worse, since the decay rate depends in some cases on the value of γ, which can vary considerably for functions in L^2. There is no simple condition, analogous to $\hat{u}_0(0) \ne 0$, to impose on an L^2 function which forces γ to assume a preferred value. For this reason we have not attempted to prove decay estimates under assumptions like these on the initial data. Instead we have focussed on the essential parameters which actually determine the sharp decay rate.

Smoothing Effects. Recall that we are assuming that $\bar{v}_-$ and v_+, where $v_\pm = v_0^{\checkmark} \pm i v_1^{\checkmark}$, satisfy the standard assumptions (see the end of section 1.5). It follows from the Hausdorff-Young inequality that

$$\frac{1}{s} < 1 + \gamma_0 < \frac{1}{s} + \lambda_0^0, \qquad \text{implies} \qquad u_0 \in L^s, \qquad 2 \le s \le \infty,$$

where $u_0 = |D|^{\gamma_0}(v_0 + \mathcal{H}v_1)$. We will not attempt here to characterize the pairs γ_0, λ_0^0 where $u_0 \in L^s$ for $1 \le s < 2$. It seems likely (and we will henceforth assume) that the same inequality $\frac{1}{s} < 1 + \gamma_0 < \frac{1}{s} + \lambda_0^0$ is a sufficient condition when $1 \le s < 2$. When $s = 1$ and $v_1 \equiv 0$, $\gamma_0 = 0$ is obviously sufficient, since $\lambda_0^0 > 0$ and v_0 has compact support. As usual, let $u(x,t)$ denote the solution of $u_t - \rho |D|^{r-1} u_x = 0$ with initial data u_0.

The sharpness of the pointwise estimates in Theorem 5.1.1 allow us to give upper bounds on the magnitude of the "dispersive smoothing effect" in L^s, i.e. if $u_0 \in L^s$ then $|D|^{\gamma - \gamma_0} u(\cdot,t) \in L^s$ for $t \ge 1$. In particular, there exist initial data $u_0 \in L^s$, $1 \le s \le \infty$, satisfying our assumptions, as well as both of the conditions $\sin(\frac{\pi}{2}\gamma)\hat{v}_0(0) \pm \cos(\frac{\pi}{2}\gamma)\hat{v}_1(0) \neq 0$, such that if $\gamma - \gamma_0 > r(\frac{1}{2} - \frac{1}{s})$ then $|D|^{\gamma-\gamma_0}u(\cdot, t) \notin L^s$ for any $t \geq 1$. The solution does not fail to be smooth, but it decays in space too slowly to be a member of the same L^s space that u_0 was in. Also, if $1 \leq s < \infty$ we can give a lower bound on $\gamma - \gamma_0$: the "smoothing effect" in L^s cannot hold for any $\gamma - \gamma_0 < 0$.

Within the family of solutions described by Theorem 5.1.1 the dispersive smoothing effect in L^s is characterized by $0 \leq \gamma - \gamma_0 \leq r(\frac{1}{2} - \frac{1}{s})$. The effect vanishes when $s = 2$ as expected. When $s = \infty$ we say the solution "gains $r/2$ derivatives in L^∞". A smoothing effect in L^∞ of this magnitude has not previously been noted in the literature, and we do not know if it will hold for general $u_0 \in L^\infty$. Note that nothing is special about $t \geq 1$; we could have proved all our results for all $t \geq \epsilon$, where $\epsilon > 0$ is fixed.

Of course smoothing effects can be measured in many ways. For example, one might suppose that $u_0 \in L^{s_0}$ and ask whether $|D|^{\gamma-\gamma_0}u(\cdot, t) \in L^s$ when $t \geq 1$. Theorem 5.1.1 shows that if this $L^{s_0} - L^s$ "smoothing" effect holds then

$$\frac{1}{s} - \frac{1}{s_0} \leq \gamma - \gamma_0 \leq r\left(\frac{1}{2} - \frac{1}{s}\right) + \frac{1}{s} - \frac{1}{s_0}.$$

A partial converse can be proved by using L^{s_1} estimates of $|D|^{\gamma-\gamma_0}G$, where $(r\rho t)^{-1/r}G(x(r\rho t)^{-1/r})$ is the fundamental solution, and Young's convolution inequality, under the assumption

$$\frac{1}{s} - \frac{1}{s_0} < \gamma - \gamma_0 < r\left(\frac{1}{s_0} - \frac{1}{2} - \frac{1}{s}\right) + \frac{1}{s} - \frac{1}{s_0}.$$

These two restictions on $\gamma - \gamma_0$ agree (except for borderline cases) when $s_0 = 1$.

Another dimension of the smoothing effect is how the L^s norm of $|D|^{\gamma-\gamma_0}u(\cdot, t)$ decays or grows as $t \to \infty$. This information can be extracted from Theorem 5.2.1. For example, consider $L^1 - L^\infty$ smoothing, where $v_1 \equiv 0$, and $r = 3$. Thus we assume $0 \leq \gamma_0 < \lambda_0^0$, so that we expect $u_0 \in L^1$. If $\gamma = \gamma_0 + \frac{1}{2}$, then $1 + \gamma \geq \frac{3}{2}$ so that case (3) in Theorem 5.2.1 never applies, and we find that $\| |D|^{1/2}u(\cdot, t)\|_{L^\infty} = O(t^{-1/2})$. This of course agrees with what is known. However, we can now assert by Theorem 5.2.2 that this decay rate is sharp. As another example, consider $L^\infty - L^\infty$ smoothing when $r = 3$. We assume $0 < 1 + \gamma_0 < \lambda_0^0$, so that $u_0 \in L^\infty$. Suppose $\gamma = \gamma_0 + \frac{3}{2}$. This implies we are in case (1) of Theorem 5.2.1 and that $\| |D|^{3/2}u(\cdot, t)\|_{L^\infty} = O(t^{-1/2})$. Fewer numbers of derivatives of the solution however will in general decay more slowly, at rates which depend on γ. An example of growth as $t \to \infty$ is $r = 3$, $\gamma_0 = \gamma = 0$, $u_0 = v_0$, $\lambda_0^0 = 2$. This falls into case (1): $\|u(\cdot, t)\|_{L^1} = O(t^{1/2})$, despite the fact that $\int_{-\infty}^\infty u(x, t)\, dx = \int_{-\infty}^\infty u_0(x)\, dx$ for all t.

Smoothing effects for linear and nonlinear dispersive effects have been studied very extensively. We refer to Kenig, Ponce Vega [KPV], Ben-Artzi Treves [bAT], Craig and Goodman [CG], and Colin [Col] for further discussion and references. We have made considerably more restrictive assumptions on u_0 than has been customary in the literature until now. The payoff has been the detailed knowledge of the solution for $t \geq \epsilon > 0$ that we have now achieved. Also, the results in the literature invariably control the solution on the entire time interval $(0, \infty)$, whereas we only do so on intervals like (ϵ, ∞). However, the question about the degree to which those results could be improved is frequently left open.

One smoothing result in the literature which is clearly optimal in some sense is due to Kenig, Ponce, and Vega [KPV]. It states that the solution of the initial-value problem

$$u_t - \rho |D|^{r-1} u_x = 0,$$
$$u(x, 0) = u_0(x)$$

satisfies the following *equality* for all $x \in \mathbb{R}$

$$\int_{-\infty}^{\infty} |\,|D|^{(r-1)/2} u(x,t)|^2 \, dt = c_r \int_{-\infty}^{\infty} |u_0(y)|^2 \, dy,$$

where $c_r > 0$ is a constant. But does this mean that for each fixed x the quantity $|D|^{(r-1)/2} u(x,t)$ decays as $t \to \infty$ like an L^2 function of t and no better (in general)? The above equality cannot answer this question. But using our results we can answer it definitively. When x is a constant, one should look in the inner region as $t \to \infty$, where we have seen that the solution decays like $O(t^{-(1+\gamma)/r})$. On the other hand, u_0 with parameters γ_0, λ_0^0 is in L^2 if $-\frac{1}{2} < \gamma_0 < -\frac{1}{2} + \lambda_0^0$. We define $\gamma = \gamma_0 + (r-1)/2$. Then $(1+\gamma)/r = 1/2 + (\gamma_0 + 1/2)/r$. Thus $|D|^{(r-1)/2} u(x,t)$ decays as $t \to \infty$ like $O(t^{-1/2-(\gamma_0+1/2)/r})$, which can be made to decay as slowly as any L^2 function of t with power-like decay by choosing γ_0 sufficiently near $-1/2$. In this respect the equality of Kenig Ponce and Vega makes an optimal statement about the large-time behavior of the solution.

Using similar arguments one can evaluate the large-time optimality of various other smoothing estimates described in the above references, such as Strichartz estimates, $L_t^p - L_x^q$ estimates, estimates in weighted Sobolev spaces, etc..

5.3 Asymptotic Balance as $t \to \infty$

In this section we will develop some principles for evaluating the leading-order large-time behavior of solutions of the nonlinear dispersive-dissipative equation

$$u_t + \partial_x P(u) + Q(D)u + iR(D)u = 0,$$
$$u(x, 0) = u_0(x),$$

introduced in section 5.1. In contrast to [D2], we will focus here on the case where the most important terms as $t \to \infty$ are u_t and $iR(D)u$. The principle of self-similarity, which was sufficient for that work, must be replaced by more general ideas.

First of all, we have found it necessary to consider various asymptotic regions, since the solution of the homogeneous linear dispersive equation has been found to be quite region dependent. The inner region had a preferred variable, namely $\xi = xt^{-1/r}$. The outer region had the preferred variable $b = xt^{-1}$. It was also found to be helpful to introduce an intermediate variable $\zeta = xt^{-\beta}$ where $\frac{1}{r} < \beta < 1$. If we relax this restriction on the variable β, we find that we can refer to an asymptotic region by a range of values for β: $\beta \in [0,1)$ is the inner region, $\beta \in (\frac{1}{r}, \infty)$ is the outer region, $\beta \in (\frac{1}{r}, 1)$ is the overlap region. (We are assuming $r \geq 2$ because when $1 < r < 2$ there are more regions to consider.)

Because of the variation in the size of the solution in the different asymptotic regions, we found it necessary to abandon the practice of measuring errors in spatial norms such as L^s. Instead we used weighted L^∞ estimates, where the weights reflected the actual behavior of the errors being estimated.

Suppose $u(x,t)$ is a function. Suppose for each $\beta \geq 0$ we define

$$d^{\pm}(\beta; u) \stackrel{\text{def}}{=} \sup\{d \in \mathbb{R} \mid \limsup_{t \to \infty} t^d |u(\zeta t^{\beta}, t)| < \infty \text{ for all } \pm\zeta > 0\}$$

$d^+(\beta; u)$ and $d^-(\beta; u)$ are called the *decay functions* of u. If u is understood, or if both of the signs $\pm$ are intended, we write $d^{\pm}(\beta; u) = d^{\pm}(\beta) = d(\beta; u) = d(\beta)$. The decay functions give us the asymptotic power-law decay rate of u in all the various asymptotic regions indexed by the number β. However, $d(\beta)$ exists even when the behavior of $u(\zeta t^{\beta}, t)$ is not precisely power-law decay; logarithmic modifications are allowed as well. In most cases of solutions of linear dispersive equations we will have $d(\beta) < \infty$. An example of a case where this is not true is for linearized KdV-type equations where $r = 4J - 1$, $J \geq 1$ an integer, for $b > 0$ when $\sin(\frac{\pi}{2}\gamma)v_0(x) = \cos(\frac{\pi}{2}\gamma)v_1(x)$ for all $x \in \mathbb{R}$. In that special but important case the solution decays exponentially as $\xi \to \infty$, and two such exponentially decaying factors involving different values of ζ are not asymptotically equivalent as $t \to \infty$. We will say $d(\beta) = \infty$ in such a case. This possibility does not ruin the general utility of the decay function concept.

Suppose that $u(x,t)$ solves the above displayed evolution equation, and that $T_1(u), T_2(u)$ are two of the terms of that equation, evaluated at the solution u. If an asymptotic region is defined by an interval $[\beta_1, \beta_2]$ together with a choice of the sign $\pm x > 0$ of x, we say the term $T_1(u)$ is *asymptotically dominant* in the region $[\beta_1, \beta_2]$ over the term $T_2(u)$ if $d^{\pm}(\beta; T_1(u)) < d^{\pm}(\beta; T_2(u))$ for all $\beta \in [\beta_1, \beta_2]$. If the asymptotic region is everything, i.e. $\beta \in [0, \infty)$ and both signs $x < 0$ and $x > 0$ are included, then we say $T_1(u)$ is *globally asymptotically dominant* over $T_2(u)$, and write $T_1(u) \gg T_2(u)$. If both $d(\beta; T_1(u)) = \infty$ and $d(\beta; T_2(u)) = \infty$ then in order to decide which term is dominant one must look directly at the exponentially decaying factors involved. Thus a separate analysis must be made in this case. We say the terms $T_1(u)$ and $T_2(u)$ are *asymptotically balanced* in the region $[\beta_1, \beta_2]$, $\pm x > 0$, and write $T_1(u) \sim T_2(u)$, if $d^{\pm}(\beta; T_1(u)) = d^{\pm}(\beta; T_2(u))$ for all $\beta \in [\beta_1, \beta_2]$. It is possible for two terms to be asymptotically balanced in one region, and yet one asymptotically dominates the other in a different region (see Example 3 below). At a given value of β there must be at least two terms which are balanced with each other and more slowly decaying than the other terms. We say these dominant terms constitute the leading-order *asymptotic balance* for that solution in the region defined by the number β. The concept of asymptotic balance is useful as a means of classifying asymptotic behavior. We do not mean to imply however that terms that decay more rapidly than those forming the asymptotic balance have no influence on the form of the leading-order large-time asymptotic behavior.

We are interested in delineating and understanding the case where the leading-order asymptotic balance is between the u_t term and the dispersive term $iR(D)u$. We specialize to the case of KdV-type dispersive terms, $iR(D)u = -\rho|D|^{r-1}u_x$. A similar analysis can be performed for Schrödinger-type equations. First we will compute the decay functions for an arbitrary solution of the homogeneous linear dispersive equation.

Lemma 5.3.1 (Decay Functions for u). *Suppose $u(x,t)$ is the solution of the equation $u_t - \rho|D|^{r-1}u_x = 0$ with initial data $u_0 = |D|^{\gamma_0}(v_0 + \mathcal{H}v_1)$. Suppose, as usual, that $\gamma_0 > -1$ and the functions v_0, v_1 are such that $\bar{v}_-$ and v_+ (see Theorem 4.4.1) satisfy the standard assumptions, with decay parameters $\lambda_0^0(\bar{v}_-)$ and $\lambda_0^0(v_+)$ respectively (see Theorem 5.1.1 for a definition of these parameters). Define $\lambda_0 = \min\{\lambda_0^0(\bar{v}_-), \lambda_0^0(v_+)\}$. Then the decay functions for u are described as follows.*

(1) *Suppose $\sin(\frac{\pi}{2}\gamma_0)\hat{v}_0(0) + \cos(\frac{\pi}{2}\gamma_0)\hat{v}_1(0) \neq 0$. If $-1 < \gamma_0 \leq -\frac{1}{2}$ then the decay function for $x \leq 0$ and $\beta \geq 0$ is given by*

$$d^-(\beta; u) = \begin{cases} \frac{1+\gamma_0}{r} & 0 \leq \beta < \frac{1}{r}, \\ \beta(1 + \gamma_0) & \frac{1}{r} \leq \beta < \infty. \end{cases}$$

If $\gamma_0 \geq -\frac{1}{2}$ define

$$\beta_0 = \begin{cases} \infty & \lambda_0 \leq r(\frac{1}{2} + \gamma_0), \\ \frac{\lambda_0 - (\frac{1}{2} + \gamma_0)}{\lambda_0 - r(\frac{1}{2} + \gamma_0)} & \lambda_0 > r(\frac{1}{2} + \gamma_0). \end{cases}$$

Then the decay function for $\gamma_0 \geq -\frac{1}{2}$, $x \leq 0$ and $\beta \geq 0$ is given by

$$d^-(\beta; u) = \begin{cases} \frac{1+\gamma_0}{r} & 0 \leq \beta < \frac{1}{r}, \\ \frac{1+\gamma_0}{r} - (\beta - \frac{1}{r})\frac{(1+\gamma_0-\frac{r}{2})}{r-1} & \frac{1}{r} \leq \beta < 1, \\ \frac{1}{2} + (\beta - 1)\frac{(\lambda_0 + \frac{r}{2} - 1 - \gamma_0)}{r-1} & 1 \leq \beta < \beta_0, \\ \beta(1 + \gamma_0) & \beta_0 \leq \beta < \infty. \end{cases}$$

(2) *Suppose $\sin(\frac{\pi}{2}\gamma_0)\hat{v}_0(0) - \cos(\frac{\pi}{2}\gamma_0)\hat{v}_1(0) \neq 0$. Then the decay function for $x \geq 0$ and $\beta \geq 0$ is given by*

$$d^+(\beta; u) = \begin{cases} \frac{1+\gamma_0}{r} & 0 \leq \beta < \frac{1}{r}, \\ \beta(1 + \gamma_0) & \frac{1}{r} \leq \beta < \infty. \end{cases}$$

Remark. The number $\beta_0 \geq 1$ is a transition point in the outer region ($x < 0$), which is finite for especially smooth initial data (λ_0 sufficiently large). If $\beta >$

β_0 then the contributions of the contour $\tilde{C}$ dominate; if $\beta < \beta_0$ the contributions of the contour C_0 dominate. It illustrates the fact that one can increase the rate of spatial decay of the solution by increasing λ_0 only up to a certain point, namely $\lambda_0 = r(\frac{1}{2} + \gamma_0)$. The number β_0 and the value of the decay function $d^-(\beta)$ for $\beta > \beta_0$ is sensitive to whether or not the condition $\sin(\frac{\pi}{2}\gamma_0)\hat{v}_0(0) + \cos(\frac{\pi}{2}\gamma_0)\hat{v}_1(0) \neq 0$ holds. The value of the decay function $d^+(\beta)$ for $\beta > 1/r$ is sensitive to whether or not the condition $\sin(\frac{\pi}{2}\gamma_0)\hat{v}_0(0) - \cos(\frac{\pi}{2}\gamma_0)\hat{v}_1(0) \neq 0$ holds. These conditions exclude some interesting special cases. The modifications of these formulae in these special cases are not difficult to work out, and we will omit this. (Sometimes we cannot avoid an infinite hierarchy of non-vanishing conditions.) Our concern here is with the "generic" case.

Proof. Theorem 5.1.1. $\square$

The nonlinear dispersive-dissipative equation we wrote down earlier is a special case of the inhomogeneous linear dispersive equation

$$u_t + iR(D)u = f,$$
$$u(x,0) = u_0(x),$$

where $f(x,t)$ is some reasonable function. This solution has the well-known integral representation

$$u(t) = u_H(t) + u_I(t) \stackrel{\text{def}}{=} e^{-iR(D)t}u_0 + \int_0^t e^{-iR(D)(t-\tau)}f(\tau)\,d\tau.$$

The leading-order large-time asymptotics of this solution depend on the behavior of $f(\tau)$. One can arrange that an arbitrary function u satisfies this equation by choosing u_0 and f appropriately. Consequently, we only wish to consider pairs (u_0, f) such that $d(\beta, f) > d(\beta; R(D)u)$ for all $\beta \geq 0$, i.e. $R(D)u \gg f$. A solution u with $R(D)u \gg f$ is said to be (globally) *asymptotically dispersive*. More generally, if $[\beta_1, \beta_2]$, $\pm x > 0$, defines an asymptotic region, then we say u is *asymptotically dispersive in the region* $[\beta_1, \beta_2]$ if $d^\pm(\beta, f) > d^\pm(\beta; R(D)u)$ for all $\beta \in [\beta_1, \beta_2]$. The inhomogeneous term f does not contribute to the leading-order asymptotic balance in an asymptotically dispersive solution. However, we must be careful not to jump to conclusions about the nature of the leading-order large-time asymptotics of asymptotically dispersive solutions.

We will say that the leading-order large-time asymptotics of u are *free* if there exists an initial datum U_0, together with the corresponding solution $U(t) = e^{-iR(D)t}U_0$ of the homogeneous linear dispersive equation, such that for all $\beta \geq 0$ we have

$$d(\beta; u - U) > d(\beta; U).$$

This means that u and U have the same leading-order large-time asymptotics in every asymptotic region. In this case one also says that f is *short range* and that the solution u is *asymptotically free*. The definition of asymptotic freedom in a particular region is an obvious modification of the above. Suppose we also have the additional relations:

$$d(\beta; u_t - U_t) > d(\beta; U_t),$$
$$d(\beta; R(D)(u - U)) > d(\beta; R(D)U),$$

for all $\beta \geq 0$. In such cases we say u is asymptotically free *in the strong sense*. Note that $d(\beta; U_t) = d(\beta; R(D)U)$. Then it follows that the solution u is asymptotically dispersive. It is easy to construct examples of asymptotically dispersive solutions with free asymptotics.

Example 1. Suppose $R(D) = \frac{1}{2}D^2$, v_0 satisfies the standard assumptions, with decay parameters λ_n, $n = 0, 1, \ldots$, i.e. $\sup_{k \in \mathbb{R}}(1 + |k|)^{n+\lambda_n}|\hat{v}_0^{(n)}(k)| < \infty$, $\hat{v}_0(0) \neq 0$, and

$$u(x, t) = \frac{1}{\sqrt{2\pi t}}\hat{v}_0(b) \exp\left[\frac{i\xi^2}{2} - \frac{i\pi}{4}\right],$$

where, as usual, $b = xt^{-1}$, and $\xi = xt^{-1/2}$. Then $u(x, t)$ satisfies

$$f \stackrel{\text{def}}{=} u_t - \frac{i}{2}u_{xx} = O\left(t^{-5/2}(1 + |b|)^{-2-\lambda_2}\right)$$

for $t \geq 1$. Since both u_t and u_{xx} are $O(t^{-3/2}(1 + |\xi|)^2(1 + |b|)^{-\lambda_0})$ we see that u is asymptotically dispersive. In this case we have $U_0 = v_0$, $U(t) = e^{-\frac{i}{2}D^2 t}v_0$, and the asymptotics of u are free since $u(x, t)$ coincides with the leading-order large-time asymptotics of $U(x, t)$.

If the leading-order asymptotics of an asymptotically dispersive solution fail to be free, we will call them *modified*. In such cases f is said to be *long range*. Solutions with modified leading-order asymptotics exist.

Example 2. Suppose $R(D)$, v_0, and λ_n are as in Example 1 above, and $\lambda_0 > 1/2$. Suppose

$$u(x, t) = \frac{1}{\sqrt{2\pi t}}\hat{v}_0(b) \exp\left[\frac{i\xi^2}{2} - \frac{i\pi}{4} + \frac{ib\ln(t)}{2\pi}|\hat{v}_0(b)|^2\right].$$

Then $u(x, t)$ satisfies (see [HO]) for all $t \geq 1$

$$f \stackrel{\text{def}}{=} u_t - \frac{i}{2}u_{xx}$$
$$= O\left(\frac{|b|}{t^{3/2}(1 + |b|)^{3\lambda_0}} + \frac{1}{t^{5/2}}\left[\frac{(\ln t)^2}{(1 + |b|)^{\alpha_2}} + \frac{\ln t}{(1 + |b|)^{\alpha_1}} + \frac{1}{(1 + |b|)^{\alpha_0}}\right]\right).$$

where $\alpha_0 = \lambda_2 + 2$, $\alpha_1 = \min\{2\lambda_0 + \lambda_2 + 1, \lambda_0 + 2\lambda_2 + 1\}$, and $\alpha_2 = 3\lambda_0 + 2\lambda_1$. Since both u_t and u_{xx} are $O(t^{-3/2}(1 + |\xi|)^2(1 + |b|)^{-\lambda_0})$ we see that u is asymptotically dispersive. However, the "logarithmic phase shift" present in the definition of $u(x, t)$ will cause it to deviate from the leading-order large-time asymptotics (in the outer region) of any solution of the homogeneous (linear) Schrödinger equation. Note that

$$\partial_x(|u|^2 u) = O\left(\frac{|b|}{t^{3/2}(1+|b|)^{3\lambda_0}} + \frac{1}{t^{5/2}}\left[\frac{\ln t}{(1+|b|)^{4\lambda_0+\lambda_1}} + \frac{1}{(1+|b|)^{2\lambda_0+\lambda_1+1}}\right]\right),$$

$$\partial_x(|u|^2 u) = f + O\left(\frac{1}{t^{5/2}}\left[\frac{(\ln t)^2}{(1+|b|)^{\alpha_2}} + \frac{\ln t}{(1+|b|)^{\alpha_1}} + \frac{1}{(1+|b|)^{\alpha_0}}\right]\right),$$

where $\alpha_0, \alpha_1, \alpha_2$ were defined above. Thus in the region $|x|/(\ln t)^2 \to \infty$ we have that $u_t - \frac{i}{2}u_{xx} - \partial_x(|u|^2 u)$ is asymptotically smaller than each of the three terms comprising it, i.e. u is an asymptotic solution of the Derivative Nonlinear Schrödinger equation in that region. We will return to this topic at the end of this section.

Diagnosing the leading-order large-time asymptotics. Here we intend to develop some rules for deciding what the leading-order large-time asymptotics of a given solution should look like, given the initial data. Many of these rules will be heuristic and conjectural in nature, and remain to be verified in further studies.

The General Inhomogeneous Linear Equation. A general analysis of the large-time behavior of asymptotically dispersive solutions of the inhomogeneous linear dispersive equation is beyond the scope of this work. Here we will describe some general problems to be solved.

The first problem concerns necessary and sufficient conditions on f such that it be short range, i.e. u_I is asymptotically free. Asymptotic freedom of u_I in the L^2 sense implies that $\lim_{t\to\infty}\|u_I(t) - e^{-iR(D)t}u_1\|_{L^2} = 0$ for some solution $e^{-iR(D)t}u_1$ of the homogeneous linear dispersive equation. It follows that $\lim_{t\to\infty}\|e^{iR(D)t}u_I(t) - u_1\|_{L^2} = 0$. This condition determines u_1 uniquely. But

$$u_1 \stackrel{\text{def}}{=} \lim_{t\to\infty} e^{iR(D)t}u_I(t) = \lim_{t\to\infty}\int_0^t e^{iR(D)\tau}f(\tau)\,d\tau.$$

Therefore we have that $\int_0^\infty e^{iR(D)t}f(\tau)\,d\tau$ converges as an improper integral in the L^2 sense.

Conversely, if the integral defining u_1 converges we have that

$$u_I(t) - e^{-iR(D)t}u_1 = -e^{-iR(D)t}\int_t^\infty e^{iR(D)\tau}f(\tau)\,d\tau,$$

which will decay to zero in the L^2 norm as $t \to \infty$. Hence u_I is asymptotically free, and f is short range. Hence in an L^2 context, f is short range if and only if the integral defining u_1 converges. Asymptotic freedom in the L^2 sense is most directly related to our sense of asymptotic freedom when $\beta = 1$. Our sense of asymptotic freedom appears to be more restrictive, since it involves all values of $\beta \geq 0$. Nevertheless, for the purposes of our discussion, we are willing to make the following:

Conjecture 1. u_I *is asymptotically free (in all regions) if and only if the integral defining u_1 converges.*

Now we will obtain a sufficient condition on f such that the integral defining $u_1(x)$ converges absolutely for each $x \in \mathbb{R}$. We can express $e^{iR(D)t}f_0$ as a convolution

$$[e^{iR(D)t}f_0](x) = \int_{-\infty}^{\infty} \frac{1}{(r\rho t)^{1/r}} G\left(\frac{y-x}{(r\rho t)^{1/r}}\right) f_0(y)\, dy,$$

where $G(y) = \frac{1}{\pi}\Re[e^{i\frac{\pi}{2r}} F_0(r, 1; iye^{i\frac{\pi}{2r}})]$ is the similarity profile of the fundamental solution of the linear dispersive equation. $G(y) = G(y; -\pi/2)$ in the notation of section 5.1. The asymptotics of this function are recorded in Theorem 3.1 of [SSS]; they can also be derived from our Lemma 2.1.1 as was done in section 4.4 for the example of the linearized Benjamin-Ono equation. The asymptotics lead to the estimates:

$$|G(y)| \le C(1+|y|)^{-\frac{(r-2)}{2(r-1)}}, \qquad y < 0,$$
$$|G(y)| \le C(1+|y|)^{-(r+1)}, \qquad y > 0.$$

These decay rates are sharp except for the case $y > 0$, $r \ge 3$ an odd integer, where $G(y)$ decays exponentially as $y \to \infty$. Using this convolution representation we obtain the following.

Lemma 5.3.2. *Suppose $r \ge 2$ and for some constant $C > 0$, all $\tau \ge 0$, and all $y \in \mathbb{R}$ we have that*

$$|f(y, \tau)| \le C(1+\tau)^{-a}\left[1 + \frac{|y|}{(1+\tau)^{1/r}}\right]^{-d}\left[1 + \frac{|y|}{(1+\tau)}\right]^{-g}.$$

If $a > \max\{1, \frac{3}{2} - d\frac{(r-1)}{r}\}$ and $d + g > \frac{r}{2(r-1)}$ then for every $x \in \mathbb{R}$ the integral

$$u_1(x) = \int_0^{\infty} \frac{1}{(r\rho\tau)^{1/r}} \int_{-\infty}^{\infty} G\left(\frac{y-x}{(r\rho\tau)^{1/r}}\right) f(y, \tau)\, dy\, d\tau$$

converges.

Proof. Since $f(y'+x, \tau)$ will satisfy the same type of estimate as $f(y, \tau)$ with a possibly different value of the constant C, we see that it suffices to consider the case $x = 0$. For simplicity, we set $\rho = 1/r$. Thus we must control two integrals of the form

$$\int_0^{\infty} \frac{1}{\tau^{1/r}(1+\tau)^a} \int_0^{\infty} \left[1 + \frac{y}{\tau^{1/r}}\right]^{-\sigma}\left[1 + \frac{y}{(1+\tau)^{1/r}}\right]^{-d}\left[1 + \frac{y}{(1+\tau)}\right]^{-g} dy\, d\tau,$$

where either $\sigma = \frac{r-2}{2(r-1)}$ or $\sigma = r + 1$, depending on which half of the original integral we are considering. Since $\sigma \ge 0$, the above double integral is bounded by

$$\int_0^\infty \frac{1}{\tau^{1/r}(1+\tau)^a} \int_0^\infty \left[1 + \frac{y}{(1+\tau)^{1/r}}\right]^{-\sigma-d} \left[1 + \frac{y}{(1+\tau)}\right]^{-g} dy\, d\tau$$

$$= \int_0^\infty \frac{1}{\tau^{1/r}(1+\tau)^a} \int_0^{1+\tau} \left[1 + \frac{y}{(1+\tau)^{1/r}}\right]^{-\sigma-d} \left[1 + \frac{y}{(1+\tau)}\right]^{-g} dy\, d\tau$$

$$+ \int_0^\infty \frac{1}{\tau^{1/r}(1+\tau)^a} \int_{1+\tau}^\infty \left[1 + \frac{y}{(1+\tau)^{1/r}}\right]^{-\sigma-d} \left[1 + \frac{y}{(1+\tau)}\right]^{-g} dy\, d\tau$$

$$= I_1 + I_2.$$

In the integral I_1, the quantity $[\ldots]^{-g}$ is bounded by a constant $c > 0$, so

$$I_1 \leq c \int_0^\infty \frac{1}{\tau^{1/r}(1+\tau)^a} \int_0^{1+\tau} \left[1 + \frac{y}{(1+\tau)^{1/r}}\right]^{-\sigma-d} dy\, d\tau$$

$$= c \int_0^\infty \frac{(1+\tau)^{1/r}}{\tau^{1/r}(1+\tau)^a} \int_0^{(1+\tau)^{\frac{r-1}{r}}} [1+z]^{-\sigma-d} dz\, d\tau.$$

This converges if $a > \max\{1, 1 + \frac{r-1}{r}[1 - \sigma - d]\}$. Likewise I_2 is bounded by

$$\int_0^\infty \frac{(1+\tau)^{-\frac{(r-1)}{r}(\sigma+d)}}{\tau^{1/r}(1+\tau)^a} \int_{1+\tau}^\infty \left[\frac{y}{(1+\tau)}\right]^{-\sigma-d-g} dy\, d\tau$$

$$= \int_0^\infty \frac{(1+\tau)^{1-\frac{(r-1)}{r}(\sigma+d)}}{\tau^{1/r}(1+\tau)^a} \int_1^\infty z^{-\sigma-d-g} dz\, d\tau,$$

which converges if $\sigma + d + g > 1$ and $a > 1 + \frac{r-1}{r}[1 - \sigma - d]$. $\square$

Another important problem concerns the decay parameters γ_1, λ_1 associated to u_1, when it exists. γ_1 is defined to be the supremum of all γ, and λ_1 is defined to be the supremum of all λ, where (γ, λ) is a pair such that

$$|\hat{u}_1(k)| \leq C \frac{|k|^\gamma}{(1+|k|)^\lambda}$$

for some $C > 0$ and all $k \in \mathbb{R}$. Although it seems unlikely that u_1 will always satisfy the same conditions that we have imposed on u_0 (for example we might have $\hat{u}_1(k) \sim c_\pm |k|^{\gamma_1} \ln(1/|k|)^m$ as $\pm k \to 0^+$), we expect that the parameters γ_1 and λ_1 will continue to play an important role in the leading-order asymptotics of $e^{-iR(D)t}u_1$. The problem is how to compute γ_1 and λ_1 if f is given.

Another problem is the determination of necessary and sufficient conditions on f such that u_I is an asymptotically dispersive solution. More generally, we seek necessary and sufficient conditions on (u_0, f) such that $u = u_H + u_I$ is an asymptotically dispersive solution.

Finally there is the general problem of describing the large-time asymptotics of u_I in all asymptotic regions when u_I is asymptotically dispersive but f is long-range.

The Linear Dispersive-Dissipative Equation. We would like to understand when, if ever, solutions of the linear dispersive-dissipative equation

$$u_t + Q(D)u + iR(D)u = 0$$

are asymptotically dispersive. In this case we suppose $f = -Q(D)u = -\nu|D|^q u$, where $\nu > 0$. We suppose $q > r$ since we know [D2] that when $q \leq r$ the solutions are not asymptotically dispersive. To do this rigorously requires the calculation and comparison of the decay functions $d(\beta; R(D)u)$ and $d(\beta; Q(D)u)$, where u solves the above equation. This is not totally infeasible, but we can also get some idea of the situation using just our accumulated knowledge of the solution of the homogeneous linear dispersive equation. We will limit ourselves here to the latter approach.

Suppose that the solution u is asymptotically free, i.e. there is a solution U of $U_t + iR(D)U = 0$ such that $d(\beta; u - U) > d(\beta; U)$ for all $\beta \geq 0$. Suppose further that u is asymptotically free in the strong sense, i.e.

$$d(\beta; Q(D)(u - U)) > d(\beta; Q(D)U)),$$
$$d(\beta; R(D)(u - U)) > d(\beta; R(D)U)),$$

for all $\beta \geq 0$. Does it follow from this that the solution u is asymptotically dispersive? To answer this we need to compute the decay functions $d(\beta; R(D)U)$ and $d(\beta; Q(D)U)$.

Lemma 5.3.3 (Decay Functions for $iR(D)U$). *Suppose $U_t + iR(D)U = 0$ and $U(x,0) = U_0(x)$, where $U_0 = |D|^\gamma(V_0 + \mathcal{H}V_1)$, $\gamma > -1$, and the functions V_0, V_1 satisfy the conditions stated in Lemma 5.3.1 for v_0, v_1. Let λ be the decay parameter, defined as λ_0 was in Lemma 5.3.1. Then the decay functions for $|D|^r \mathcal{H}U$ are described as follows.*

(1) *Suppose* $-\sin[\frac{\pi}{2}(\gamma + r)]\hat{V}_1(0) + \cos[\frac{\pi}{2}(\gamma + r)]\hat{V}_0(0) \neq 0$. *Define*

$$\beta_0^{(r)} = \begin{cases} \infty & \lambda \leq r(\frac{1}{2} + \gamma + r), \\ \frac{\lambda - (\frac{1}{2} + \gamma + r)}{\lambda - r(\frac{1}{2} + \gamma + r)} & \lambda > r(\frac{1}{2} + \gamma + r). \end{cases}$$

Then the decay function for $x \leq 0$ and $\beta \geq 0$ is given by

$$d^-(\beta; R(D)U) = \begin{cases} \frac{1+\gamma+r}{r} & 0 \leq \beta < \frac{1}{r}, \\ \frac{1+\gamma+r}{r} - (\beta - \frac{1}{r})\frac{(1+\gamma+\frac{r}{2})}{r-1} & \frac{1}{r} \leq \beta < 1, \\ \frac{1}{2} + (\beta - 1)\frac{(\lambda - 1 - \gamma - \frac{r}{2})}{r-1} & 1 \leq \beta < \beta_0^{(r)} \\ \beta(1 + \gamma + r) & \beta_0^{(r)} \leq \beta < \infty. \end{cases}$$

(2) *Suppose* $\sin[\frac{\pi}{2}(\gamma + r)]\hat{V}_1(0) + \cos[\frac{\pi}{2}(\gamma + r)]\hat{V}_0(0) \neq 0$. *Then the decay function for $x \geq 0$ and $\beta \geq 0$ is given by*

$$d^+(\beta; R(D)U) = \begin{cases} \frac{1+\gamma+r}{r} & 0 \leq \beta < \frac{1}{r}, \\ \beta(1 + \gamma + r) & \frac{1}{r} \leq \beta < \infty. \end{cases}$$

Proof. $|D|^r \mathcal{H}U$ is the solution of the same equation that U satisfies with the initial data $|D|^{\gamma+r}(-V_1 + \mathcal{H}V_0)$. This datum has the same decay parameter λ as U_0 (the non-degeneracy conditions in the standard assumptions are satisfied). We apply Lemma 5.3.1 with $\gamma_0 = \gamma + r$. □

Lemma 5.3.4 (Decay Functions for $Q(D)U$). *Make the same assumptions as in the previous lemma. The decay functions for $|D|^q U$ are described as follows.*

(1) *Suppose* $\sin[\frac{\pi}{2}(\gamma + q)]\hat{V}_0(0) + \cos[\frac{\pi}{2}(\gamma + q)]\hat{V}_1(0) \neq 0$. *Define*

$$\beta_0^{(q)} = \begin{cases} \infty & \lambda \leq r(\frac{1}{2} + \gamma + q), \\ \frac{\lambda - (\frac{1}{2} + \gamma + q)}{\lambda - r(\frac{1}{2} + \gamma + q)} & \lambda > r(\frac{1}{2} + \gamma + q). \end{cases}$$

Then the decay function for $x \leq 0$ and $\beta \geq 0$ is given by

$$d^-(\beta; Q(D)U) = \begin{cases} \frac{1+\gamma+q}{r} & 0 \leq \beta < \frac{1}{r}, \\ \frac{1+\gamma+q}{r} - (\beta - \frac{1}{r})\frac{(1+\gamma+q-\frac{r}{2})}{r-1} & \frac{1}{r} \leq \beta < 1, \\ \frac{1}{2} + (\beta - 1)\frac{(\lambda+\frac{r}{2}-1-\gamma-q)}{r-1} & 1 \leq \beta < \beta_0^{(q)} \\ \beta(1+\gamma+q) & \beta_0^{(q)} \leq \beta < \infty. \end{cases}$$

(2) *Suppose* $\sin[\frac{\pi}{2}(\gamma + q)]\hat{V}_0(0) - \cos[\frac{\pi}{2}(\gamma + q)]\hat{V}_1(0) \neq 0$. *Then the decay function for $x \geq 0$ and $\beta \geq 0$ is given by*

$$d^+(\beta; Q(D)U) = \begin{cases} \frac{1+\gamma+q}{r} & 0 \leq \beta < \frac{1}{r}, \\ \beta(1+\gamma+q) & \frac{1}{r} \leq \beta < \infty. \end{cases}$$

Notice that when $q > r$ we have $d^+(\beta; Q(D)U) > d^+(\beta; R(D)U)$ for all $\beta \geq 0$, and $d^-(\beta; Q(D)U) > d^-(\beta; R(D)U)$ for all $0 \leq \beta < 1$. But $d^-(1; Q(D)U) = d^-(1; R(D)U) = 1/2$. Therefore these same relations hold for $d(\beta; Q(D)u)$ and $d(\beta; R(D)u)$. Thus it is not the case that $iR(D)u \gg Q(D)u$, i.e. the solution u is not asymptotically dispersive.

Thus if the solution were asymptotically dispersive, it could not be asymptotically free in the strong sense. Our argument does not eliminate the possibility that u has modified asymptotics, or that u fails to be asymptotically dispersive in the outer regions. Further investigation of the asymptotics of u will be pursued elsewhere.

The Nonlinear Dispersive Equation. Finally we come to the diagnosis of the leading-order large-time asymptotics of solutions of the nonlinear dispersive equation

$$u_t + \partial_x(u^p) + iR(D)u = 0,$$

where $p \geq 2$ is an integer. We will assume in our discussion that $iR(D) = \rho|D|^r \mathcal{H}$, but there are analogous results for Schrödinger-type equations.

First we will derive necessary conditions on the initial data such that the solution is asymptotically free in the strong sense. The derivation is not entirely rigorous, nevertheless at the end we will conjecture that the conditions we derive are in fact necessary. So assume u is asymptotically free in the strong sense, i.e. there is a solution U of $U_t + iR(D)U = 0$ such that for all $\beta \geq 0$ we have that

$$d(\beta; u - U) > d(\beta; U)),$$
$$d(\beta; u_t - U_t) > d(\beta; U_t)),$$
$$d(\beta; u_x - U_x) > d(\beta; U_x)),$$
$$d(\beta; R(D)(u - U)) > d(\beta; R(D)U)).$$

It follows that for all $\beta \geq 0$ we have

$$d(\beta; \partial_x(u^p - U^p)) > d(\beta; \partial_x(U^p)),$$

and therefore that the solution u is asymptotically dispersive, i.e. $iR(D)u \gg \partial_x(u^p)$. From this it follows that $iR(D)U \gg \partial_x(U^p)$. To draw out the implications of this we must first compute the decay functions of $\partial_x(U^p)$.

Lemma 5.3.5 (Decay Functions for $\partial_x(U^p)$). *Adopt the assumptions and notation of Lemma 5.3.3. Then the decay functions for $\partial_x(U^p)$ are described as follows.*

(1) *Suppose $\sin(\frac{\pi}{2}\gamma)\hat{V}_0(0) + \cos(\frac{\pi}{2}\gamma)\hat{V}_1(0) \neq 0$. Define*

$$\beta_0' = \begin{cases} \infty & \lambda \leq r(\frac{3}{2} + \gamma), \\ \frac{\lambda - (\frac{3}{2} + \gamma)}{\lambda - r(\frac{3}{2} + \gamma)} & \lambda > r(\frac{3}{2} + \gamma). \end{cases}$$

If $-1 < \gamma \leq -\frac{1}{2}$ then the decay function for $x \leq 0$ and $\beta \geq 0$ is given by

$$d^-(\beta; \partial_x(U^p))$$
$$= \begin{cases} \frac{1 + (1+\gamma)p}{r} & 0 \leq \beta < \frac{1}{r}, \\ \frac{1 + (1+\gamma)p}{r} + (\beta - \frac{1}{r})\left[(1+\gamma)p - \frac{1 + r(\frac{1}{2}+\gamma)}{r-1}\right] & \frac{1}{r} \leq \beta < 1, \\ \frac{1}{2} + (1+\gamma)(p-1) + (\beta - 1)\left[\frac{\lambda + \frac{r}{2} - \gamma - 2}{r-1} + (1+\gamma)(p-1)\right] & 1 \leq \beta < \beta_0' \\ \beta[1 + (1+\gamma)p] & \beta_0' \leq \beta < \infty. \end{cases}$$

If $\gamma \geq -\frac{1}{2}$ then define also

$$\beta_0 = \begin{cases} \infty & \lambda \leq r(\frac{1}{2} + \gamma), \\ \frac{\lambda - (\frac{1}{2} + \gamma)}{\lambda - r(\frac{1}{2} + \gamma)} & \lambda > r(\frac{1}{2} + \gamma). \end{cases}$$

For $\gamma \geq -\frac{1}{2}$, $x \leq 0$, and $\beta \geq 0$, the decay function is given by

$d^-(\beta; \partial_x(U^p))$

$$
= \begin{cases}
\frac{1+(1+\gamma)p}{r} & 0 \leq \beta < \frac{1}{r}, \\[2mm]
\frac{1+(1+\gamma)p}{r} - (\beta - \frac{1}{r})\frac{1+(1+\gamma-\frac{r}{2})p}{r-1} & \frac{1}{r} \leq \beta < 1, \\[2mm]
\frac{p}{2} + (\beta - 1)\frac{p(\lambda+\frac{r}{2}-1-\gamma)-1}{r-1} & 1 \leq \beta < \beta_0 \\[2mm]
\beta_0(1+\gamma)p - \frac{\beta_0-1}{r-1} + (\beta - \beta_0)\left[\frac{\lambda+\frac{r}{2}-\gamma-2}{r-1} + (1+\gamma)(p-1)\right] & \beta_0 \leq \beta < \beta_0' \\[2mm]
\beta[1 + (1+\gamma)p] & \beta_0' \leq \beta < \infty.
\end{cases}
$$

(2) *Suppose $\sin(\frac{\pi}{2}\gamma)\hat{V}_0(0) - \cos(\frac{\pi}{2}\gamma)\hat{V}_1(0) \neq 0$. Then the decay function for $x \geq 0$ and $\beta \geq 0$ is given by*

$$
d^+(\beta; \partial_x(U^p)) = \begin{cases}
\frac{1+(1+\gamma)p}{r} & 0 \leq \beta < \frac{1}{r}, \\[2mm]
\beta[1 + (1+\gamma)p] & \frac{1}{r} \leq \beta < \infty.
\end{cases}
$$

Proof. A tedious but straightforward calculation based on Lemma 5.3.1. $\square$

Remarks. (1) As can be seen from the above, if U_0 is very smooth, i.e. $\lambda > r(\frac{3}{2} + \gamma)$, then the decay function $d^-(\beta; \partial_x(U^p))$ is quite complicated, but it is nevertheless piecewise linear and continuous.

(2) By Theorem 5.1.1, an explicit function with the same decay functions as $\partial_x(U^p)$ is given by

$$
\frac{1}{t^{\frac{1+(1+\gamma)p}{r}}}\left[\frac{1}{(1+|\xi|)^{2+\gamma}} + [1 - \mathrm{sgn}(b)]\frac{(1+|\xi|)^{\frac{1}{r-1}(2+\gamma-\frac{r}{2})}}{(1+|b|)^{\frac{\lambda}{r-1}}}\right] \times
$$
$$
\times \left[\frac{1}{(1+|\xi|)^{(1+\gamma)(p-1)}} + [1 - \mathrm{sgn}(b)]\frac{(1+|\xi|)^{\frac{p-1}{r-1}(1+\gamma-\frac{r}{2})}}{(1+|b|)^{\frac{\lambda(p-1)}{r-1}}}\right].
$$

We have made strong assumptions on the initial datum U_0 of the solution U in order to compute these decay functions. The decay parameters γ, λ for U_0 can be defined in a more general manner as γ_1, λ_1 were for u_1 above. Our strong assumptions can probably be relaxed significantly without changing the decay functions. We will therefore assume that the decay functions for all the quantities defined in terms of U are as given in Lemmata 5.1, 5.3, and 5.5, where γ, λ have the more general meanings.

Now we can derive conditions on γ, λ which are equivalent to the relation: $iR(D)U \gg \partial_x(U^p)$.

Lemma 5.3.6. *Adopt the assumptions of Lemma 5.3.3.*

(1) *If $iR(D)U \gg \partial_x(U^p)$ then $p + \gamma(p-1) > r$ and $\lambda - \gamma \geq 1 - \frac{r}{2} - \frac{r-1}{p-1}$.*

(2) *If* $p + \gamma(p-1) > r$ *and* $\lambda - \gamma \geq 1 - \frac{r}{2} - \frac{r-1}{p-1}$ *and the four conditions*

$$\sin(\tfrac{\pi}{2}\gamma)\hat{V}_0(0) \pm \cos(\tfrac{\pi}{2}\gamma)\hat{V}_1(0) \neq 0,$$

$$\pm \sin[\tfrac{\pi}{2}(\gamma + r)]\hat{V}_1(0) + \cos[\tfrac{\pi}{2}(\gamma + r)]\hat{V}_0(0) \neq 0,$$

are satisfied, then $iR(D)U \gg \partial_x(U^p)$.

Proof. (1) $p + \gamma(p-1) > r$ follows from consideration of the inner region. If the condition $\lambda - \gamma \geq 1 - \frac{r}{2} - \frac{r-1}{p-1}$ were not true then $\lambda < r(\frac{1}{2} + \gamma) < r(\frac{1}{2} + \gamma + r)$. When the non-vanishing condition of Lemma 5.3.5(1) is true, $\beta = \beta_0$ is the transition region where the contributions (for $x < 0$) of the contour $C_0(b)$ and $\tilde{C}(b)$ are balanced. If that non-vanishing condition is not satisfied then the contribution of the contour $\tilde{C}(b)$ gets smaller while the contribution of $C_0(b)$ stays the same; so the transition value of β gets even larger. Since $\beta_0 = \infty$, the contribution of the contour $\tilde{C}(b)$ is negligible, and the decay function in Lemma 5.3.5(1) is correct without the non-vanishing condition. A similar argument shows that the decay function in Lemma 5.3.3(1) is correct without the need to impose the non-vanishing condition for that part. Even though $d^-(1; \partial_x(U^p)) > d^-(1; R(D)U)$ we have that $(d^-)'(\beta; \partial_x(U^p)) < (d^-)'(\beta; R(D)U)$ if $\beta > 1$ so that for sufficiently large β we will have $d^-(\beta; \partial_x(U^p)) < d^-(\beta; R(D)U)$. This contradicts the assumption $iR(D)U \gg \partial_x(U^p)$. Thus the condition $\lambda - \gamma \geq 1 - \frac{r}{2} - \frac{r-1}{p-1}$ must be true.

(2) A careful case by case comparison of two piecewise linear continuous functions. $\square$

Remarks. (1) The condition $p + \gamma(p-1) > r$ was mentioned in section 5.1 as a candidate condition for "asymptotically weak nonlinearity" on the basis of the hypothesis of self-similarity. The term "asymptotically weak nonlinearity", although precise enough to handle cases where self-similarity was the rule, is ambiguous here. It splits into two distinct properties of the solution u: "asymptotically dispersive", and "asymptotically free". However we have already remarked that the self-similar hypothesis is unreliable, and here the necessity of imposing the regularity condition $\lambda - \gamma \geq 1 - \frac{r}{2} - \frac{r-1}{p-1}$ is a consequence of its failure.

(2) The four non-vanishing conditions assumed in the above theorem are needed in order to precisely determine the decay functions for large values of β. We will call any situation where one of these four quantities vanishes an *anomaly*. There are anomalous situations where $d(\beta; \partial_x(U^p)) < d(\beta; R(D)U)$ for some $\beta \geq 0$, even though $p + \gamma(p-1) > r$ and $\lambda - \gamma \geq 1 - \frac{r}{2} - \frac{r-1}{p-1}$. (For example, if $p = r = 2$, $0 < \gamma < 1$, $V_0 = -\tan(\frac{\pi}{2}\gamma)V_1$, and $\hat{V}_1(0) \neq 0$, then $d^+(1; UU_x) = 3 + 2\gamma < d^+(1; \mathcal{H}U_{xx}) = 5 + \gamma$.) However, much of the previous work would be considered anomalous by this definition. For example, if $r = 3$, $\gamma = 0$, and $V_1 \equiv 0$, then all four quantities vanish! Thus the word anomaly is not intended to connote unimportance. Such important special cases must be considered separately. Our results provide the tools to analyze these cases.

The conditions $p+\gamma(p-1) > r$ and $\lambda-\gamma \geq 1-\frac{r}{2}-\frac{r-1}{p-1}$ involve the parameters γ, λ associated to U_0, a function which is generally not easily computable from the initial data u_0. However if one grants Conjecture 1 that whenever u is asymptotically free we have that the integral

$$u_1 \overset{\text{def}}{=} -\int_0^\infty e^{iR(D)\tau} \partial_x [u(\tau)^p] \, d\tau$$

converges to a function u_1 having parameters γ_1, λ_1, and $U_0 = u_0 + u_1$, then we can relate γ, λ to γ_0, λ_0 and γ_1, λ_1 as follows:

$$\gamma = \min\{\gamma_0, \gamma_1\}, \qquad\qquad \lambda - \gamma = \min\{\lambda_0 - \gamma_0, \lambda_1 - \gamma_1\}.$$

Therefore it follows that $p + \gamma_0(p-1) > r$ and $\lambda_0 - \gamma_0 \geq 1 - \frac{r}{2} - \frac{r-1}{p-1}$.

Conjecture 2. *If the solution u of the nonlinear dispersive equation with initial data u_0 is asymptotically free in the strong sense then the conditions $p + \gamma_0(p-1) > r$ and $\lambda_0 - \gamma_0 \geq 1 - \frac{r}{2} - \frac{r-1}{p-1}$ hold.*

The condition $\lambda_0 - \gamma_0 \geq 1 - \frac{r}{2} - \frac{r-1}{p-1}$ is a very weak smoothness condition on the initial data u_0. It would be remarkable if global solutions could be proved to exist under weaker smoothness conditions than this one. Hence the condition $p + \gamma_0(p-1) > r$ is the primary one.

Now we will discuss two examples from the literature which reinforce this conjecture.

Example 3 (MKdV). $u_t - 6u^2 u_x + u_{xxx} = 0$, $r = 3$, $p = 3$, $\gamma_0 = 0$, u_0 in the Schwartz class (the equation has no real-valued soliton solutions). In this example, $p + \gamma_0(p-1) = r$, so we will compare the leading-order large-time asymptotics of u in the outer region $(x < 0)$ to those for a solution U of the homogeneous linearized KdV equation (see section 4.6), i.e.

$$U(x,t) \sim \frac{1}{\sqrt{3\pi t}} \left|\frac{b}{3}\right|^{-1/4} R(b) \cos\left[2\left|\frac{\xi}{3}\right|^{3/2} - \frac{\pi}{4} + \phi(b)\right].$$

The functions $R(b)$ and $\phi(b)$ (as usual $b = x/t$ and $\xi = xt^{-1/3}$) are related to the initial data by the rule $R(b)e^{i\phi(b)} = \mu_0(|\frac{b}{3}|^{1/2}) + i\eta_0(|\frac{b}{3}|^{1/2})$, where μ_0 and η_0 are defined in terms of U_0 as in section 4.4. We have assumed that $\gamma = 0$ and $V_1 \equiv 0$ for clarity. But according to Deift and Zhou [DZ1], the leading-order large-time asymptotics for u in this region (their region II) are of the form

$$u(x,t) \sim \frac{1}{\sqrt{3\pi t}} \left|\frac{b}{3}\right|^{-1/4} R_N(b) \cos\left[2\left|\frac{\xi}{3}\right|^{3/2} - \frac{\pi}{4} + \phi_N(b) - \frac{R_N(b)^2}{2\pi} \ln(t)\right].$$

The functions $R_N(b)$ and $\phi_N(b)$ are explicitly expressed in terms of the reflection coefficient of the initial data, and hence depend on that data in a nonlinear manner. Although bearing a close formal resemblance to our outer expansion for the linearized KdV equation, the asymptotics of u differ from it in a significant way, namely the logarithmic phase shift. This phase shift implies that u is not asymptotically free. It is interesting to note that $d^-(\beta; u^2 u_x) > d^-(\beta; u_{xxx})$ if $\frac{1}{3} < \beta \leq 1$ (in particular). So in the region $\frac{1}{3} < \beta \leq 1$ the solution is asymptotically dispersive but not asymptotically free. (Here and below we assume that the asymptotic results of Deift and Zhou remain valid when differentiated with respect to x up to three times.)

In the inner region the situation is mostly the same. The leading-order asymptotics of U are given by (assuming $\gamma = 0$ and $V_1 \equiv 0$)

$$U(x, t) \sim \frac{1}{(3t)^{1/3}} \hat{U}_0(0) \operatorname{Ai}\left(\frac{x}{(3t)^{1/3}}\right).$$

However, in the same region (their region IV) Deift and Zhou show that

$$u(x, t) \sim \frac{1}{(3t)^{1/3}} w\left(\frac{x}{(3t)^{1/3}}\right),$$

where w is the particular solution of the second Painlevé equation $w''(\xi) = \xi w(\xi) + 2w(\xi)^3$ with monodromy parameters given explicitly [DZ1] in terms of the reflection coefficient of u_0 evaluated at zero. This clearly does not agree with the asymptotics of any solution of the homogeneous linearized KdV equation. In fact, u is not even asymptotically dispersive in the inner region, since $u^2 u_x \sim u_{xxx}$ for $0 \leq \beta \leq 1/3$.

Example 4 (KdV). $u_t - 6uu_x + u_{xxx} = 0$, $r = 3$, $p = 2$, $\gamma_0 = 0$, u_0 in the Schwartz class (we consider the solution containing no solitons; non-vanishing solutions of this kind have $\int_{-\infty}^{\infty} u_0(x)\, dx > 0$, [Sch]). In this example $p + \gamma_0(p - 1) < r$, and there are many parallels with the previous example. The leading-order large-time asymptotics (see Deift, Venakides, Zhou [DVZ]) in the region $\frac{1}{3} < \beta \leq 1$, $x < 0$, (their region IV) again has a logarithmic phase shift, so u is not asymptotically free. In the inner region $0 \leq \beta \leq \frac{1}{3}$ the asymptotics are again self-similar, where the profile $w = v' + v^2$ and v is a solution of Painlevé II. The new feature here is the "collisionless shock region", which generically appears in the vicinity of $\beta = 1/3$. In contrast to the previous example, which could be viewed as being on the borderline, this example has features which are very dissimilar to linear asymptotics.

Next we inquire into sufficient conditions for the solution to be asymptotically free. By Conjecture 1 this means we want sufficient conditions so that the integral defining u_1 converges. First we will derive conditions on a solution U of the homogeneous linear equation such that the integral $-\int_0^{\infty} e^{iR(D)\tau} \partial_x [U(\tau)^p]\, d\tau$ converges.

Lemma 5.3.7. *Suppose $U(t) = e^{-iR(D)t} U_0$, $\gamma > -1$, and $\lambda > 0$ are as in Lemma 5.3.3. Suppose*

$$p > \max\{3, \tfrac{r-1}{1+\gamma}, \tfrac{2+\gamma}{1+\gamma}\},$$

$$\lambda > \max\{1 + \gamma - \tfrac{r}{2} + \tfrac{1}{p}(1 + \tfrac{r}{2}), \tfrac{1}{p-1}[1 + p(1 + \gamma - \tfrac{r}{2}) - r(1 + \gamma)],$$

$$1 + (p + r - pr)(1 + \gamma)\}.$$

Then the integral $\int_0^\infty e^{iR(D)\tau} \partial_x[U(\tau)^p]\,d\tau$ converges for each value of $x \in \mathbb{R}$.

Proof. Use Lemma 5.3.2 and remark (2) after Lemma 5.3.5. □

We are interested in a sufficient condition for the solution u to be asymptotic to U, where $\gamma = \gamma_0$ and $\lambda = \lambda_0$. We impose the (conjectural) necessary condition $p + \gamma_0(p - 1) > r$. We assume that λ_0 is sufficiently large so that the condition on λ_0 in Conjecture 2 and the three conditions on $\lambda = \lambda_0$ in Lemma 5.3.7 are satisfied. The only other condition of Lemma 5.3.7 which does not follow from these assumptions is $p > 3$. If $p > 3$ and if u has decay functions equal to those of $u_H(t) = e^{-iR(D)t}u_0$, then u_1 will exist. Let γ_1, λ_1 be the parameters for this u_1. It is not obvious but plausible that these agree with the parameters of

$$u_2 \stackrel{\text{def}}{=} -\int_0^\infty e^{iR(D)\tau} \partial_x[u_H(\tau)^p]\,d\tau,$$

which are in principle computable from the initial data u_0. If in addition to the above we have $\gamma_1 > \gamma_0$ and $\lambda_1 - \gamma_1 > \lambda_0 - \gamma_0$ then by Lemma 5.3.1 we have that $u_H \gg u_I$, where

$$u_I(t) \stackrel{\text{def}}{=} -\int_0^t e^{-iR(D)(t-\tau)} \partial_x[u(\tau)^p]\,d\tau.$$

Therefore iteration in the integral equation $u = u_H + u_I$ makes sense, because $u_H + u_I$ will have the same behavior as that assumed of u. Since iteration in the integral equation is the basic proof method, we believe these conditions are sufficient.

Conjecture 3. *Suppose u is a global solution of the nonlinear dispersive equation with initial data u_0 having parameters γ_0, λ_0 satisfying the conditions of Lemma 5.3.1. Suppose $p + \gamma_0(p - 1) > r$, $p > 3$, and λ_0 is sufficiently large. Suppose that the parameters γ_1, λ_1 of u_2 (defined above) satisfy $\gamma_1 > \gamma_0$ and $\lambda_1 - \gamma_1 > \lambda_0 - \gamma_0$. Then u is asymptotically free, i.e. it has the same leading-order large-time asymptotics in all regions as $U(t) = e^{-iR(D)t}U_0$, and $U_0 = u_0 + u_1$ also has parameters γ_0, λ_0.*

Unfortunately, we do not know how to express γ_1, λ_1 efficiently in terms of the initial data u_0, hence the conditions $\gamma_1 > \gamma_0$ and $\lambda_1 - \gamma_1 > \lambda_0 - \gamma_0$ are in a somewhat unsatisfactory form, with which we must be content at the present time. If the parameters of u_2 do not agree with those of u_1, then the idea behind our conjecture fails, and we do not know of a substitute.

On a more optimistic tack, if $p > \max\{4, \frac{r+1}{1+\gamma_0}\}$ then it is not hard to convince oneself that generically $\gamma_1 = 1$. For this purpose note that

$$\lim_{k \to 0} \frac{\hat{u}_1(k)}{ik} = -\int_0^\infty [u(\tau)^p]\widehat{}(0)\,d\tau$$

and, assuming u behaves (in regard to decay functions) like u_H, Theorem 5.2.2 shows that this integral converges absolutely precisely when $p > \max\{4, \frac{r+1}{1+\gamma_0}\}$.

It would be an anomaly if this integral converged to the value zero; for real valued solutions it is impossible if p is even. For $p > \max\{4, \frac{r+1}{1+\gamma_0}\}$ the condition $\gamma_1 > \gamma_0$ becomes $\gamma_0 < 1$. If $p = \max\{4, \frac{r+1}{1+\gamma_0}\}$ it is likely that $\gamma_1 = 1$, and therefore $\hat{u}_1(k)$ does not have power-like behavior as $k \to 0$. Consequently, we expect that the relation $p = \max\{4, \frac{r+1}{1+\gamma_0}\}$ describes a transition point in the decay properties of certain norms of $u_I(t)$.

It is interesting to inspect the results in the literature in light of this conclusion. For example, when $r = 3$, $\gamma_0 = 0$, we expect that $\gamma_1 = 1$ when $p > 4$. But a change when $p = 4$ in the behavior of $\hat{u}_1(k)$ as $k \to 0$ may partly explain why Christ and Weinstein [CW] prove two different theorems concerning decay estimates for solutions of the generalized KdV equation: Proposition 4.1 for the case $p > 4$, and Proposition 5.1 covering (in particular) the case $p = 4$. For another example, the restriction $p > \frac{r+1}{1+\gamma_0}$ appears when $\gamma_0 = 0$, $r \geq 4$, in the linear scattering results of Kenig Ponce and Vega [KPV]. Possibly this is because their estimates of u_I are sufficiently tight so as to detect the transition which takes place at $p = \frac{r+1}{1+\gamma_0}$. The main other case, which is not covered in these two papers, is the generalized Benjamin-Ono equation $u_t + \partial_x(u^p) - \rho \mathcal{H} u_{xx} = 0$. Sidi Sulem and Sulem [SSS] prove decay estimates and linear scattering for small solutions of this equation (and others) when $p \geq 4$ (an integer). Their proof detects no difference between $p = 4$ and $p > 4$ because of the evident "looseness" in their estimates of u_I.

When $p = 4$ and $\gamma_0 = 0$ it is plausible that γ_1 is a continuous function of $r \in [2, 4)$. We have already asserted that $\gamma_1 = 1$ when $2 \leq r \leq 3$. Probably γ_1 decreases as $r \to 4^-$. We do not know where, or if, it happens that $\gamma_1 = 0$. If this happens for some $3 < r' < 4$, then it could still be true that u is asymptotically free for $r' < r < 4$, but one would have to reformulate the conjecture under the assumption that $\gamma = \gamma_1$ instead of $\gamma = \gamma_0$. However, in such a case γ_1 would be even more difficult to determine from the initial data. We include these speculations to underline the central role that the parameters γ_1, λ_1 play in the study of the leading-order large-time asymptotics of u.

There is some evidence in the literature to suggest that asymptotic freedom holds even in situations where the condition $p + \gamma_0(p - 1) > r$ is nearly not satisfied. For example, consider the linear scattering result (Theorem 2.14) of Kenig Ponce Vega [KPV2] which they proved for solutions of the "critical power" generalized KdV equation $u_t + \partial_x(u^5) + \rho u_{xxx} = 0$. They showed that if $\|u_0\|_{L^2}$ is sufficiently small, then this equation has a global solution u, and there exists a unique function $U_0 \in L^2$ such that $\lim_{t \to \infty} \|u(t) - e^{-iR(D)t}U_0\|_{L^2} = 0$. The point is that if u_0 satisfies the assumptions we have made on our initial data, then $u_0 \in L^2$ implies that $\gamma_0 > -1/2$. So when $p = 5$ we have $p + \gamma_0(p - 1) > 5 - \frac{1}{2} \cdot 4 = 3 = r$. We can get as close to violating the condition $p + \gamma_0(p-1) > r$ as we want using L^2 initial data. Similar results were obtained by Choi [Ch] for solutions of the equation $u_t + \partial_x(u^p) + u_{xxxxx} = 0$. When $p \geq 5$ and $|D|^{\frac{1}{2} - \frac{4}{p-1}} u_0$ is sufficiently small in L^2 then Choi proved that a global solution u exists and a unique function U_0 exists so that $\lim_{t \to \infty} \||D|^{\frac{1}{2} - \frac{4}{p-1}}[u(t) - e^{-iR(D)t}U_0]\|_{L^2} = 0$.

In order for our type of initial data to satisfy Choi's assumption, we would need $\gamma_0 > \frac{4}{p-1} - 1$. Therefore $p + \gamma_0(p-1) > p + (\frac{4}{p-1} - 1)(p-1) = 5 = r$. So initial data satisfying Choi's assumptions can come as close as we want to violating the condition $p + \gamma_0(p-1) > r$. Yet some sort of asymptotic freedom holds in these cases. It is not obvious, but plausible, that the full strength of asymptotic freedom (as we have defined it) holds for our type of initial data satisfying the assumptions of Conjecture 3.

Finally, let us consider what happens when $p = 3$. Conjecture 3, and the analogous result for Schrödinger-type equations, exclude this case for good reason. Consider for $r = 2$, $\gamma_0 = 0$, the Derivative Nonlinear Schrödinger (DNLS) equation $u_t - \partial_x(|u|^2 u) - \frac{i}{2}u_{xx} = 0$. Hayashi and Ozawa [HO] have shown that global (necessarily complex-valued) solutions of this equation exist with modified asymptotics (in an L^2 sense). The leading-order large-time asymptotics of these solutions in the region $\beta \geq \epsilon > 0$ are given by the function in Example 2 of this section (the function was also denoted by u in that example). They also prove that any global solution satisfying certain reasonable regularity and decay criteria cannot be asymptotically free. The error between the solution and its leading-order large-time asymptotic approximation is controlled in the H^2 norm (asymptotic freedom in the strong sense), hence these solutions seem to be asymptotically dispersive (at least in the outer region). As we saw in Example 2, the approximate solution is definitely asymptotically dispersive in all regions.

This example raises the question of whether suitably small global solutions of the generalized Benjamin-Ono equation $u_t + \partial_x(u^3) - \rho \mathcal{H} u_{xx} = 0$ have behavior similar to solutions of the DNLS equation. As we have seen in Examples 3 and 4, logarithmic phase shifts are present in real-valued cases, so there is no reason why modified asymptotics should be confined to the DNLS case. If the solutions of this generalized Benjamin-Ono equation are not asymptotically free, then this would explain why no one has ever proved a linear scattering result covering this case!

5.4 Asymptotic Behavior as $|x| \to \infty$

As a final application of our results we investigate the rate of spatial decay of solutions of the inhomogeneous linear dispersive equation

$$y_t - \rho|D|^{r-1}y_x = \partial_x(f_0 + \mathcal{H}f_1),$$
$$y(x,0) = 0,$$

where $f_0(x,t)$, $f_1(x,t)$ are compactly supported in x for each $t > 0$. We will not attempt to be rigorous, since a careful study of solutions of the inhomogeneous linear dispersive equation is beyond the scope of this work. Our goal here is to gain what understanding we can from our asymptotic results without proving a theorem. Since we are not being rigorous we should say that our real intent is to try to understand the large $|x|$ behavior of solutions of the nonlinear dispersive equation $u_t + \partial_x(u^p) - \rho|D|^{r-1}u_x = 0$. In this we are motivated by some papers of Iório [I1], [I2], and Abdelouhab [Abd]. In these works solutions of the initial-value problem for this nonlinear dispersive equation are obtained in weighted

Sobolev spaces, which allow for a range of different spatial decay rates. The most interesting result was for the Benjamin-Ono equation ($p = q = 2$) in that Iório proved that any solution $u \in C([0,T], H^4 \cap L_4^2)$ must vanish identically, where H^4 is the usual Sobolev space of functions $u(x)$ whose Fourier transform $\hat{u}$ is in L_4^2, and where L_4^2 consists of those functions $v(x)$ such that $(1+|x|)^4 v(x)$ is in L^2 over all $\mathbb{R}$. Hence nontrivial solutions satisfying $u(x,t) \sim c_{\pm}|x|^{-a}$ as $\pm x \to \infty$ do not exist if $a > 4.5$. In a companion result, Iório proved that solutions u of the Benjamin-Ono equation in $C([0,T], H^3 \cap L_3^2)$ exist, but must satisfy $u(t)\hat{}(0) = 0$ for all $t \in [0,T]$. This result seems to allow for the existence of nontrivial solutions with decay rate $a = 4.5$. But we ask, are there really any solutions with this decay rate? What is the maximum decay rate possessed by a solution of the Benjamin-Ono equation? We cannot give a conclusive answer here, but the asymptotic calculations presented in this section suggest that no solution of Benjamin-Ono decays more rapidly than $|x|^{-4}$ as $|x| \to \infty$; c.f. [HKO].

This limitation in the rate of spatial decay of solutions is not present in all nonlinear dispersive equations. For example, Kato [K] has shown that solutions of the Korteweg-de Vries equation ($p = 2, r = 3$) exist in $C([0,\infty), H^{2s} \cap L_s^2)$ for all $s \geq 1$. Abdelouhab [Abd] has shown that when the dispersive symbol $R(k)$ is a smooth function, then there is no limit to the rate of spatial decay of solutions of nonlinear dispersive equations. Our calculations confirm that this unlimited decay rate of solutions only happens when $r \geq 3$ is an odd integer. For the other values of $r \geq 2$ the decay rate of solutions is limited as in the Benjamin-Ono case; no solutions decay more rapidly than $|x|^{-2-r}$ as $|x| \to \infty$.

We are interested in the calculations in this section for another reason beyond the question we raised above. In the previous section we raised the question about the relative asymptotic sizes of u_H and u_I, where u_H solves the homogeneous linear dispersive equation, and u_I solves an inhomogeneous linear dispersive equation. For all $r \geq 2$ it is possible by choosing γ_0 and $\lambda_0^0 - \gamma_0$ sufficiently large to cause u_H to decay as $|x| \to \infty$ at an arbitrarily rapid rate. However, we will see that y (as well as u_I) is limited in its spatial decay rate (when r is not an odd integer) irrespective of u_0. Thus it is quite possible for the asymptotic contribution of u_I to be significantly larger than that of u_H.

To begin our analysis, we will try to justify our assumptions on the inhomogeneous term $f = \partial_x(f_0 + \mathcal{H}f_1)$. The decomposition $f_0 + \mathcal{H}f_1$ is a device to obtain a slowly decaying function from two rapidly decaying functions f_0, f_1. This is because the Hilbert transform $\mathcal{H}f_1$ is generally slowly decaying. In order to see this, we present the following.

Lemma 5.4.1. *Suppose $n \geq 0$ is an integer and $v(x)$ is a function such that its Fourier transform satisfies $\hat{v} \in W^{n,1}(\mathbb{R})$, i.e. $\hat{v} \in L^1$ and its nth distributional derivative $\hat{v}^{(n)} \in L^1$. Then*

$$(\mathcal{H}v)(x) = \frac{1}{\pi} \sum_{j=1}^{n} \frac{[x^{j-1}v(x)]\hat{}(0)}{x^j} + o(|x|^{-n}), \qquad as \ |x| \to \infty.$$

Remark. Under weaker assumptions on v the Hilbert transform $\mathcal{H}v$ can have much more complicated asymptotic behavior. For such details, see chapter six of Wong [W]. The above lemma will be sufficient for our purposes.

Proof. It suffices to show that the Fourier transform of

$$(-ix)^n(\mathcal{H}v)(x) - \frac{(-i)^n}{\pi}\sum_{j=1}^{n}[x^{j-1}v(x)]^\wedge(0)x^{n-j}$$

(as a tempered distribution) is an L^1 function. The Fourier transform of this quantity is

$$\left(\frac{d}{dk}\right)^n(\mathcal{H}v)^\wedge(k) - \frac{(-i)^n}{\pi}\sum_{j=1}^{n}[x^{j-1}v(x)]^\wedge(0)2\pi i^{n-j}\delta^{(n-j)}(k),$$

where $\delta(k)$ is the Dirac delta distribution. Using the product rule for differentiating a product of absolutely continuous functions (or more specifically, Lemma 5.2.2 of [D1]) one finds that all the delta functions cancel, and one is left with $-i\,\text{sgn}(k)\hat{v}^{(n)}(k)$, which is in L^1 by assumption. $\square$

In all of the following arguments, we assume our functions are smooth enough so that the above lemma applies. Thus in order for $\mathcal{H}f_1$ to decay faster than $1/x$, certain moments of f_1 must vanish. If $v = v_0 + \mathcal{H}v_1$ where v_0 and v_1 have compact support and are (at least) in H^1 then it is not difficult to see that v^p can be decomposed in the same fashion, i.e. $v^p = f_0 + \mathcal{H}f_1$, where f_0, f_1 have compact support and are (at least) in H^1. In fact, using the product rule for the Hilbert transform we have

$$(v_0 + \mathcal{H}v_1)(w_0 + \mathcal{H}w_1) = v_0w_0 + v_0\mathcal{H}w_1 + w_0\mathcal{H}v_1 + v_1w_1 + \mathcal{H}(v_1\mathcal{H}w_1 + w_1\mathcal{H}v_1).$$

The sum of the first four terms on the right has compact support and belongs to H^1 since v_0, v_1, w_0, w_1 do. Also, $v_1\mathcal{H}w_1 + w_1\mathcal{H}v_1$ has compact support and lies in H^1. If $\hat{v}_1(0) \neq 0$ and $\hat{w}_1(0) \neq 0$ then by the lemma $v_0 + \mathcal{H}v_1$ and $w_0 + \mathcal{H}w_1$ will decay like $O(|x|^{-1})$ as $|x| \to \infty$, and so the product $(v_0 + \mathcal{H}v_1)(w_0 + \mathcal{H}w_1)$ will decay like $O(|x|^{-2})$ as $|x| \to \infty$. It is not hard to see that the 0th moment of $v_1\mathcal{H}w_1 + w_1\mathcal{H}v_1$ vanishes, as required by the lemma. In a similar way, if $v = O(|x|^{-m})$ as $|x| \to \infty$ ($m > 0$ is a real number), then $v^p = f_0 + \mathcal{H}f_1 = O(|x|^{-pm})$ as $|x| \to \infty$. Hence if $j_0 = ((pm - 1))$, i.e. the smallest integer greater than or equal to $pm - 1$, and if $\hat{f}_1 \in W^{j_0,1}$, then $[x^{j-1}f_1]^\wedge(0) = 0$ for $j = 1, \ldots, j_0$.

On the other hand, if $v^p = f_0 + \mathcal{H}f_1$ then there is no particular reason why $\hat{f}_0(0) = \int_{-\infty}^{\infty} v(x)^p\, dx$ should vanish, and in general it will not. When v is real-valued and $p \geq 2$ is an even integer, $\hat{f}_0(0)$ cannot vanish unless $v \equiv 0$. Since we want our conclusions to apply to solutions of the nonlinear dispersive equation, we want $f_0 + \mathcal{H}f_1$ to have properties appropriate to the pth power of some function v. Hence we will assume $f_0(t)^\wedge(0) \neq 0$ for all $t \geq 0$ and $[x^j f_1(x,t)]^\wedge(0) = 0$ for $j = 0, 1, \ldots, j_0 - 1$.

When $v(x,t)$ solves the linear or nonlinear dispersive equation, then for $t > 0$ it will not be the case that $v = v_0 + \mathcal{H}v_1$ where v_0 and v_1 have compact support. Hence it could be objected that our assumption on $f = f_0 + \mathcal{H}f_1$ is too restrictive to apply to solutions of the nonlinear dispersive equation. Strictly speaking this is true. However, even though v may not have such a decomposition into compactly supported components, it is still quite possible that it can be approximated by such a construction in such a way that the leading-order asymptotics of $y(x,t)$ as $|x| \to \infty$ are independent of whether $f = \partial_x v^p$ or $f = \partial_x(f_0 + \mathcal{H}f_1)$ with f_0, f_1 of compact support. We believe this is plausible, since as we will see, the leading-order large $|x|$ asymptotics of $y(x,t)$ are quite insensitive to whether or not f_0, f_1 have compact support. The important quantities are the moments, which we can arrange by approximation to have the correct values. So this is how we propose that the calculations of this section be interpreted when thinking about solutions of the nonlinear dispersive equation. It is obviously a nontrivial matter to prove that this approximation can be done effectively, but this proof is beyond the scope of this work.

Now that our assumptions on f_0 and f_1 are clear, we can begin the analysis of the large x asymptotics of

$$y(t) = \int_0^t e^{-iR(D)(t-\tau)} f(\tau)\, d\tau.$$

Define $v(\cdot, \sigma; \tau) = e^{-iR(D)\sigma} f(\tau)$ for $\sigma \geq 0$. In chapters one through four we have gained a detailed picture of the asymptotic behavior of this solution of the homogeneous linear dispersive equation. Its initial data is $f(\tau) = \partial_x[f_0(\tau) + \mathcal{H}f_1(\tau)] = |D|^1[f_1(\tau) - \mathcal{H}f_0(\tau)]$, so $\gamma = 1$ for this solution. We cannot apply Theorem 5.1.1 or Lemma 5.3.1 to ascertain the large $|x|$ behavior of $v(x, s; \tau)$ because $\sin(\frac{\pi}{2})f_1(\tau)\hat{}(0) \pm \cos(\frac{\pi}{2})f_0(\tau)\hat{}(0) = 0$. However, if f_0 and f_1 are sufficiently smooth, it follows from Theorems 3.2.1A and B that the largest contribution to the asymptotics when $x = \zeta\sigma^\beta$, β sufficiently large, comes from the contour $\tilde{C}(-x/\sigma)$. A calculation based on Theorem 1.4.9 yields then the following expansion:

$$v(x, \sigma; \tau) = \frac{-1}{\pi} \sum_{j=0}^{N-1} \sum_{n=0}^{n_j - 1} \frac{\Gamma(2 + j + nr)(\rho\sigma)^n}{j!n!|x|^{2+j+nr}} (-1)^{\alpha n + (1-\alpha)j}.$$

$$\cdot \left\{ \cos[\tfrac{\pi}{2}n(r-1)]i^j f_1(\tau)^{\hat{}(j)}(0) + (-1)^\alpha \sin[\tfrac{\pi}{2}n(r-1)]i^j f_0(\tau)^{\hat{}(j)}(0) \right\}$$

$$+ O(|x|^{-2-N}).$$

In the above, $\alpha \in \{0, 1\}$, and $(-1)^\alpha x \leq 0$; also for each $j \geq 0$ we define n_j to be the smallest integer such that $j + n_j(r - \frac{1}{\beta}) \geq N$. $N \geq 1$ is an integer; we assume f_0 and f_1 are sufficiently smooth so that the asymptotic contributions of the other contours are contained in the above error term. Because of our assumptions, when $n = 0$ the first non-vanishing term is with $j = j_0 = (\!(pm - 1)\!)$

$$\frac{-1}{\pi} \frac{\Gamma(2 + j_0)(-1)^{(1-\alpha)j_0}}{j_0! |x|^{2+j_0}} [x^{j_0} f_1(\tau)]^{\hat{}}(0).$$

Notice that this term does not depend on σ, whereas all the terms with $n \geq 1$ do depend on σ, and vanish as $\sigma \to 0^+$. The above term represents the asymptotic behavior expected of $f(\tau)$. Strictly speaking, Theorem 1.4.9 only yields the above expansion when $\sigma \geq 1$, or more generally when $\sigma \geq \epsilon > 0$. However, the above observations make it quite plausible that this expansion also holds for all $0 < \sigma < 1$ as well, and hence for all $\sigma > 0$. We have not proved this, but for the sake of our discussion we will assume that this is the case.

When $r \neq 3, 5, 7, \ldots$, the above term will not always be the leading-order term of $v(x, \sigma; \tau)$ as $|x| \to \infty$. For such r the term with $(j, n) = (0, 1)$ is important:

$$\frac{-1}{\pi} \frac{\Gamma(2+r)\rho\sigma}{|x|^{2+r}} \sin[\tfrac{\pi}{2}(r-1)]f_0(\tau)\widehat{\ }(0).$$

This term will dominate as $|x| \to \infty$ when $r < j_0$.

We will suppose that the leading-order asymptotics of $y(x, t)$ can be obtained by a formal integration in τ in these two leading-order terms.

$$
\begin{aligned}
y(x, t) \sim &\ \frac{-1}{\pi} \frac{\Gamma(2+j_0)(-1)^{(1-\alpha)j_0}}{j_0! |x|^{2+j_0}} \int_0^t [x^{j_0} f_1(\tau)]\widehat{\ }(0)\, d\tau \\
&+ \frac{-1}{\pi} \frac{\Gamma(2+r)\rho}{|x|^{2+r}} \sin[\tfrac{\pi}{2}(r-1)] \int_0^t (t-\tau) f_0(\tau)\widehat{\ }(0)\, d\tau.
\end{aligned}
$$

This is the expression for the leading-order asymptotics of $y(x, t)$ as $|x| \to \infty$, $t > 0$ fixed, or as $x = \zeta t^\beta$, $t \to \infty$, $(-1)^\alpha \zeta < 0$ fixed, β sufficiently large, that we have been seeking.

Let us now consider some of the implications of this formula in some special cases. First of all, consider the case $r = p = 2$ and suppose this formula is applicable to the asymptotics of (recall from section 5.3)

$$u_I(t) = -\int_0^t e^{-iR(D)(t-\tau)} \partial_x[u(\tau)^p]\, d\tau,$$

where u is a real-valued solution of the Benjamin-Ono equation $(iR(D)u = \rho|D|^2 \mathcal{H}u)$. As in section 5.3 define $u_H(t) = e^{-iR(D)t}u_0$. The results of section 4.4 show that if $u_0 = |D|^{\gamma_0}(v_0 + \mathcal{H}v_1)$ is sufficiently smooth we can arrange for $u_H(x, t)$ to decay like $O(|x|^{-m})$ for any real number $m > 0$. Let us consider what happens for some small values of m and for $t > 0$ fixed.

   (1) $m = 1$, $j_0 = 1$. This sort of decay results when $\gamma_0 = 0$ and $\hat{v}_1(0) \neq 0$. It is consistent to suppose $u = O(|x|^{-1})$ as $|x| \to \infty$, because then the above formula indicates that $u_I = O(|x|^{-3})$. The first line of the above formula is the dominant one.

   (2) $m = 2$, $j_0 = 3$. This sort of decay results in the anomalous case where $\gamma_0 = 0$, $\hat{v}_1(0) = 0$ and $\hat{v}_1'(0) \neq 0$. It is consistent to suppose $u = O(|x|^{-2})$ as $|x| \to \infty$, because then the above formula indicates that $u_I = O(|x|^{-4})$. Now the second line of the above formula is the dominant one (as is true for (3)-(5) below).

(3) $m = 3$, $j_0 = 5$. This sort of decay results in the anomalous case where $\gamma_0 = 0$, $\hat{v}_0(0) \neq 0$, $\hat{v}_1(0) = 0$ and $\hat{v}_1'(0) = 0$. This is the most rapid decay we can get for u_H when $\gamma_0 = 0$. It is consistent to suppose $u = O(|x|^{-3})$ as $|x| \to \infty$, because then the above formula indicates that $u_I = O(|x|^{-4})$.

(4) $m = 4$, $j_0 = 7$. This sort of decay results in the anomalous case where $\gamma_0 = 1$, $\hat{v}_1(0) \neq 0$, $\hat{v}_0(0) = 0$ and $\hat{v}_0'(0) = 0$. This is the most rapid decay we can get for u_H when $\gamma_0 = 1$. It is consistent to suppose $u = O(|x|^{-4})$ as $|x| \to \infty$, because then the above formula indicates that $u_I = O(|x|^{-4})$. Here the contributions u_H and u_I decay at the same rate. This case corresponds to Iório's theorem that a solution $u \in C([0, T], H^3 \cap L_3^2)$ of the Benjamin-Ono equation must satisfy $u(t)\hat{}(0) = 0$ for $t \in [0, T]$.

(5) $m = 5$, $j_0 = 9$. This is the non-anomalous decay rate for u_H when $\gamma_0 = 4$, and can also happen in anomalous ways when $\gamma_0 = 2$ or 3. However, it is not consistent to suppose that $u = O(|x|^{-5})$, because u_I cannot decay faster than $O(|x|^{-4})$, as the above formula indicates. So if it were true that $u = O(|x|^{-5})$ as $|x| \to \infty$, then

$$\int_0^t (t - \tau) f_0(\tau)\hat{}(0)\, d\tau = \int_0^t (t - \tau) \int_{-\infty}^{\infty} u(x, t)^2\, dx\, d\tau = 0, \qquad t > 0,$$

implying that $u \equiv 0$. This corresponds to Iório's theorem that a solution $u \in C([0, T], H^4 \cap L_4^2)$ of the Benjamin-Ono equation must vanish identically.

For similar reasons, we expect that any sufficiently smooth solution u of Benjamin-Ono decaying like $O(|x|^{-4-\epsilon})$, $\epsilon > 0$, must vanish identically. As in (5) above, it is clearly possible for the Benjamin-Ono equation to possess solutions where u_H decays more rapidly than u_I; This can happen when $-(-1)^\alpha x \to \infty$, $t > 0$ fixed, for anomalous initial data satisfying

$$\sin(\tfrac{\pi}{2}\gamma_0) i^j \hat{v}_0^{(j)}(0) + (-1)^\alpha \cos(\tfrac{\pi}{2}\gamma_0) i^j \hat{v}_1^{(j)}(0) = 0, \qquad j = 0, 1.$$

whenever $\gamma_0 > 1$.

In [KPV2] and [Ch] linear scattering results were proved for all $r < \tilde{p} = p + \gamma_0(p - 1)$, $p \geq 5$, $r = 3, 5$, but for smaller values of p such results are known only for values of r significantly bounded away from $\tilde{p}$. Thus it is interesting to see if anything unusual happens in the relation between u_H and u_I as $r \to \tilde{p}^-$. Consider the example $p = 4$ and $2 < r < \tilde{p} = 4 + 3\gamma_0$, $r \neq 3$. We assume $u_0 \in L^1$; thus $\gamma_0 \geq 0$ and if $\gamma_0 = 0$ then $\hat{v}_1(0) = 0$. Under these assumptions we have $u_H = O(|x|^{-m})$ with $m > 1$ ($m \geq 1 + \gamma_0$), hence $j_0 = (\!(4m - 1)\!) \geq (\!(3 + 4\gamma_0)\!)$. Therefore, provided $\gamma_0 \in \{0\} \cup (\tfrac{1}{4}, \tfrac{1}{3}] \cup (\tfrac{1}{2}, \tfrac{2}{3}] \cup (\tfrac{3}{4}, \infty)$ and $\int_0^t [x^{j_0} f_1(\tau)]\hat{}(0)\, d\tau$ grows only like a power of t, the the second term in the above formula for the asymptotics of $y(x, t)$ is dominant for all $r < 4 + 3\gamma_0$ when $x = \zeta t^\beta$, $t \to \infty$, β sufficiently large. The above restriction on the values of γ_0 implies that $r < j_0$ for all $r < 4 + 3\gamma_0$. As before we suppose that these asymptotics for $y(x, t)$ apply to u_I, where u solves the nonlinear dispersive equation. Assuming that u has the same sharp L^4 decay rate as u_H, we have by Theorems 5.2.1 and 5.2.2 that

$$f_0(\tau)\hat{\,}(0) = \int_{-\infty}^{\infty} u(x,\tau)^4\, dx$$

$$= \begin{cases} O((1+\tau)^{-1}) & 2 < r < 3 + 4\gamma_0, r \neq 3, \\ O((2+\tau)^{-1}\ln(2+\tau)) & r = 3 + 4\gamma_0, r \neq 3, \\ O((1+\tau)^{-(3+4\gamma_0)/r}) & 3 + 4\gamma_0 < r < 4 + 3\gamma_0. \end{cases}$$

Therefore

$$\int_0^t (t-\tau)f_0(\tau)\hat{\,}(0)\, d\tau = \begin{cases} O(t\ln(t)) & 2 < r < 3 + 4\gamma_0, r \neq 3, \\ O(t[\ln(t)]^2) & r = 3 + 4\gamma_0, r \neq 3, \\ O(t^{2-(3+4\gamma_0)/r}) & 3 + 4\gamma_0 < r < 4 + 3\gamma_0, \end{cases}$$

as $t \to \infty$. Thus $y(x,t)$ and hence $u_I(x,t)$ will be asymptotically larger than $v(x,t;\tau)$. But if we compare $y(\zeta t^\beta, t)$ to $u_H(\zeta t^\beta, t)$ we find even as $r \to 4 + 3\gamma_0$ that $u_H(\zeta t^\beta, t)$ is asymptotically larger than $y(\zeta t^\beta, t)$ if β is sufficiently large and $\gamma_0 < 1 + r$. If β is not so large, our estimates for $y(\zeta t^\beta, t)$ can be asymptotically larger than those of $u_H(\zeta t^\beta, t)$, but this could be meaningless, since for smaller values of β we might not be looking at the dominant terms. Hence we do not observe any tendency for $u_I(\zeta t^\beta, t)$ to become asymptotically larger than $u_H(\zeta t^\beta, t)$ as $r \to 4 + 3\gamma_0$ when β is sufficiently large and $\gamma_0 < 1 + r$.

When we compare the two examples we have discussed so far we seem to have a conflict of principles. In the first example we found situations when $\gamma_0 > 1$ where $u_H \ll u_I$, i.e. u_H was asymptotically smaller than u_I, as $|x| \to \infty$, $t > 0$ fixed. In the second example we found that $u_H \gg u_I$ when $\gamma_0 < r + 1$ and β is sufficiently large. These findings can be reconciled if we note that the examples we found in the first case where $u_H \ll u_I$ had $\gamma_0 \geq 3$ or were for highly anomalous initial data. In the second case we mostly avoided anomalous initial data (except when $\gamma_0 = 0$). It stands to reason that anomalies can introduce complications into the study of asymptotic behavior. This is why we have tried to define and avoid them as much as possible. Thus the principles are not really in conflict.

As a final example, consider the case $r = 3$, or more generally $r \geq 3$ an odd integer. Returning to the asymptotic expansion we derived for $v(x,\sigma;\tau)$, we find that $\sin[\pi 2n(r-1)] = 0$ for all $n \geq 0$, and hence there is no obstruction to having $v(x,\sigma;\tau)$ or $y(x,t)$ decay at an arbitrarily rapid rate as $|x| \to \infty$. The results of Kato [K] and Abdelouhab [Abd] mentioned at the beginning of this section show that this behavior extends to solutions of the nonlinear dispersive equation when $r \geq 3$ is an odd integer.

In conclusion we remark that the non-rigorous results of this section point up the need to obtain careful rigorous asymptotic expansions of the solution $y(x,t)$ of the inhomogeneous linear dispersive equation analogous to the results we have presented for the homogeneous linear dispersive equation. Hopefully, we have also convinced the reader that considerable understanding of the solutions of the nonlinear dispersive equation can be gained by the application of the results proved in chapters one through four.

REFERENCES

[Abd]   L. Abdelouhab, *Nonlocal Dispersive Equations in Weighted Sobolev Spaces*, Diff. Int. Eq. **5** (1992), 307–338.

[AblSe1] M. J. Ablowitz, H. Segur, *Asymptotic solutions of the Korteweg-de Vries equation*, Stud. Appl. Math. **57** (1977), 13–44.

[AblSe2] M. J. Ablowitz, H. Segur, *Solitons and the inverse scattering transform*, SIAM Studies in Applied Mathematics, 4, SIAM, Philadelphia, 1981.

[AS]   M. Abramowitz, I. A. Stegun (editors), *Handbook of Mathematical Functions*, Dover, New York, 1964.

[AM]   M. M. Agrest, M. S. Maksimov, *Theory of Incomplete Cylindrical Functions and their Applications*, Springer Verlag, Berlin, 1971.

[Ahl]   L. V. Ahlfors, *Complex Analysis*, McGraw-Hill, New York, 1979.

[ABS]   C. J. Amick, J. L. Bona, and M. E. Schonbek, *Decay of Solutions of some Nonlinear Wave Equations*, J. Diff. Eq. **81** (1989), 1–49.

[bAT]   M. Ben-Artzi, F. Treves, *Uniform Estimates for a Class of Evolution Equations*, J. Funct. Anal. **120** (1994), 264–299.

[BG]   C. A. Berenstein, R. Gay, *Complex Variables, An Introduction*, Springer-Verlag, New York, 1991.

[Bi]   P. Biler, *Asymptotic Behaviour in Time of Solutions to Some Equations Generalizing The Korteweg-de Vries-Burgers Equation*, Bull. Polish Acad. Sci. Math. **32** (1984), 275–282.

[B]   N. Bleistein, *Uniform asymptotic expansions of integrals with many nearby stationary points and algebraic singularities*, J. Math. Mech. **17** (1967), 533–559.

[BH]   N. Bleistein, R. A. Handelsman, *Asymptotic expansions of Integrals*, Holt, Rinehart and Winston, New York, 1975.

[BDP]   J. L. Bona, F. Demengel, K. S. Promislow, *Fourier Splitting and Dissipation of Nonlinear Dispersive Waves*, preprint.

[BL1]   J. L. Bona, L. Luo, *Decay of the solutions to nonlinear, dispersive equations*, Diff. Int. Eq. **6** (1993), 961–980.

[BL2]   J. L. Bona, L. Luo, *More results on the decay of the solutions to nonlinear, dispersive wave equations*, Discrete and Continuous Dynamical Systems **1** (1995), 151–193.

[BL3]   J. L. Bona, L. Luo, *The effect of dissipation to some generalized nonlinear, dispersive wave equations*, preprint.

[BPW1]   J. L. Bona, K. S. Promislow, C. E. Wayne, *On the asymptotic behavior of solutions to nonlinear, dispersive, dissipative wave equations*, J. Math and Computers in Simulation **37** (1994), 264–277.

[BPW2]   J. L. Bona, K. S. Promislow, C. E. Wayne, *Higher-order asymptotics of decaying solutions of some nonlinear, dispersive, dissipative wave equations*, Nonlinearity **8** (1995), 1179–1206.

[Caz]   T. Cazenave, *An Introduction to Nonlinear Schrödinger Equations*, Textos de Métodos Matemáticos, 22, Instituto de Matemática- UFRJ, Rio de Janeiro, 1989.

[CFU]   C. Chester, B. Friedman, F. Ursell, *An extension of the method of steepest descents*, Proc. Camb. Phil. Soc. **53** (1991), 599–611.

[Ch]   Y. Choi, *Well-Posedness and Scattering Results for Fifth Order Evolution Equations*, Ph.D. Thesis, University of Chicago, Chicago, Illinois, 1994.

[CW]   F. M. Christ, M. I. Weinstein, *Dispersion of small amplitude solutions of the generalized Korteweg-de Vries equation*, J. Funct. Anal. **100** (1991), 87–109.

[Col]   T. Colin, *Smoothing effects for dispersive equations via a generalized Wigner transform*, SIAM J. Math. Anal. **25** (1994), 1622–1641.

[Con]   P. Constantin, *Decay Estimates for Schrödinger Equations*, Commun. Math. Phys. **127** (1990), 101–108.

[CG]   W. Craig, J. Goodman, *Linear dispersive equations of Airy type*, J. Diff. Eq. **87** (1990), 38–61.

[DZ1]   P. Deift, X. Zhou, *A steepest descent method for oscillatory Riemann-Hilbert problems. Asymptotics for the MKdV equation*, Annals of Math. **137** (1993), 295–368.

[DZ2]   P. Deift, X. Zhou, *Long-time asymptotics for integrable systems. Higher order theory.*, Commun. Math. Phys. **165** (1994), 175–191.

[DVZ]   P. Deift, S. Venakides, X. Zhou, *The Collisionless Shock Region for the Long-Time Behavior of Solutions of the KdV Equation*, Comm. Pure Appl. Math. **47** (1994), 199–206.

[D1]   D. B. Dix, *Temporal Asymptotic Behavior of Solutions of the Benjamin-Ono-Burgers Equation*, J. Diff. Eq. **90** (1991), 238–287.

[D2]   D. B. Dix, *The Dissipation of Nonlinear Dispersive Waves: The Case of Asymptotically Weak Nonlinearity*, Comm. P. D. E. **17** (1992), 1665–1693.

[D3]   D. B. Dix, *Large-time behavior of solutions of Burgers' equation*, Proc. Roy. Soc. Edinburgh A.

[EVZ]   M. Escobedo, J. L. Vazquez, E. Zuazua, *Asymptotic behavior and source-type solutions for a diffusion convection equation*, Arch. Rational. Mech. Anal. **124** (1993), 43–65.

[EZ]   M. Escobedo, E. Zuazua, *Large Time Behavior for Convection-Diffusion Equations in $\mathbb{R}^N$*, J. Funct. Anal. **100** (1991), 119–161.

[F]   M. V. Fedoryuk, *Metod Perevala (The Saddle-Point Method)*, Nauka, Moscow, 1977. (Russian)

[HKO]   N. Hayashi, K. Kato, T. Ozawa, *Dilation method and smoothing effects of solutions to the Benjamin-Ono equation*, Proc. Roy. Soc. Edinburgh **126A** (1996), 273–285.

[HO]   N. Hayashi, T. Ozawa, *Modified wave operators for the derivative nonlinear Schrödinger equation*, Math. Ann. **298** (1994), 557–576.

[I1]   R. J. Iório, *On the Cauchy Problem for the Benjamin-Ono Equation*, Comm. P. D. E. **11** (1986), 1031–1081.

[I2]   R. J. Iório, *KdV, BO and friends in weighted Sobolev spaces*, Lecture Notes in Math **1450** (1990), Springer Verlag, Berlin, 104–121.

[K]   T. Kato, *On the Cauchy Problem for the (generalized) Korteweg-de Vries equations*, Studies in Applied Mathematics, Advances in Mathematics Supplementary Studies, vol. 8, Academic Press, 1983, pp. 93–128.

[KPV]   C. E. Kenig, G. Ponce, L. Vega, *Oscillatory Integrals and Regularity of Dispersive Equations*, Indiana Univ. Math. J. **40** (1991), 33–69.

[KPV2]   C. E. Kenig, G. Ponce, L. Vega, *Well-posedness and scattering results for the generalized Korteweg-de Vries equation via the contraction principle*, Comm. Pure Appl. Math. **46** (1993), 527–620.

[KC]   J. Kevorkian, J. D. Cole, *Perturbation Methods in Applied Mathematics*, Springer-Verlag, New York, 1981.

[N1]   P. I. Naumkin, *Asymptotic behavior of solutions to nonlinear equations with dissipation for large x and t*, Stud. Appl. Math. **87** (1992), 45–60.

[N2]   P. I. Naumkin, *On the asymptotic behavior for large time of the solutions of nonlinear equations in the case of maximal order*, Differential Equations **29** (1993), 926–929.

[NS1]   P. I. Naumkin, I. A. Shishmarev, *Asymptotic behavior, for large time values, of solutions of the Korteweg-de Vries with dissipation*, Differential Equations **29** (1993), 253–263.

[NS2]   P. I. Naumkin, I. A. Shishmarev, *Asymptotic relationship as $t \to \infty$ between solutions to some nonlinear equations I, II*, Differential Equations **30** (1994), 806–814, 1329–1340.

[NS3]   P. I. Naumkin, I. A. Shishmarev, *Nonlinear nonlocal equations in the theory of waves*, Translations of Mathematical Monographs 133, American Math. Soc., Providence RI, 1994.

[N]   A. C. Newell, *Solitons in Mathematics and Physics*, SIAM, Philadelphia, 1985.

[oDT]   A. B. Olde Daalhuis, N. M. Temma, *Uniform Airy-Type Expansions of Integrals*, SIAM J. Math. Anal. **25** (1994), 304–321.

[O1]   F. W. J. Olver, *Asymptotics and Special Functions*, Academic Press, New York, 1974.

[O2]   F. W. J. Olver, *Error Bounds for Stationary Phase Approximations*, SIAM J. Math. Anal. **5** (1974), 19–29.

[PV]   G. Ponce, L. Vega, *Nonlinear small data scattering for the generalized Korteweg-de Vries equation*, J. Funct. Anal. **90** (1990), 445–457.

[Sch]   P. C. Schuur, *Asymptotic Analysis of Soliton Problems*, Lecture Notes in Math **1232** (1986), Springer Verlag, Berlin.

[SSS]   A. Sidi, C. Sulem, P. L. Sulem, *On the Long Time Behaviour of a Generalized KdV Equation*, Acta Appl. Math. **7** (1986), 35–47.

[Wa]   C. E. Wayne, *Invariant manifolds for Parabolic Partial Differential Equations on Unbounded Domains*, Arch. Rat. Mech. Anal. (to appear).

[W]   R. Wong, *Asymptotic Approximations of Integrals*, Academic Press, Boston, 1989.

[Z1]   L. Zhang, *Asymptotic property for the solution to the generalized Korteweg-de Vries equation*, Acta Math. Appl. Sinica (English Series) **10** (1994), 377–386.

[Z2]   L. Zhang, *Decay estimates for the solutions of some nonlinear evolution equations*, J. Diff. Eq. **116** (1995), 31–58.

SUBJECT INDEX

Lecture Notes in Mathematics

For information about Vols. 1–1479
please contact your bookseller or Springer-Verlag

9 783540 634348